AF389498

Space for Sustainability: Using Data from Earth Observation to Support Sustainable Development Indicators

Space for Sustainability: Using Data from Earth Observation to Support Sustainable Development Indicators

Editors

Stephen Morse
Richard Murphy
Ana Andries

MDPI • Basel • Beijing • Wuhan • Barcelona • Belgrade • Manchester • Tokyo • Cluj • Tianjin

Editors

Stephen Morse	Richard Murphy	Ana Andries
University of Surrey	University of Surrey	University of Surrey
UK	UK	UK

Editorial Office
MDPI
St. Alban-Anlage 66
4052 Basel, Switzerland

This is a reprint of articles from the Special Issue published online in the open access journal *Actuators* (ISSN 2076-0825) (available at: https://www.mdpi.com/journal/sustainability/special_issues/space_sus).

For citation purposes, cite each article independently as indicated on the article page online and as indicated below:

LastName, A.A.; LastName, B.B.; LastName, C.C. Article Title. *Journal Name* **Year**, *Volume Number*, Page Range.

ISBN 978-3-0365-4265-2 (Hbk)
ISBN 978-3-0365-4266-9 (PDF)

Cover image courtesy of Ana Andries.

Contents

About the Editors

Stephen Morse

Stephen Morse currently holds the position of Chair in Systems Analysis for Sustainability at the Centre for Environment and Sustainability at the University of Surrey, Guildford, UK. He followed higher education degree programmes at the University of Wales (BSc), University of Reading (MSc and Ph.D.) and University of Southampton (Ph.D.) and is a Fellow of the Royal Geographical Society (FRGS), the Royal Society of Biology (FRSB) and the High Education Academy (FHEA). Steve has a background in applied ecology and the environment, and before becoming an academic, he worked for 10 years in rural development in Nigeria for agencies such as the World Bank and nongovernmental organisations. His research and teaching interests are broad, spanning both the natural and social sciences but especially the interface between them. These interests include methods for the assessment of sustainability (e.g., indicators and indices) in order to help guide intervention. He has also helped pioneer a number of participatory methodologies for sustainability assessment. He has been involved in research and sustainable development projects across Europe, the Mediterranean, South America, Africa and Asia. Steve has authored 170 journal papers and 20 books, with the latter including 'The Rise and Rise of Indicators: Their History and Geography' (2019; Routledge). He has also jointly edited four major collections, including the 'Routledge Handbook of Sustainability Indicators and Indices' (2018; Routledge).

Richard Murphy

Richard Murphy is Professor of Life Cycle Assessment and a Fellow of IOM3 and The Royal Society of the Arts. His research interests span life cycle sustainability assessment, bio-based materials and product systems (including bioplastics and biofuels), biodegradation and the use of satellite and Earth Observation data for sustainable development. He has a background in biological sciences with a BSc in Botany with Zoology from King's College London and a Ph.D. in Pure and Applied Biology from Imperial College London. He has undertaken postdoctoral work in New Zealand and the Netherlands and has been a Reader in Plant Sciences at Imperial College London prior to joining the University of Surrey in 2013. Richard is a past President of the Institute of Wood Science (now part of IOM3), a past member of DEFRA's Hazardous Substances Advisory Committee, and BSI committees. He is a Founder and Director of LCAworks Ltd., was Chief Scientific Officer of Mycologix Ltd and has advised the UK Climate Change Committee on LCA for bioenergy systems. He has received funding from UK Research Councils, the European Commission, international agencies and organisations and has published over 150 journal papers and articles.

Ana Andries

Ana Andries is an early-stage researcher, currently working as a Postdoctoral Research Fellow at the Centre for Environment and Sustainability at the University of Surrey, Guildford, UK. She obtained her higher education degree at the University of Alexandru Ioan Cuza (BSc and MSc), University of South Wales (MSc) and University of Surrey (Ph.D.) and is also a member of the Royal Cartography Society (BCS). Ana has an academic background in environmental geography, sustainability, and conservation, and before starting her PhD in assessing the potential of Earth Observation (EO) to support SDG indicators, she spent several years working as GIS Quality Control for Rural Payment Agency (RPA), office of DEFRA, inside European Union projects. Her research interests cover SDGs, EO satellite data, corruption, education, poverty, inequality, biodiversity, conservation biology and soil health. In exploring these interests, Ana has worked with mixed research methods, including semi-structured interviews and secondary data analysis (e.g., processing satellite data, statistical techniques, geoprocessing analysis). Ana published six peer-reviewed articles during her Ph.D.

Preface to "Space for Sustainability: Using Data from Earth Observation to Support Sustainable Development Indicators"

Global progress toward living sustainably is now urgently needed. Actions for sustainability are typically informed through the use of indicator-based frameworks that encompass diverse attributes of the environmental, social and economic dimensions of 'sustainability'. Reporting on such indicators is embedded in frameworks such as the United Nations Sustainable Development Goals (SDGs), with the primary responsibilities for reporting borne by national and local governments. Additionally, many businesses and public bodies (e.g., universities, health services) are increasingly under internal and external pressure to similarly report via these sustainability indicators, especially as part of the SDGs, and such reporting is of increasing interest to investors and the financial services sectors from a risk and assurance perspective. However, the use of these indicator-based frameworks involves many challenges, and one of the most significant of these is the challenge of acquiring sufficient, timely and good-quality data to populate these indicators via 'conventional' methods (e.g., surveys at the local, national or corporate level) as this is often expensive and time consuming. Many developing regions, in particular, suffer from a lack of resources or established systems for such data collection and, indeed, this is also proving to be challenging for more developed economies. One approach to address this issue of data provision for indicators of sustainable development (SD) is the use of Earth Observation (EO). EO-based data, geospatial information and 'big data' can support the population of sustainability indicators at all scales, and the integration of these sources is a step forward in advancing the well-being of our societies. While EO-derived data have been used for many years to assess important issues such as deforestation and changes in land use, their use to address more socioeconomic issues (e.g., inequality, poverty, corruption, health care) within SD remains limited. Nonetheless, EO tools and technologies are developing rapidly with an expanding range of capabilities, resolutions, frequency, data power, accuracy, etc., and this is anticipated to continue into the foreseeable future. The papers in this book set out some of the frontiers regarding the use of EO data for SD indicators, and given the rapid progress in the field, it provides a timely and welcome milestone in the journey.

Stephen Morse, Richard Murphy, and Ana Andries
Editors

 *sustainability*

Article

Using Data from Earth Observation to Support Sustainable Development Indicators: An Analysis of the Literature and Challenges for the Future

Ana Andries [1,*], Stephen Morse [1], Richard J. Murphy [1], Jim Lynch [1] and Emma R. Woolliams [2]

[1]   Centre for Environment and Sustainability, University of Surrey, Guildford GU2 7XH, UK; s.morse@surrey.ac.uk (S.M.); rj.murphy@surrey.ac.uk (R.J.M.); j.lynch@surrey.ac.uk (J.L.)

[2]   Climate and Earth Observation Group, National Physical Laboratory, Teddington TW11 0LW, UK; emma.woolliams@npl.co.uk

*   Correspondence: a.andries@surrey.ac.uk

Abstract: The Sustainable Development Goals (SDG) framework aims to end poverty, improve health and education, reduce inequality, design sustainable cities, support economic growth, tackle climate change and leave no one behind. To monitor and report the progress on the 231 unique SDGs indicators in all signatory countries, data play a key role. Here, we reviewed the data challenges and costs associated with obtaining traditional data and satellite data (particularly for developing countries), emphasizing the benefits of using satellite data, alongside their portal and platforms in data access. We then assessed, under the maturity matrix framework (MMF 2.0), the current potential of satellite data applications on the SDG indicators that were classified into the sustainability pillars. Despite the SDG framework having more focus on socio-economic aspects of sustainability, there has been a rapidly growing literature in the last few years giving practical examples in using earth observation (EO) to monitor both environmental and socio-economic SDG indicators; there is a potential to populate 108 indicators by using EO data. EO also has a wider potential to support the SDGs beyond the existing indicators.

Keywords: earth observation; SDGs; indicator type; data challenges

Citation: Andries, A.; Morse, S.; Murphy, R.J.; Lynch, J.; Woolliams, E.R. Using Data from Earth Observation to Support Sustainable Development Indicators: An Analysis of the Literature and Challenges for the Future. *Sustainability* **2022**, *14*, 1191. https://doi.org/10.3390/su14031191

Academic Editor: Jamal Jokar Arsanjani

Received: 11 December 2021
Accepted: 17 January 2022
Published: 21 January 2022

Publisher's Note: MDPI stays neutral with regard to jurisdictional claims in published maps and institutional affiliations.

1. Introduction

In 2012, the United Nations (UN) Conference on Sustainable Development (Rio + 20) was held in Rio de Janeiro; 193 member states agreed on a new and comprehensive framework called the Sustainable Development Goals (SDGs). The SDG framework was intended to be an integrated development agenda to 2030 that would apply equally to developed and developing countries and address all three dimensions of sustainable development (environment, social and economic) and their interlinkages [1].

The SDGs were ratified by the UN General Assembly at the Sustainable Development Summit in New York on 25 September 2015. All signatory countries committed to monitoring their progress towards the 17 goals at the heart of the SDGs by assessing past and current conditions at national and sub-national levels. To achieve this, the UN created a framework of 169 targets and 232 unique indicators that are meant to frame national agendas and policies up to 2030 [2].

The Inter-Agency Expert Group for the SDGs (IAEG-SDGs) established three working groups responsible for formulating these indicators and targets, as well as methods for 'populating' the indicators with appropriate data and communication and coordination with all partners (international organisations, civil society, governments, academia, and the private sector). The IAEG-SDGs Global Indicator Framework was officially adopted by the UN Statistical Commission in March 2017 [3] and, initially, the IAEG-SDGs classified the indicators into three different tiers based on how well established the methodology was at that time and the data available to allow the population of the indicators: Tier I (well-established methodology and data are widely available), Tier II (well-established methodology but data are not collected regularly by the countries), and Tier III (no established methodology to collect the required data and data are not available). Each of the SDG indicators has a set of metadata guidelines to aid interpretation and transferability; these include indicator definition (objective and purpose), computational methodology (refers to how the indicator is computed and disaggregated into multiple sub-indicators) and sourcing of data (the main source of data, collecting method, the frequency of data collection, data providers and data availability) [4].

The IAEG-SDGs regularly reviews the Global Indicator Framework to add more indicators, alter existing ones and, if required, update the status of the tier categorisation of the indicator (see Table 1).

Table 1. SDG indicator framework and tier classification update [3,5].

Date of SDG Indicator Framework Update	Number of Indicators	Indicator Amendments	Indicator Tier Classification
March 2016 (original framework)	232	N/A	93 Tier I 66 Tier II 68 Tier III 5 multiple tiers
December 2020	231	36 changes in which 14 indicators were proposed to replace existing ones, 8 indicators were revised, 8 proposals were made for additional indicators, and 6 indicators were recommended for removal	231 130 Tier I 97 Tier II 4 multiple tiers
March 2021	231	No modifications	No modifications

To populate the IAEG-SDG indicators at the country level, many different types of data are required; these are mostly derived from eight sources: census data (CD), household surveys (HS), agricultural survey (AS), administrative data (AD), Civil registration (CR), economic statistics (ES), geospatial data (GD), and other environmental data (Env) (Table 2) [6]. However, data alone are not enough, as they must be transformed (e.g., data combined to generate a single indicator), analysed, interpreted, and communicated to those who would make use of the indicators. Therefore, the UN noted a requirement to *"intensify efforts to strengthen statistical capacities of developing countries and least developed countries are among the particular cases that need special attention in this regard"* ([2], paragraph 74). This includes a requirement to ensure access to quality and disaggregated data, including geospatial and earth observation (EO)-derived data [2]. Environmental data can include in situ observations, ground-based or aircraft remote sensing or satellite remote sensing. Here we use the broad term earth observation (EO) to describe predominantly satellite-based remote sensing. This term often also includes remote sensing from both crewed and autonomous aircraft where similar imaging techniques are used.

Geospatial data play a crucial role for many of the SDG indicators, as well as for disaggregated analysis of socioeconomic SDG indicators [6–10]. For instance, converting household surveys into geospatial data can facilitate disaggregation and analysis by spatial characteristics, e.g., proximity to roads or levels of urban development [6]. In recent years, there have been efforts to integrate the geospatial approach into the national statistics offices (NSOs) of many countries [11]. Likewise, geospatial tools can easily manipulate large datasets such as those derived from EO satellites, aerial photography (AP), lidar, information and communications technologies (ICT) (e.g., social media, mobile, and crowdsourcing data). The processing to derive information products from these different sources is increasingly dependent on artificial intelligence and machine learning. These tools can manipulate large data sets and the artificial intelligence methods are often referred to collectively as 'big data' or the 'data revolution' [6–8]. Processing and analysing large datasets in real-time can reveal patterns, trends, and interactions, thereby deriving information and insights on human behaviours and wellbeing, and helping to target aid interventions to help vulnerable people [12].

Table 2. Data type and challenges associated.

Data Source	Data Type	Description	Data Challenges
Traditional national statistical data	Census data (CD)	CD is a systematic recording of information, focusing mainly on the population and takes place at least once every 10 years (varies between countries); NSOs are responsible for collecting these data.	Challenges in obtaining quality data - lack of data accessibility and availability, data quality, data continuity, transparency, and accountability - lack of funding/financing Challenges in data processing - difference in methodology - limited human and technical capacity (e.g., skills and training, adequate staff) - lack of developed infrastructure to support networking, high-performance computing, and the use of digitalisation
	Household surveys (HS)	HS provides demographic and socio-economic data.	
	Agricultural survey (AS)	AS covers information about land use and ownership, operator characteristics, production practices, crop yields and productivity, income, and expenditures.	
	Administrative data (AD)	ADs are collected by government departments and include information about welfare, tax, health, and educational record systems.	
	Civil registration (CR)	CR is a collection of data that records main events in a person's life (such as birth, marriage, divorce, adoption, and death). The primary responsibility for collecting such data is typically attributed to different Ministries and National Statistical Offices (NSOs).	
	Economic statistics (ES)	ES typically include data on the labour force, tax returns, trade statistics, etc. The scope of these data are to measure the financial performance of economic agents and can include estimates of Gross Domestic Product, Gross National Income, national poverty levels, household income, labour force participation and employment status, and economic losses from disasters.	
Non-traditional data	Geospatial data (GD)	GD refers to all data from previous categories that are located with environmental data, that include geolocation as coordinates and topology, allowing the data to be illustrated geographically. Environmental data can be derived from readily available imagery from satellites, airborne, drones, etc. at different spatial resolutions that are coupled with geospatial tools. Satellite imagery is currently used to populate environmental indicators linked to agriculture, biodiversity, forestry and land cover and use change.	Challenges in using EO data - EO does not provide statistical indicators by default and requires expert processing and analysis - analysis prone to software errors and human misinterpretation - incompatibility with current statistical methodology - cloud coverage limits availability, particularly in rainforest areas Challenges in obtaining EO data - limited technical capacity and EO/software skills in NSOs - limited hardware and software infrastructure in NSOs - lack of understanding of EO data value prevents investment - reluctance to change the current method
	Environmental data (Env)	Other environmental data refers to ground technologies or surveys. Many environmental indicators also include real-time monitoring of conditions, such as air quality in urban areas or water supply.	

However, of all these elements, the one area that is perhaps still in its infancy is the use of EO technology to populate the IAEG-SDG indicators (herein referred to as the SDG indicators) particularly for those indicators in the social and economic domains of sustainable development. Using EO technology for social and economic domains is an area of active research; much progress has been made in recent years as the availability of satellite data and tools to manipulate them are more readily available. Notwithstanding this progress, much remains to be done; the time is now right to review the current state-of-the-art and chart a path for the future.

Therefore, the purpose of this paper is to provide a conceptual review [13] of the literature focusing on efforts to use EO technology in populating the SDG indicators. The paper structure is presented in Figure 1 and it begins with an analysis of the literature of data challenges (including cost) behind populating the SDG indicators before moving on to explore the potential of using EO derived data. As part of this, it is necessary to give the reader a sense of how EO technology has progressed to date and the authors have included Supplementary Materials which set out some of the technical capabilities of EO-based instruments, EO data portals and data platforms currently available. Our review highlights the main applications of satellite imagery in sustainability and later links to peer-reviewed articles and best practices of populating SDG indicators. The paper applies a framework first developed by Andries et al. [14,15] designed to assess the potential of EO for the SDG indicators and expands it by considering the classification of indicators into the various indicator types based on the sustainable development pillars. This analysis was designed to identify the areas where more work is needed in the use of EO data for indicators, but also to understand the field that has been covered in recent years.

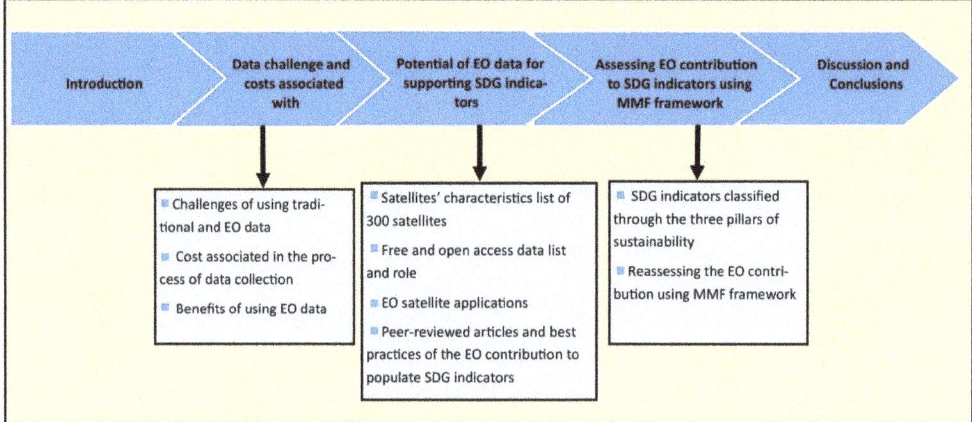

Figure 1. Paper structure.

2. Data Challenges and Costs

Results from the Voluntary National Reviews (VNRs) 2019 [16], a process by which each country assesses progress made in achieving the SDGs every year, showed that most developing countries failed to populate the SDG indicators due to a lack of data. However, even when data are available, countries can encounter other challenges as highlighted in Table 2 [17].

Developing countries tend to have relatively weak statistical institutions with poor governance, constraints in obtaining quality data, a lack of time series [11], and often have large variations in data collection and presentation methodologies across space and time [18]. Kindornay et al. [19] highlighted other issues, such as lack of investment in staff, infrastructure, and tools (e.g., computers and software), low human capacity, highly fragmented statistical systems, and inadequate funding, most of these issues are under political influence; thus, data may even be misreported and suppressed for political reasons [20]. All of these issues can make indicator comparisons unreliable [11,16–18,21,22] (see Table 2).

Regarding EO-derived data, a survey conducted by the Intergovernmental Group on Earth Observations (GEO) [23], with 72 respondents from GEO's member countries, revealed a series of challenges that cover organisational, technical capacity and data accessibility issues for NSOs to obtain the necessary data (Table 2). However, according to the UN Global Working Group on Big Data survey, 60% of interviewees (within 93 countries) noted that the main benefits of using big data from satellite imagery and other sources are "faster and more timely effective statistics", "modernisation of the statistical production process", generating "new products and services", and "cost reductions" [24]. There have been challenges regarding the temporal resolution (revisit times) of EO satellites when these data are used for certain sensitive applications; however, in recent years there has been a shift from multi-day revisit times to daily revisit times [25], particularly through the establishment of small satellite constellations such as CubeSats, SkySat, and Vivid-I, which are able to close this temporal resolution gap [26]. In addition, cloud coverage, particularly in tropical areas, and satellite sensor failure can result in data gaps; this may limit their applicability. However, algorithms have been developed and are now often easily accessible on online computational platforms and portals (e.g., Google Earth Engine), which can partially fill in such data gaps [27,28].

Given recent developments in satellite programmes, many have noted that there is a need to invest in geospatial departments in NSOs, ensuring adequate human capacity with relevant skills, equipment, and software and policy frameworks to support the collection and use of EO data; countries also need to be open to reform in their institutional and regulatory context [29]. For example, the Applied Sciences Programme within the National Aeronautics and Space Administration (NASA) Earth Science Division has been implementing free training on the practical use of satellite data, covering the application of EO to disaster management, water quality management, health care (e.g., malaria early warning), fisheries management, air quality monitoring, and wildfires [30,31].

Jerven [32] made one of the first estimations of the financial costs of populating the SDG indicators in the policy paper *'Benefits and Costs of the Data for Development Targets for the Post-2015 Development Agenda'*, later elaborated in [33]. In both reports, Jerven estimated a global cost of US $254 billion for populating the SDG indicators over the 2016–2030 period, taking into account the need for the population census, living standards measurement studies, demographic and health surveys, core welfare indicator questionnaires, and multiple indicator cluster surveys. However, Demombynes and Sandefur [34] provided a much lower estimated cost of US $4.5 billion to populate the SDG indicators over the same period. However, these estimates do not consider the use of EO and 'big data', despite the UN "Data for Development" report (2015) suggestions that *"cost savings might result from the use of new technologies"* and should result in better quality and efficiency, and that the new technologies *"may yield lower costs in the long run, but in the near term, they are likely to require new investments"* ([6] p. 29). The report estimated a total cost of populating the SDG indicators to be US$902 to US$941 million per annum for all data types across 77 developing countries, where US$787 million would be attributed to traditional national statistical data and US$114 million to EO data (Table 2).

EO derived data are increasingly being recognised as a promising resource for tackling the challenges involved in data collection for indicators and the Integrated United Nations Committee of Experts on Global Geospatial Information Management (UN-GGIM) and GEO play key roles in realising this promise. These agencies are working with statistical communities, at both global and national levels, to support them on how they could benefit from the use of EO data to monitor SDGs. As a result, NSOs in countries such as South Africa, the United Kingdom, Nigeria, Philippines, Sweden [27], USA [35], Colombia, Senegal, Sierra Leone, Kenya, Tanzania and Ghana [36] have started exploring the use of EO satellite imagery and geospatial frameworks, especially to address indicators of SDGs 1 (End Poverty), 2 (Zero Hunger), 6 (Clean water and sanitation), 11 (Sustainable cities), 13 (Climate Action), and 15 (Life on Land). EO data can also provide opportunities to measure the indicator at the subnational level [35].

3. Potential of Earth Observation (EO) for Supporting SDG Indicators

The era of observing the Earth through satellites began in 1957 with the launch of the first satellites (Sputnik-1 and Sputnik-2) by the Soviet Union followed soon after (1958 to 1959) by NASA launching the Explorer satellite series. While images of the Earth had been captured by sub-orbital craft in the 1940s, the first orbital satellite image of the Earth was captured by Explorer 6 in 1959 [37]. Technology progressed rapidly and, in 1972, Landsat 1 provided the first images with a resolution of 80 m, followed by the European Space Agency (ESA) that has observed the Earth from space since the launch of its first meteorological mission in 1977 and later by European Remote Sensing (ERS) and Envisat missions satellites [38]. Today the main space agencies provide high-resolution satellite data at no cost to the user, including Landsat from NASA with 30 m and ESA Sentinel 2 with up to 10 m resolutions.

There are currently 260 operational EO satellites (having full functionality), 333 non-operational (usually historical satellites that have stopped sending data for several reasons) and approximately 200 satellites are in development (this includes satellites that have been approved and planned to be launched).

However, it needs to be noted here that a single satellite may carry several instruments (sensors) that provide EO data. A list of important current satellites, non-operational satellites and satellites in development (along with their sensors) is given in Supplementary Material Table S1 (which contains a list of 300 satellites, where 162 are operational, 116 non-operational, and 22 in development), based on data provided in three sources: WMO-Oscar [39], EO satellite portal [40] and CEOS [41]. The list covers the satellites' characteristics (e.g., date and lifetime, type of instrument, spectral coverage), capabilities (temporal and spatial resolution, primary mission, imaging capability, data accessibility), and main applications. The satellites and instruments in Table S1 have been selected based on three criteria:

- they have been or are used in a wide range of programmes in sustainable development by organisations such as GEO, ESA, NASA, JAXA, etc.
- the instruments on board the satellites have provided data for more than 6 consecutive months
- the instruments on the satellites can offer time-series data from a series of similar satellites (e.g., Landsat, Sentinel, NOAA, GOES, METEOSAT, HY).

EO satellites can be passive (measuring the light reflected from or the thermal energy emitted by the Earth) or active (measuring the interaction of energy emitted by the satellite with the Earth, e.g., through radar). Passive EO satellite sensors provide data at various spectral, spatial, and temporal resolutions. The spectral resolution refers to the wavelengths of radiation that can be detected by the instrument and, for example, in the visible and shortwave infrared regions, the information in different spectral bands can provide a spectral 'signature' for different land cover types (e.g., vegetation, soil, water, buildings). Spatial resolution (typically reported as the number of metres covered by one dimension of a single pixel) is a measure of the observable detail in an image, and modern optical sensors provide multispectral and panchromatic (PAN) imagery at much finer spatial resolutions than seen in previous decades [42] (although the chosen resolution for a modern mission is a compromise between resolution and coverage.) Different space agencies, companies, and other organizations categorise the spatial resolution of satellite images differently; thus, 'high resolution' can have a variety of meanings. In general, the categories used for images are classified as shown in Table 3. Satellite data can also be categorised by the type of orbit the satellite is in, whether it is in a high geostationary orbit (so it always observes the same locations on Earth) or in a low Earth orbit (so it tracks the surface as it orbits).

Table 3. Spatial resolution of satellites in EO [43].

Spatial Resolution	Examples of Satellites	Scale
Coarse resolution (>1000 m)	e.g., GCOM, Envisat, Aeolus, etc.	Global and regional
Medium resolution (100 m to 1000 m)	e.g., MODIS, AVHRR, Sentinel-3 OLCI, etc.	Global and regional
Fine resolution (5 m to 100 m)	e.g., Landsat 5–8 TM, Sentinel-2, SPOT 5, DMC etc.,	Regional and local
Very high resolution (VHR) (<5 m)	e.g., Rapid Eye, WorldView, Pleiades, SkySat, SuperView, etc.	Local

Free and open access data play a key role in enabling the discovery, retrieval, and manipulation of data to monitor the planet [44]. For instance, NASA and the U.S. Geological Society (USGS) have provided free Landsat imagery since 2008 [45]; ESA has released free data (e.g., ERS, Envisat, Meteosat) since 2010; now, all data from the operational Sentinel missions are freely available via the Copernicus Programme [46].

Open data can be freely used, re-used, and redistributed by anyone, whereas commercial data, such as VHR imagery from commercial satellites, require the user to purchase the data and to follow a license agreement. Commercial satellite imagery can be expensive depending on the size of the area of interest, spatial resolution and number of observation dates required. Rudd et al. [47] provided an example of using commercial satellite data for precision agriculture, as this application benefited from higher spatial and temporal resolutions than space agency sensors can provide; however, they considered these data expensive for their application. Comparisons of the cost-effectiveness of EO-derived data versus other sources of data are still relatively rare in the literature; nevertheless, examples are provided by [48–52].

The findings of these studies suggest that using EO-derived data can be more cost advantageous compared with the use of other data; however, the savings depend on the scale of the project and what other types of data are available. For example, Bruzelius et al. [52] conducted a cost-effectiveness analysis for their approach in detecting the health-care service buildings for an area in south-eastern Liberia; the computational and data cost involved was USD 12.30; if the method was applied for the entire country, the cost would be approximately USD 141. These low prices show that EO-derived data can provide information at a substantially reduced cost compared to the traditional monitoring costs, which determines SDG indicators accessible for all nations, thus providing an opportunity for *"leaving no one behind"* [51].

There are numerous web portals and services that enable the discovery, access, and use of EO data and derived information products. These can be classified into three main categories:

- EO data portals
- EO processing, visualisation, and cloud computing platforms (where the term 'platform' in this paper refers to those online resources layer which provides the 'back-end' functionality of EO satellite images, rather than the engineering use of "platform" used to contrast the "satellite" with its "sensor")
- EO derived thematic products and services [53,54]

There are around 25 EO data portals that provide free (to the user) and open satellite data at medium spatial resolution (e.g., Sentinel and Landsat) and historical VHR data (these are listed in Supplementary Material Table S2) [55–80]. Free and open access EO data are generally provided by taxpayer-funded national and international space agencies and are provided for the public good. They play a major role in expanding the spectrum of new users and applications [81]. There are numerous commercial suppliers (e.g., PlanetLab, Maxar Technologies, Iceye, Earth-I, EarthBlox, Pixera, Surrey Satellite Technologies, etc.) that offer EO satellite data and/or products for a cost or a subscription plan tailored to the clients' needs.

EO processing and visualisation platforms provide images that have passed through several 'levels' of processing, integration, and analysis of raw EO data. Corrections at these levels include radiometric calibration, atmospheric correction, and the derivation of products from the satellite data (such as land classification maps); these EO-based products are often further processed and combined with other geo-referenced socio-demographic, economic and environmental data. The new cloud-based infrastructure services are becoming more important as they enable users to access, store, and analyse large volumes of EO data without having to download (and in some cases process) the raw data [53,81]. These cloud platforms provide 'analysis ready data' (ARD) which are normally geometric, orthorectified and radiometrically calibrated [24]. They can include integrated technologies such as application programming interfaces (API) and web services to provide a more complete solution for big EO data management and analysis. Currently, there are around 14 major platforms that include big EO data management and analysis which offer mainly free data access with commercial infrastructure, but use different storage systems, and access interfaces of large EO data sets (listed in Supplementary Material Table S3) [53]. Among these platforms, Google Earth Engine (GEE) combines a multi-petabyte catalogue of satellite imagery and geospatial datasets with planetary-scale analytical capabilities and is available for scientists, researchers and developers to detect changes, map trends, and to quantify differences on the Earth's surface [28]. However, one barrier to the use of such platforms is the need for computing skills and facilities to download and manipulate large datasets.

In terms of EO-derived thematic products and services, ESA, NASA, and other partners have created dedicated portals with advanced visualisation tools and services where users can directly access spectral indices such as NDVI, as well as products derived using advanced and complex algorithms that assess land degradation, land cover change, water quality, etc. These free-to-the-user higher-level products require less expertise to manipulate the EO data and generate products but are more susceptible to data misinterpretation as a result of accessibility with a lower expertise level. These products provide quantitative data to develop a baseline, assess trends, and address SDG indicators (a selected list of these thematic products is presented in Table S4) [82–88].

Such portals and platforms integrated with big data provide several benefits, as they provide a consistent, standardised product that has been developed by experts, that can be used by a wide range of users without those users needing an expert understanding (of the instruments) but lead to greater misinterpretation with assumptions made about the sensor performance [89,90]. One of the first systematic data cubes to be established was the Australian Geoscience Data Cube [91], which provides standardised data in a common format on a range of processing platforms and with common pre-processing. These analysis-ready data initiatives have great potential for NSOs; however, there has not yet been the investment to establish how best to use the data at the national level [24]. There can also be a misconception that such data can be considered reliable simply due to the volume of data available; however, this is not necessarily the case [24]. Efforts are being made through satellite communities such as CEOS CARD4L [92] to provide some consistent validation and evaluation of such products, although those initiatives are still at an early stage and quality statements should be cautiously interpreted.

It is critical for users of EO information to understand and consider the key characteristics (spectral, temporal, and spatial resolutions, and calibration uncertainties) of the satellites and their applications to determine which are the most suitable to address specific needs [37,93,94]. Some applications require data over relatively short periods such as monitoring oil spills [95], forest fires [96], and sea ice motion [97], while, for understanding long term trends of environmental and climatological issues, long time series of images, and multi-decadal stability of the data set, which implies radiometric accuracy, are essential [97]. Other applications are in between, needing seasonal imaging especially for crop identification [98], phenology [99], and wetland monitoring [100]. Table 4 illustrates a representative range of EO applications with strong relevance to various aspects of sustainable development, although it should be noted that this is by no means an exhaustive list.

Table 4. Example EO satellite applications with relevance to the SDGs.

Main Application	Classification	Example EO Applications
Environmental	Terrestrial	Land use and cover change [85] Biodiversity and habitat assessment [101,102] Inland water resources [103] Forestry- Deforestation and Afforestation [104] Reforestation [99] Forest fire [105] Agriculture (crop mapping) [98] Hydrology [106] Mining and mineral exploitation [107] Species growth (phenology) [108] Soil Moisture [109] Soil organic carbon [110] Snow cover and glaciers [111] Species identification [112]
	Ocean	Sea-surface temperatures [113] Ocean colour and algae blooming [114] Sea levels [115] Algae blooming [116] Floating plastics [117] Sea-Ice [97] Marine and coastal environments [118] Species identification [119]
	Atmospheric	Weather forecasting [120] Radiation, evapotranspiration [121]
	Climate system	Air quality and greenhouses gases [122] Carbon Dioxide [123] Ozone [124] Nitrogen dioxide [125] Methane [126] Nitrogen dioxide [127] Sulphur dioxide [128]
Socio-economic	Disaster management	Oil spill [95] Disaster risk and damage assessment [129] Geohazard risks [130] Flooding [131]
	Corruption	Illegal logging [132] Favouritism [133] Illegal fisheries [134] Inflated GDP [135]
	Transport	Ship tracking [136] Transportation infrastructure [137] Smart transport and logistics [138]
	Socio-Economic development	Global population density [139] Quality of life [140] Ethnic minorities development index [141] Poverty [142] Economic growth and GDP estimation socio-economic activities [143] Urbanisation dynamics [144] Regional inequality [145] Urban planning [146]
	Energy	Electricity consumption [147] Access to electricity [148]
	Humanitarian	Human rights [149] Natural disasters, structural damage assessment, and population estimation in settlements in conflict [150] Forced labor [151]
	Health	Incidence of breast cancer [152] Response for Vector-Borne Diseases [153] Access to health services [52]

Several peer-reviewed articles and case studies published since 2017 have explored the contribution of satellite imagery towards the SDG indicators and targets [14,15,27,35,36, 90,154–158]. Some examples are provided in Table 5.

Table 5. Published assessments of EO data use in support of SDG indicators and targets.

Title	Type	Year	EO Data Contribution on SDG Indicators
Earth observation in service of the 2030 Agenda for Sustainable Development [154]	Peer review	2017	First review of the EO contribution for SDGs, focusing on the role of GEO and GEOSS work and the actual use of EO in support of the SDGs. Presents the major GEO projects that address indicators from SDG 2, SDG 6, SDG 15. Also, it discussed the importance of capacity building, data access, and global collaboration with NSOs and custodian agencies.
Satellite Earth Observation in support of the Sustainable Development Goals [36] EO4SDGs Initiative [27]	Report	2018	The report presented best practices of using EO data for several indicators that cover SDG 2, SDG 6, SDG 11, SDG 14, and SDG 15.
		2019	EO4SDGs Initiative is launching a series of pilot projects to apply and test uses of EO to support the assessment and tracking of the SDGs, including integration with national statistical accounts for the indicators.
Maturity Matrix Framework (MMF) 1.0 [14]	Framework	2018	MMF 1.0 is an analytical framework that is based on 3 premises (methods of processing EO data, requirement of non-EO information, and level of completeness). A total of 80 peer reviews and reports were systematically reviewed under MMF 1.0 to seek the potential of EO data to support SDG indicators. The MMF 1.0 was applied to all SDG indicators and found 84 indicators that were classified into three categories: weak support from EO, indirect measure by EO data and high potential of EO data directly to populate the SDG indicator.
Maturity Matrix Framework (MMF) 2.0 [15]	Framework	2019	MMF 2.0 is similar to the previous framework, but it evolved into 6 premises after expert interviews were conducted with 38 specialists in EO and sustainability.
A Review of the Sustainability Concept and the State of SDG Monitoring Using Remote Sensing [157]	Review	2020	The scope of this review was to summarise the work and best practices of using EO for supporting the SDG indicators.
Compendium of Earth Observation contributions to the SDG Targets and Indicators [156]	Compendium/Framework	2020	ESA Compendium presents case studies of using EO satellite-derived data for a total of 34 indicators which were assessed against a framework that discriminates the indicators into two categories; 17 indicators can be directly measured, and 17 indicators indirectly informed by EO data across 29 targets and 11 goals.
EnviroAtlas [35]	Review	2020	EnviroAtlas is a web-based, interactive map of environmental and socio-economic data relevant to the SDGs. It provides multi-resolution spatial indicators and introduced proxy indicators related to the SDGs. The paper presents multiple examples of using EO for 47 targets across most goals.
Maxar company (formerly known as DigitalGlobe) [158]	Review	2021	The report summarised the case studies that address all SDGs.

GEO and its space-agency arm CEOS has set out how EO can support the SDG indicators through their EO4SDG initiative [154]; more recently, they have published a report detailing the achievements of the programme, examples of application and further steps [155]. In particular, they noted successes in collaboration with UNEP in assessing SDG Indicators 6.6.1 (spatial extent of water-related ecosystems) and 6.3.2 (ambient water quality), with UNCCD on Indicator 15.3.1 (proportion of degraded land per total land), and with the UN Habitat on Indicator 11.7.1 (average share of the built-up area of cities that is open space for public use) which enabled these four indicators to be recognised as conceptually clear with internationally accepted methodologies. GEO suggests that EO data have a role to play in monitoring progress with 71 out of the 169 Targets and 30 out of the 232 indicators either directly or indirectly [154,155].

More recently, as part of the GEO EO4SDG initiative, in collaboration with the ESA, an analysis has been completed regarding the contribution of EO for the Global Indicator Framework [156]. In this analysis, the authors sought to identify those indicators where EO could reasonably contribute to the development or implementation of indicator methodology and based this on two main criteria which they refer to as "readiness" and "adequacy". A "traffic light" system of red, amber, and green colours was then applied across a suite of 34 selected indicators to flag EO relevance.

A further example of an analysis of the potential of EO for the SDG indicators is provided by EnviroAtlas, developed by the United States Environmental Protection Agency and other partners. EnviroAtlas provides a collection of web-based, interactive maps of environmental and socio-economic data that support numerous SDG targets and indicators that cover goals 6, 11, 15 and 4.

The increasing trend of launching constellations of nano and microsatellites (operates between 60 and 100 small satellites) into space has been clear and well established in the last two decades, aiming for higher spatial resolution and near real-time measurements and continuous monitoring due to the daily revisit times [159]. In addition, there is an increasing willingness to a partnership between the EO companies and custodian agencies exploring the use of EO data to address SDGs. For instance, Planet Labs is working with UN agencies, Kongsberg Satellite Services (KSAT) and Airbus to help measure the SDG indicators by offering free access to VHR satellite, (<5 m) for 64 countries involved in monitoring tropical rainforests. This partnership also helped support an interactive platform called "Global Forest Watch", (an open-source web application that monitors global forests in near real-time) [160].

Another example of an EO company engaging with the SDGs is the Maxar company [158]. Engstrom et al. [161] investigated whether high-resolution satellite imagery from Maxar Technology can accurately and affordably estimate the economic well-being of 1291 Sri Lankan villages from space using machine learning and VHR imagery focused on proxy indicators, such as car counts, building density, green space, and more, which have all been used in models to predict the spatial variability of poverty across urban areas. Maxar VHR satellite data have also been used to map areas at high risk for malaria, based on population density and proximity to standing water. This has helped health workers determine the required number of life-saving mosquito nets and the exact locations to use insecticides [162]. Another example is the use of EO and mobile phone data to help understand the disparity in transportation use by gender and the role it plays in inequality (SDG 5) in Santiago, Chile. The GovLab and its partners DigitalGlobe, UNICEF together with other organisations, established the first-ever baseline study of the urban mobility experiences of women and girls, in which VHR satellite data play a role in human mobility patterns [163–165].

Among many case studies that address the more environmental SDGs, ESA presented an example using satellite data and algorithms targeted at detecting illegal waste sites [36,166]. They identified 207 sites classified as suspicious across one test area of 7000 km^2 in the UK [167]. These sites are often associated with lower income and high-rate crime areas and, thus, can help address indicators within SDG targets 12.4, 6.3 and 11.6 [36].

4. Assessing EO Contribution to SDGs Indicators Using a Maturity Matrix Framework (MMF) Approach

In 2017 and 2019, Andries et al. [14,15] presented analyses of published literature and other reports on the applicability of EO satellite data for SDG indicators using a structured MMF based on six premises (uncertainty assessment, directness, completeness, requirement for non-EO information, practicability, and cost-effectiveness analysis). The indicators were allocated an aggregate assessment score (entitled Maturity Matrix Score (MMS)) based on the premises detailed in [15] and their overall potential for being populated with data from EO sources. Two versions of the MMF framework were produced and, using the latest version (MMF 2.0), applied to all 232 SDG indicators some 80 of them were determined to have some potential for being populated with data by EO. Of these, EO was found to provide weak support (MMS between 1 and 2) for 25 indicators, partial support (MMS values between 2 and 4) for 40 and strong support (MMS values between 4 and 5) for 15 indicators. The remaining 152 indicators had no evidence that EO could be used (at the time of the research).

The SDG indicators have not been formally classified in terms of the three pillars of sustainability (economic, social, and environmental) by the IAEG-SDGs; there have been, perhaps surprisingly, few attempts to do this reported in the literature. Some exceptions are provided by Paoli and Addeo [168] on the SDGs, Tremblay et al. [169] for the SDG targets and Cochran et al. [35] for the SDG indicators. Such a classification of indicators can be useful from a policy point of view [170] and can help to identify priorities for early action, understand implementation challenges, detect data gaps and indicate the trade-offs among SDGs [168].

A widely used taxonomy of indicators is based on the three pillars of sustainability that includes environmental, economic and social indicators (sometimes institutions are listed as a fourth dimension) [171]. Social indicators deal with broad categorical concerns of well-being (e.g., OECD's survey "How is life?" [172]), values (e.g., human values [173]), agency (demographic and health surveys [174] and inequality (often calculated via Gini coefficients) [175]). Economic indicators are statistics that allow an analysis of the economic performance of a country and are classified into two main categories: economic structure (e.g., gross domestic product (GDP), investment share in GDP, the balance of trade in goods and services) and consumption and production patterns (e.g., consumer price index (CPI), producer price index (PPI), household consumption, etc.) [171]. Environmental indicators are based on the four main Earth systems: atmosphere, hydrosphere, lithosphere, biosphere. Some environmental indicators are based on the human (anthropogenic) impact on the four main Earth spheres. The processes of human interaction and the Earth-system response to that interaction, such as pollution flows and natural resources stock (e.g., greenhouse gas emissions—CO_2, SO_x, NO_x, methane), use of water resources, river quality, wastewater treatment, land-use changes, protected areas, use of nitrogenous fertilizers, use of forest resources, trade in tropical wood, threatened species, fish catches, waste generation, municipal waste, industrial accidents). Lastly, partnership indicators measure the effectiveness of cooperation between two or more organisations [176].

The authors classified the SDG indicators into five main categories, as follows: social, economic, environmental, the comprehensive indicators—socio-economic and socio-environmental—and partnership; the results are presented in Figure 2. The assessment resulted in 94 social indicators that mainly related to social inclusion and inequality reduction (SDG 5 and SDG 10), poverty eradication (SDG 1), quality education (SDG 4), sustaining good health and well-being (SDG 3), and promoting peaceful and inclusive societies (SDG 16). There are 14 economic indicators mostly represented by GDP (in its various forms) that can be found in SDG 2, SDG 8, and SDG 10. There are 27 SDG environmental indicators that encompass issues of water scarcity, water resource management (SDG 6), climate change (SDG 13), loss and degradation of biodiversity and ecosystems services (SDG 6, SDG 13, SDG 14, SDG 15), deforestation (SDG 15). Most partnership indicators (14 out of 16) occur in SDG 17, as the goal focuses on progress made in strengthening global partnership and in improving financial governance. The last category, comprehensive SDG indicators, cover 96 indicators, of which 30 have socio-environmental features and 66 include socio-economic dimensions of sustainability. Therefore, the SDG indicator framework has an imbalance across the three pillars of sustainability and is mainly focused on the socio-economic aspects with only 27 environmental and 30 socio-environmental indicators.

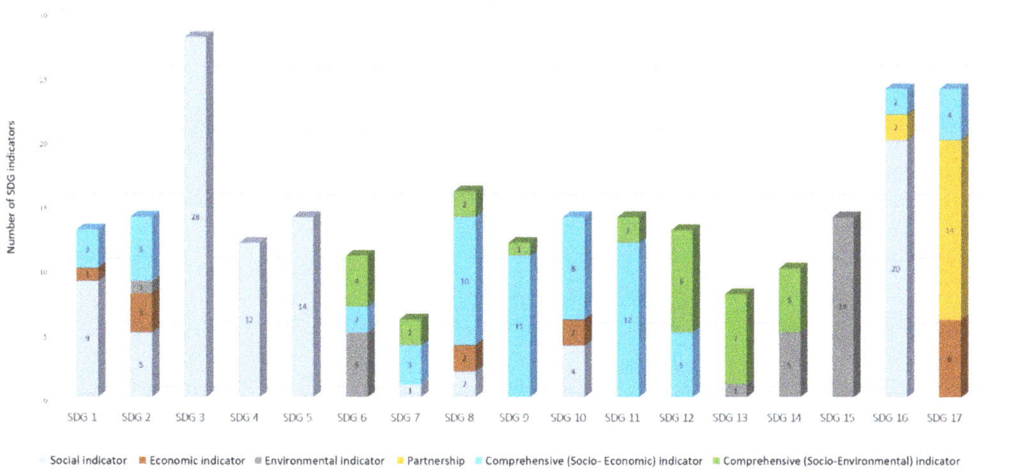

Figure 2. SDG Indicator type based on the three dimensions of sustainability (Authors' assessment).

Based on the MMF 2.0 [15] summarised above, we have re-evaluated all SDG indicators using the latest best practice case studies and approaches available in the literature and reports up to mid-2021 and have applied the categories presented in Figure 2. The results (see Figure S1) suggest that 108 SDG indicators could be populated, at least in part, by using data from EO. These comprise 19 indicators with weak support from EO, 67 with partial support from EO, and 22 with strong EO support. The remaining 139 indicators had no evidence (yet) (see Figure 3) that EO data could be used for their support; however, this is a rapidly evolving field in which future approaches to using EO data for social indicators may be developed.

These results have been sorted further in Figure 3, based on the indicator types of Figure 2. The findings suggest that a strong contribution of EO data is predominantly within the environmental type indicators; however, that was not exclusive, as several social-economic indicators (e.g., indicators 1.4.2, 1.5.1, 1.5.2., 2.1.1, 11.5.1, 11.6.2) can be also measured directly by EO data. For instance, indicator 1.5.2. (direct economic loss attributed to disasters in relation to global GDP) can be directly measured by EO datasets; Pham et al. [177] have shown an approach of automatic detection of building damage of an earthquake, and EO optical and radar imagery can be used to gather useful information on the economic loss on other disasters (e.g., flooding, transport infrastructure, fires, landslides, etc.) at different stages. Satellite data of very high spatial and temporal resolutions play a critical role in disaster management.

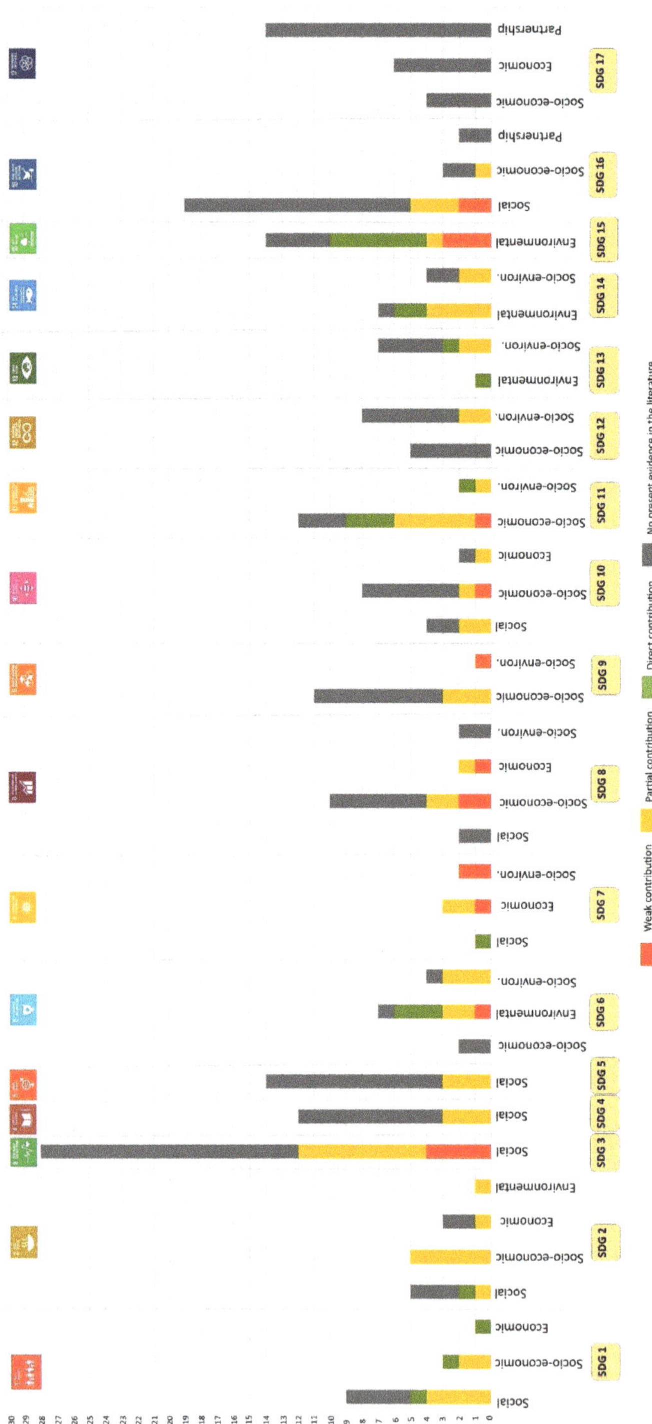

Figure 3. SDG indicators type and the EO satellite contribution (based on MMF 2.0) (Authors' assessment).

Moreover, indicators with partial support from EO are spread amongst all indicator types, reflecting the potential of EO satellite-derived data to play a role in calculating the SDG indicator via a proxy indicator. For example, SDG indicators 16.1.1 (number of victims of intentional homicide per 100,000 population, by sex and age) cannot be measured directly by satellite data. However, Patino et al. [178] investigated the influence of the urban layout (e.g., impervious surface percentage, type of roofs, land cover types) on homicide rates using VHR image and integrated census data for socio-economic variables; thus, they found associations of higher homicide rates with more heterogeneous and disordered urban layouts. Likewise, EO data have been used to monitor conflicts (bombing, military presence, war) by using historical records of the areas that changed over time and humanitarian efforts, for example, by spotting refugee camps and estimating the number of refugees from space [149]; all these applications can be linked to SDG indicator 16.1.2 (conflict-related deaths per 100,000 population, by sex, age and cause).

A weak contribution of EO data to supporting the indicators occurs primarily among indicators within the more socio-economic category. By its nature, SDG indicator 8.3.1 (proportion of informal employment in total employment, by sector and sex) cannot be supported directly by EO data; nevertheless, Ghosh et al. [179] explored the potential for estimating the formal and informal economy for Mexico using known relationships between the spatial patterns of nighttime satellite imagery and economic activity. They developed regression models between spatial patterns of nighttime imagery and adjusted official gross state product (AGSP) for the USA states. These regression parameters derived from the regression models of the USA were 'blindly' applied to Mexico to estimate the estimated gross state income (EGSI) at the sub-national level and the estimated gross domestic income (EGDI) at the national level. Comparison of the EGDI estimate of Mexico against the official gross national income (GNI) estimates suggested that the magnitude of Mexico's informal economy and the inflow of remittances are 150% larger than their existing official estimates in the GNI.

Comparison of the MMF results from 2019 to 2021 (see Supplementary Material Figures S1 and S2) reveals 30 new indicators over these approx. 2 years where there is published evidence that EO can provide some support. This suggests that researchers are increasingly finding ways to use EO data in social and economic indicators. It also reflects the amendments agreed by IAEG-SDG in December 2020 that introduced new indicators and revised or removed others.

Figure 4 illustrates these 30 indicators with new EO-based methods, categorised into their type and role of the EO contributions. Many of the social indicators that have new EO-based methods are related to education. For instance, in one of the new approaches, Yazdani et al. [180] noted that EO data can provide appropriate reliability, accuracy, and convenience for identifying rural schools in Liberia via machine learning approaches; based on their approach, UNICEF [181] launched an initiative to map every school in the world, thus contributing to achieving SDG 4 (quality education). In addition, the usage of night-time satellite data can help monitor aspects of education facilities, such as access to electricity [182,183] and drinking water [184], which are closely linked to SDG 4.a.1 indicator (schools with basic service as electricity and internet). Furthermore, Andries et al. [185] used VHR satellite data and enrolment data to calculate the primary school classroom area per pupil in rural Nigeria. They found that over 70% of the 1900 schools evaluated could be classified as overcrowded (according to Nigerian government metrics), this being highly correlated with poverty and literacy rates. Andries et al. [185] also analysed their area per pupil result via the MMF approach, which characterised this EO support as partial for the SDG indicators 4.1.2. (Completion rate in primary education, lower secondary education, upper secondary education) and 4.6.1 (Proportion of population in each age group achieving at least a fixed level of proficiency in functional (a) literacy and (b) numeracy skills, by sex). While area per pupil is not an SDG indicator, it potentially provides a useful contribution since overcrowded schools influence SDG indicators, such as completion of primary education and absenteeism, and can lead to low literacy and numeracy rates. The study is a proof of concept that was further validated in situ by Andries et al. [186].

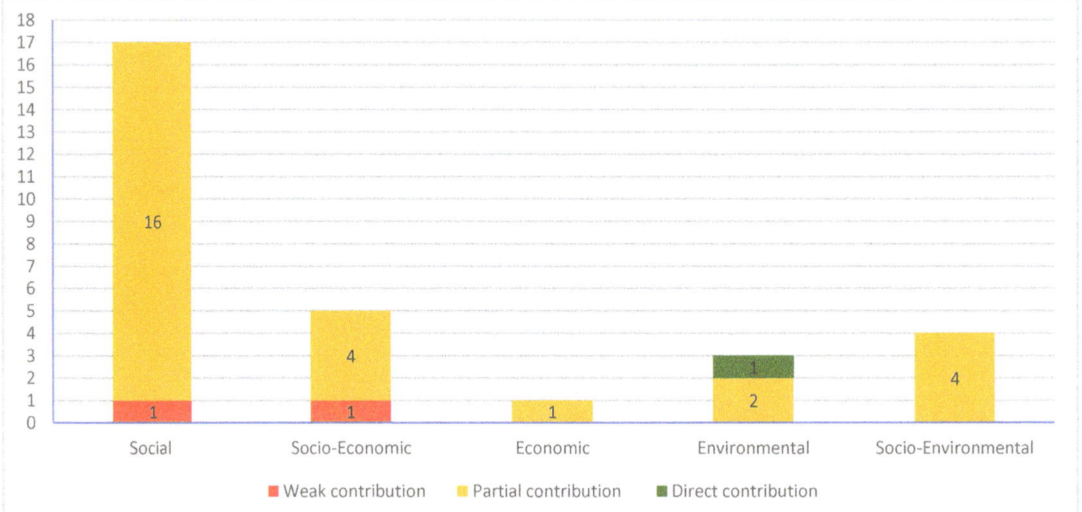

Figure 4. EO contributions for SDG indicators against the MMF 2.0 and the latest publications (Authors' assessment).

The good health and well-being goal (SDG 3) is another social goal where new methods have been derived to evaluate indicators using EO data [187–189]. For example, Bruzelius et al. [52] used EO images and machine learning to assess health services in remote villages, thus providing information that partially contributes to indicator 3.8.1 (coverage of essential health services).

The latest developments in satellites continue to help to support the more environmental-based SDG indicators. One of the latest SDG indicators introduced at the IAEG-SDG December 2020 update [5] is 13.2.2 (total greenhouse gas emissions per year). EO data can help validate or complement this indicator through the wide availability of satellites; however, EO is not currently discussed in the methodological guidelines [190]. Satellites (e.g., Gases Observing Satellite (GOSAT), Orbiting Carbon Observatory-2 (OCO-2), TanSat) have been used to measure atmospheric column-averaged concentrations of the key greenhouse gases CO_2 and CH_4 at the global level. These can be used to complement the current methods, using flux inversions to provide estimates of natural fluxes of CO_2 and CH_4. ESA's Greenhouse Gases Climate Change Initiative [191] has created consistent and quality-assured products with rigorous uncertainty analysis based on these satellites, which are available via the Copernicus Climate Data Store [192]. More recent public (e.g., Sentinel 5P) [193,194] and private (e.g., GHGSat) [126,195] missions provide more detailed spatial information and can begin the process of determining emission fluxes (rather than concentrations of mixed gases) for the larger emissions (city air pollution and oil and gas fugitive emissions) [196,197]. Therefore, the current data from these satellites can be translated into actionable information to achieve the goals of COP 26 and SDG 13 (climate action) [198]. Furthermore, there are new missions in development that will further increase our understanding and direct monitoring of GHG emissions (e.g., Microcarb). An area of current research is the exploration of the potential of other medium and high resolution platforms, not initially designed for atmospheric retrievals, such as Sentinel 2, PRISMA, WorldView-3, to fill the spatial and temporal gaps in mapping methane plumes from point emitters [197].

As can be seen in Supplementary Material Figure S2, 11 indicators that had weak support from EO, have 'upgraded' to providing partial or strong support from EO due to new publications with new EO-based methods. For instance, indicator 7.1.1 (proportion of population with access to electricity) had an MMS of 2.5 (amber) in the original MMF 2.0 dashboard analysis, based on the project India Lights [199]. This project was a collaboration between Development Seed, the World Bank and the University of Michigan that used the DMSP images to extract the light output of individual villages to show those that had access to electricity. However, in this latest review, we increased this indicator's MMS score from 2.5 (amber) to 4 (green) based on a joint project (by Facebook, The Energy Sector Management Assistance Program (ESMAP) at the World Bank, the KTH Royal Institute of Technology, the World Resources Institute (WRI), and the University of Massachusetts Amherst) that developed a predictive model for mapping medium-voltage distribution using night-time satellite data, a MODIS land cover dataset and geospatial products (e.g., different type of roads, railways, political boundaries). The increased score came from the new work on assessing uncertainty (with more validation against ground-truth data) and improved practicability premises (with the method now integrated into the end-user's decision-making in several countries [200] with the possibility to replicate the model (using the model documentation and codes) to other countries [201].

5. Discussion

The SDG framework, with its targets and indicators, stands as the most comprehensive, universal, and ambitious plan to end poverty, promote prosperity and people's well-being and protect the environment, and has been agreed to by 193 world leaders spanning the developed and developing worlds. The achievement of the SDGs relies prominently on the availability and use of relevant, reliable, and timely data to measure progress, inform policies and target those areas that need improvement. Traditional data sources have administrative and technical limitations, in terms of data access, standardisation and quality, lack of awareness concerning the benefits of current technologies, lack of financial resources, technology and skills gaps, geographical constraints (in terms of developing space projects), and coverage gaps across space and time [202]. Indeed, surveys suggest that 73% of countries require assistance to upgrade existing data sources [203]. Thus, the world is currently not on track in terms of being able to populate the SDG indicators; only 44% of the SDG indicators have sufficient data for global monitoring. However, the advent of new EO technologies, either by themselves or combined with traditional data sources and methods, offers the opportunity to provide robust and timely data on a routine basis for monitoring and decision-making that supports the achievement of the SDGs.

Given a large number of satellites/instruments and their capabilities (spatial, spectral and temporal resolution) (Table S1), free and open access portals (Table S2), data cubes (Table S3), and EO thematic platforms (Table S4), there is potential to monitor every place on Earth. EO data can help save time and cost [204,205] compared to the generally more labour-intensive traditional approaches. There are some common misconceptions related to the economic cost of VHR [206,207]; performing cost-effectiveness analysis can help identify the most effective and competitive data [208,209]. For instance, Watmough et al. [51] examined the role of using satellite data to assess the household wealth by considering the spatial landscape use, size of the building, agricultural productivity, land use and cover surrounding the household as predictor variables. They also estimated the price for traditional and EO analyses. Surveying 330 households in rural areas of western Kenya would cost USD 106,500 per year, compared with USD 1750 to USD 5000 per year for 100 km^2 (covering the same 330 households) using VHR satellite data. They demonstrated that satellite data can predict household wealth with 62% accuracy compared with the traditional sampling methods and suggest that EO data be used to understand the dynamics of changes in wealth between less frequent household surveys.

Nevertheless, there are capacity issues within NSOs that need to be addressed to enable access to and use of EO data. These issues are not only about providing training to NSO staff, but also about the strategic direction in the political agreement process and cooperation between geospatial and EO communities, organisations, ministries, and stakeholders. Likewise, there is a need for investment in the official statistics community in terms of EO data manipulation, interpretation, and digitalisation.

This issue can be resolved by increasing technical capacity and infrastructure, but also via regulations designed to build and consolidate trust among different stakeholders involved in data sharing. Since 2015, GEO and its initiative EO4SDG, custodian agencies and academia increased their efforts to explore the potential of EO and geospatial data to support SDG indicators and targets; thus, in the last couple of years, there have been several reviews that assess the EO contributions for SDG indicators under various frameworks and analyses alongside practical guidance [14,15,27,35,36,90,154–158]. Despite these developments, some countries are reluctant adopters for several reasons, such as concerns about satellite data "spying" and confidentially, or because of a lack of data standardisation and/or because they do not yet have the technical and infrastructure capacity in using EO data [210]. Notwithstanding such reluctance, there are several examples of fruitful collaborations between country NSOs and geospatial agencies, such as the case of Ireland, Sweden, South Africa, Colombia, the Philippines, and the UK that integrated EO and/or geospatial data for indicators that cover SDG 6, 11, 13 and 15.

In this paper, the framework developed by Andries et al. [15] was applied to the most recent publications (up to June 2021), updating a similar analysis presented in the earlier paper. Interestingly, rapid progress in terms of the EO data usage for supporting the SDG indicators has occurred in the last 3 years, revealing an additional 30 SDG indicators that can be measured by EO data (mostly through an indirect approach), with 26 of these indicators having socio-economic features. This rapid progress in the most recent years is likely driven by the increased availability of open source data, the collaboration between various organisations, including the commercial sector, academia and GEO and other organisations that continue to provide efforts promoting this type of data at the Statistical Geospatial Integration Forum [211] and other relevant events [212].

However, there remain 139 SDG indicators for which there is currently no EO support. When classifying the SDG indicators based on their sustainability pillars in the present research, we observed that 174 (71%) of the SDG indicators are strongly oriented towards social or economic pillars, which implies some barriers in the use of EO data for indicators in those categories.

Likewise, we considered it valuable to classify the SDG indicators with regard to the well-established Driver-Pressure-State-Impact-Response (DPSIR) analysis [213–215] (a framework that describes the interactions between society, environment, and socio-economic outcomes) as a further perspective on the components of the SDG that can be monitored using EO data. This classification has been performed by Andries et al. [216] and Masó et al. [217], and both found an uneven distribution of the SDG indicators into the five DPSIR components, with a large percentage of response and state indicators (32% and 41% respectively), followed by impact (18%), driving forces (6%) and 3% for pressure indicators.

Response indicators cannot be measured directly by EO and can only rarely be assessed via proxy indicators. As an example, consider the response indicator SDG indicator 14.6.1 (degree of implementation of international instruments aiming to combat illegal, unreported and unregulated fishing). This indicator monitors the human action (the response), i.e., whether countries are enforcing international agreements about illegal fishing through management strategies, regulations and policies. EO data can support the overall aims behind this indicator through an indirect approach, as satellites sensors can monitor maritime traffic at night using ship lighting and, thus, implicitly observe fishing activities [136,218,219]. When illegal fishing is spotted from space, as Park et al. [209] have shown in their study, this would reflect poor governance and legislation in those countries (thus this is an indirect observation of the response indicator).

Earth observation (EO) could have the highest contributions (based on the wide variety of satellite sensors and capabilities) on the environmental SDG 15 (life on land). However, even here, (as shown in Figure S1) four indicators, 15.6.1 (number of countries that have adopted legislative, administrative and policy frameworks to ensure fair and equitable sharing of benefits), 15.8.1 (proportion of countries adopting relevant national legislation and adequately resourcing the prevention or control of invasive alien species), 15.a.1 ((a) official development assistance on conservation and sustainable use of biodiversity, and (b) revenue generated and finance mobilized from biodiversity-relevant economic instruments) and 15.b.1 (same as 15.a.1), cannot be supported by EO data due to their nature of the human decisions to the changes. On the other hand, pressure indicators on the environment such as wildfires, illegal logging, pollution, land-use change, light pollution, urbanisation, are readily measurable by EO; however, they are not covered in the SDG framework.

Therefore, the ability of EO data to support the existing SDG indicators is affected by the socioeconomic nature of the SDGs framework and by the conception and formulation of the indicators. At the same time, EO data can bring to the SDGs framework new opportunities which have not been yet explored and new indicators not yet listed in the formal system, but which could have applicability to help address the SDG targets. As Bell and Morse [220] have highlighted, an indicator must be measurable in the sense that data must be available at the required quality, spatial relevance and timeliness; such data must also be cost-effective. In this vein, Masó et al. [217] and Cochran et al. [35] have proposed new indicators that are deliberately EO based and that can complement the existing indicators, providing valuable data that is currently missing.

The UN and custodian agencies, GEO, researcher communities, and other stakeholders should take the opportunity to work together to create new or adapted indicators that optimise the use of the available data such as from EO. Much needs to be done; however, the potential of EO-derived data for populating the SDG indicators is here and is growing. Further research, exploration and examples of applications will help to expand this frontier of possibilities.

6. Conclusions

This paper presents the data challenges for monitoring the SDGs and how EO satellite data have been used by different communities to support the SDGs and their indicators. It has analysed the contribution of EO data from assessments carried out by various organisations and the classification of the SDG indicators based on the three pillars of sustainability. We have drawn the following main conclusions from this:

- To achieve the SDGs, inform policies and investment decisions, high-quality data (relevant, timely, reliable, and internationally comparable) for measurement and validation are needed.
- Indicators require good quality data: the lack of data and/or outdated data, particularly in developing countries, is a major constraint for monitoring the SDGs.
- Traditional data can be expensive and come with other disadvantages: new technologies, including satellite EO data and methods, can play a major role in the provision of data for many SDG indicators.
- The plethora of satellites now observing the Earth provides data at different resolutions and capabilities that have been used in many applications related to sustainable development, including in populating the SDG indicators.
- The framework of SDG indicators has an imbalance between the three pillars of sustainability and is mainly focused on the socio-economic aspects of sustainability, with only 27 environmental and 30 socio-environmental indicators out of its total of 231 unique indicators. The framework is also biased towards the "state" and "response" indicators of the DPSIR framework.
- Satellite data can be used as indirect support for many SDG indicators and has greater potential to support the SDGs beyond the existing indicators.

- Despite the efforts of various organisations in developing satellite open access platforms (e.g., data cube, EO portals and visualisation platforms), there are still constraints to using these data and services. Furthermore, NSOs in many/several developed countries are constrained in making use of EO data for populating their SDG indicators.

Supplementary Materials: The following are available online at https://www.mdpi.com/article/10.3390/su14031191/s1, Figure S1: MMF 2.0 Dashboard (version 2019); Figure S2: MMF 2.0 Dashboard (version 2021); Table S1: List of satellites and their characteristics; Table S2: Free satellite imagery platform sources; Table S3: Platforms for big EO data management; Table S4: EO thematic products.

Author Contributions: Conceptualization, A.A., S.M., R.J.M., E.R.W. and J.L.; methodology, A.A., S.M., R.J.M., E.R.W. and J.L.; software, A.A.; validation, A.A., S.M., R.J.M., E.R.W. and J.L.; formal analysis, A.A. and E.R.W.; investigation, A.A.; resources, A.A., S.M., R.J.M., E.R.W. and J.L.; data curation, A.A.; writing—original draft preparation, A.A.; writing—review and editing, A.A., S.M., R.J.M., E.R.W. and J.L.; visualization, A.A. and E.R.W.; supervision, S.M., R.J.M., E.R.W. and J.L.; project administration, S.M.; funding acquisition, S.M., R.J.M., and E.R.W. All authors have read and agreed to the published version of the manuscript.

Funding: This research was funded by Natural Environment Research Council (NERC) SCENARIO Doctoral Training Partnership, Grant/Award NE/L002566/1, CASE award partner the National Physical Laboratory (NPL) and the APC was funded through the University of Surrey.

Institutional Review Board Statement: Not applicable.

Informed Consent Statement: Not applicable.

Data Availability Statement: Not applicable.

Acknowledgments: This research was supported by the SCENARIO Doctoral Training Partnership of the UK Natural Environment Research Council (NERC) with PhD funding for the first author. We thank the National Physical Laboratory (NPL) for co-funding and supervision support as the CASE award partner for this Ph.D. research.

Conflicts of Interest: The authors declare no conflict of interest. The co-funder had a role in the design of the study, in the collection, analyses and interpretation of data, in the writing of the manuscript, and in the decision to publish the results.

References

1. UN. Rio 20+ United Nations Conference on Sustainable Development—The Future We Want. Available online: https://sustainabledevelopment.un.org/content/documents/733FutureWeWant.pdf (accessed on 19 August 2021).
2. UN. A/Res/70/1 Transforming Our World: The 2030 Agenda for Sustainable Development, 21 October 2015. Available online: https://www.un.org/en/development/desa/population/migration/generalassembly/docs/globalcompact/A_RES_70_1_E.pdf (accessed on 19 August 2021).
3. UN. Tier Classification for Global SDG Indicators (updated 29 March 2021). Available online: https://unstats.un.org/sdgs/files/Tier%20Classification%20of%20SDG%20Indicators_29%20Mar%202021_web.pdf (accessed on 18 August 2021).
4. UN. SDG indicator Metadata. Available online: https://unstats.un.org/sdgs/metadata/files/Metadata-01-01-01a.pdf (accessed on 18 August 2021).
5. UN. SDG Tier Classification. Available online: https://unstats.un.org/sdgs/iaeg-sdgs/tier-classification/ (accessed on 18 August 2021).
6. SDSN. Data for Development: A needs Assessment for SDG Monitoring and Statistical Capacity Development. Available online: https://sustainabledevelopment.un.org/content/documents/2017Data-for-Development-Full-Report.pdf (accessed on 17 August 2021).
7. Rajabifard, A. *Sustainable Development Goals Connectivity Dilemma*; CRC Press: Boca Raton, FL, USA, 2019.
8. Avtar, R.; Aggarwal, R.; Kharrazi, A.; Kumar, P.; Kurniawan, T.A. Utilizing geospatial information to implement SDGs and monitor their Progress. *Environ. Monit. Assess.* **2020**, *192*, 35. [CrossRef] [PubMed]
9. UN-GGIM. The Global Statistical Geospatial Framework. Available online: https://ggim.un.org/meetings/GGIM-committee/9th-Session/documents/The_GSGF.pdf (accessed on 17 August 2021).
10. UN-GGIM. The Territorial Dimension in SDG Indicators: Geospatial Data Analysis and its Integration with Statistical Data. Available online: https://un-ggim-europe.org/wp-content/uploads/2019/05/UN_GGIM_08_05_2019-The-territorial-dimension-in-SDG-indicators-Final.pdf (accessed on 19 August 2021).

11. OECD. Overview: What will it take for data to enable development? In *Development Co-operation Report 2017*; Data for Development, OECD Publishing: Paris, France, 2017. [CrossRef]

12. Bill-Weilandt, A.; Bonino, C.; Diakite, T.; Freitas, L.; Hauray, G. Monitoring Progress Towards the SDGs: The Proliferation of Quantification in International Development Policy and Practice. Available online: https://forccast.hypotheses.org/files/2017/06/PSIA-2016-Monitoring_Progress_Towards_the_SDGs.pdf (accessed on 18 August 2021).

13. Grant, M.J.; Booth, A. A typology of reviews: An analysis of 14 review types and associated methodologies. *Health Info Libr. J.* **2009**, *26*, 91–108. [CrossRef]

14. Andries, A.; Morse, S.; Murphy, R.; Lynch, J.; Woolliams, E.; Fonweban, J. Translation of Earth observation data into sustainable development indicators: An analytical framework. *Sustain. Dev.* **2018**, *27*, 366–376. [CrossRef]

15. Andries, A.; Morse, S.; Murphy, R.J.; Lynch, J.; Woolliams, E.R. Seeing Sustainability from Space: Using Earth Observation Data to Populate the UN Sustainable Development Goal Indicators. *Sustainability* **2019**, *11*, 5062. [CrossRef]

16. Cázarez-Grageda, K.; Zougbede, K. National SDG Review: Data Challenges and Opportunities. Available online: https://paris21.org/sites/default/files/inline-files/National-SDG_Review2019_rz.pdf (accessed on 17 August 2021).

17. Sarvajayakesavalu, S. Addressing challenges of developing countries in implementing five priorities for sustainable development goals. *Ecosyst. Health Sustain.* **2017**, *1*, 1–4. [CrossRef]

18. Lu, Y.; Nakicenovic, N.; Visbeck, M.; Stevance, A.-S. Policy: Five priorities for the UN Sustainable Development Goals. *Nature* **2015**, *520*, 432–433. [CrossRef]

19. Kindornay, S.; Bhattacharya, D.; Higgins, K. Implementing Agenda 2030: Unpacking the Data Revolution at Country Level. Available online: https://www.think-asia.org/handle/11540/6956 (accessed on 18 August 2021).

20. SciDevNet. Africa's 'Sluggish Data Collection Needs a Revolution'. Available online: https://www.scidev.net/global/news/africa-s-sluggish-data-collection-needs-a-revolution/ (accessed on 18 August 2021).

21. Bhattacharya, D.; Khan, T.; Rezbana, U.; Mostaque, L. Moving forward with the SDGs: Implementation Challenges in developing countries. *Cent. Policy Dialogue (CPD). Bangladish Civ. Soc. Think Thank.* **2016**, 1–41. Available online: https://library.fes.de/pdf-files/iez/12673.pdf (accessed on 18 August 2021).

22. SDSN. State of Development Data Funding. Available online: https://opendatawatch.com/wp-content/uploads/2016/09/development-data-funding-2016.pdf (accessed on 18 August 2021).

23. GEO. Responses to the Questionnaire on Uses of Earth Observation Data for SDG Analysis and Reporting by GEO Member Countries. Available online: https://eo4sdg.org/wp-content/uploads/2020/03/Responses-to-Questionnaire-on-the-Uses-of-Earth-Observation-Data-for-SDG-analysis-and-reporting-by-GEO-Member-Countries.pdf (accessed on 18 August 2021).

24. United Nations Satellite Imagery and Geo-Spatial Data Task Team. Earth Observations for Official Statistics Satellite Imagery and Geospatial Data Task Team Report. 2017, p. 170. Available online: https://unstats.un.org/bigdata/task-teams/earth-observation/UNGWG_Satellite_Task_Team_Report_WhiteCover.pdf (accessed on 18 August 2021).

25. Denis, G.; Claverie, A.; Pasco, X.; Darnis, J.-P.; de Maupeou, B.; Lafaye, M.; Morel, E. Towards disruptions in Earth observation? New Earth Observation systems and markets evolution: Possible scenarios and impacts. *Acta Astronaut.* **2017**, *137*, 415–433. [CrossRef]

26. Wood, D.; Stober, K.J. Small Satellites Contribute to the United Nations' Sustainable Development Goals. In Proceedings of the Conference on Small Satellites, Logan, UT, USA, 4–9 August 2018; pp. 1–9.

27. Kavvada, A.; Metternicht, G.; Kerblat, F.; Mudau, N.; Haldorson, M.; Laldaparsad, S.; Friedl, L.; Held, A.; Chuvieco, E. Towards delivering on the Sustainable Development Goals using Earth observations. *Remote Sens. Environ.* **2020**, *247*, 111930. [CrossRef]

28. Gorelick, N.; Hancher, M.; Dixon, M.; Ilyushchenko, S.; Thau, D.; Moore, R. Google Earth Engine: Planetary-scale geospatial analysis for everyone. *Remote Sens. Environ.* **2017**, *202*, 18–27. [CrossRef]

29. PARIS21. *A Road Map for a Country-Led Data Revolution*; OECD Publishing: Paris, France, 2015. [CrossRef]

30. NASA. Earth Science Applied Sciences. Available online: https://appliedsciences.nasa.gov (accessed on 17 August 2021).

31. ESA and the Sustainable Development Goals. Available online: https://www.esa.int/Enabling_Support/Preparing_for_the_Future/Space_for_Earth/ESA_and_the_Sustainable_Development_Goals (accessed on 10 October 2021).

32. Jerven, M. *Benefits and Costs of the Data for Development Targets for the Post-2015 Development Agenda*; Copenhagen Consensus Center: Tewksbury, MA, USA, 2014; p. 14.

33. Jerven, M. How Much Will a Data Revolution in Development Cost? *Forum Dev. Stud.* **2017**, *44*, 31–50. [CrossRef]

34. Demombynes, G.; Sandefur, J. Costing a data revolution. *Cent. Glob. Dev. Work. Pap.* **2014**, *14*, 1–12. [CrossRef]

35. Cochran, F.; Daniel, J.; Jackson, L.; Neale, A. Earth Observation-Based Ecosystem Services Indicators for National and Subnational Reporting of the Sustainable Development Goals. *Remote Sens Env.* **2020**, *244*, 1–111796. [CrossRef]

36. ESA. Satellite Earth Observation in Support of the Sustainable Development Goals. Available online: http://eohandbook.com/sdg/files/CEOS_EOHB_2018_SDG.pdf (accessed on 17 August 2021).

37. Khorram, S.; Frank, H.K.; van der Wiele, C.F.; Stacy, A.C.N. *Introduction: Remote Sensing*; SpringerBriefs in Space Development: Berlin/Heidelberg, Germany, 2012.

38. Freeden, W.; Nashed, M.; Sonar, T. *Handbook on Geomathematics*; Springer: Berlin/Heidelberg, Germany, 2010; Volume 2, p. 1371.

39. World Meteorological Organization-Observing Systems Capability Analysis and Review Tool (OSCAR). Available online: https://www.wmo-sat.info/oscar/satellites (accessed on 17 August 2021).

40. Directory EO Portal Missions Database. Available online: https://directory.eoportal.org/web/eoportal/satellite-missions/a (accessed on 18 August 2021).
41. CEOS Database. Available online: http://Database.Eohandbook.Com/about.Aspx (accessed on 17 August 2021).
42. Kumar, P.S.J.; Huan, T.L. *Earth Science and Remote Sensing Applications*; Springer: Berlin/Heidelberg, Germany, 2018; Volume 43.
43. Liang, S.; Wang, J. Chapter 1-A systematic view of remote sensing. In *Advanced Remote Sensing*, 2nd ed.; Academic Press: Cambridge, MA, USA, 2020; pp. 1–57.
44. Harris, R.; Baumann, I. Open data policies and satellite Earth observation. *Space Policy* **2015**, *32*, 44–53. [CrossRef]
45. Woodcock, C.E.; Allen, R.; Anderson, M.; Belward, A.; Bindschadler, R.; Cohen, W.; Gao, F.; Goward, S.N.; Helder, D.; Helmer, E.; et al. Free Access to Landsat Imagery. *Science* **2008**, *320*, 1011. [CrossRef]
46. Turner, W.; Rondinini, C.; Pettorelli, N.; Mora, B.; Leidner, A.K.; Szantoi, Z.; Buchanan, G.; Dech, S.; Dwyer, J.; Herold, M.; et al. Free and open-access satellite data are key to biodiversity conservation. *Biol. Conserv.* **2015**, *182*, 173–176. [CrossRef]
47. Rudd, J.D.; Roberson, G.T.; Classen, J.J. Application of satellite, unmanned aircraft system, and ground-based sensor data for precision agriculture: A review. In Proceedings of the 2017 ASABE Annual International Meeting, Spokane, WA, USA, 16–19 July 2017; p. 1.
48. Andries, A.; Morse, S.; Murphy, R.J.; Lynch, J.; Mota, B.; Woolliams, E.R. Can Current Earth Observation Technologies Provide Useful Information on Soil Organic Carbon Stocks for Environmental Land Management Policy? *Sustainability* **2021**, *13*, 12074. [CrossRef]
49. Allen, M. Contextual Overview of the use of Remote Sensing data Within CAP Eligibility Inspection and Control. Available online: http://www.niassembly.gov.uk/globalassets/documents/raise/publications/2015/dard/3115.pdf (accessed on 17 August 2021).
50. Sadlier, G.; Flytkjær, R.; Sabri, S.; Robin, N. *Value of Satellite-Derived Earth Observation Capabilities to the UK Government Today and by 2020*; London Economics: London, UK, 2018.
51. Watmough, G.R.; Marcinko, C.L.J.; Sullivan, C.; Tschirhart, K.; Mutuo, P.K.; Palm, C.A.; Svenning, J.C. Socioecologically informed use of remote sensing data to predict rural household poverty. *Proc. Natl. Acad. Sci. USA* **2019**, *116*, 1213–1218. [CrossRef] [PubMed]
52. Bruzelius, E.; Le, M.; Kenny, A.; Downey, J.; Danieletto, M.; Baum, A.; Doupe, P.; Silva, B.; Landrigan, P.J.; Singh, P. Satellite images and machine learning can identify remote communities to facilitate access to health services. *J. Am. Med. Inform. Assoc. JAMIA* **2019**, *26*, 806–812. [CrossRef]
53. Gomes, V.; Queiroz, G.; Ferreira, K. An Overview of Platforms for Big Earth Observation Data Management and Analysis. *Remote Sens.* **2020**, *12*, 1253. [CrossRef]
54. GEF. Earth Observation and the Global Environment Facility (GEF)-Technical Guide. Available online: https://stapgef.org/sites/default/files/2021-04/Earth%20Observation%20and%20the%20GEF%20Technical%20Guide_web.pdf (accessed on 17 August 2021).
55. USGS. Available online: http://earthexplorer.usgs.gov/ (accessed on 18 August 2021).
56. GloVis. Available online: https://glovis.usgs.gov/ (accessed on 18 August 2021).
57. NASA. Earth Data. Available online: https://earthdata.nasa.gov/earth-observation-data (accessed on 18 August 2021).
58. ESA. Available online: https://landsat8portal.eo.esa.int/portal/ (accessed on 18 August 2021).
59. ASTER. Available online: https://ssl.jspacesystems.or.jp/ersdac/GDEM/E/index.html (accessed on 18 August 2021).
60. USGS. Earth Resources Observation and Science (EROS) Centre. Available online: https://www.usgs.gov/centers/eros/science/usgs-eros-archive-products-overview?qt-science_center_objects=0#qt-science_center_objects (accessed on 18 August 2021).
61. Copernicus. Open Access Hub. Available online: https://sentinel.esa.int/web/sentinel/ (accessed on 18 January 2022).
62. SentinelHub. Sentinel Playground. Available online: https://apps.sentinel-hub.com/sentinel-playground (accessed on 18 August 2021).
63. Hub, S. EO Browser. Available online: https://apps.sentinel-hub.com/eo-browser/ (accessed on 18 August 2021).
64. NASA. Worldview. Available online: https://worldview.earthdata.nasa.gov/ (accessed on 18 August 2021).
65. NCDC. National Centers for Environmental Information. Available online: https://www.ncdc.noaa.gov/data-access/quick-links (accessed on 18 August 2021).
66. MapBox. Remote Pixel. Available online: https://search.remotepixel.ca/#3/40/-70.5 (accessed on 18 August 2021).
67. CSA. Earth Observation Data Management System. Available online: https://www.asc-csa.gc.ca/eng/satellites/radarsat/ (accessed on 18 August 2021).
68. INPE. Catálogo de Imagens. Available online: http://www.dgi.inpe.br/CDSR/ (accessed on 18 August 2021).
69. ISRO. Bhuvan. Available online: https://bhuvan.nrsc.gov.in/bhuvan_links.php (accessed on 18 August 2021).
70. NASA. Global Precipitation Measurements. Available online: https://gpm.nasa.gov/data/directory (accessed on 18 August 2021).
71. EUMETSAT. Available online: http://www.eumetsat.int/website/home/Data/DataDelivery/ (accessed on 18 August 2021).
72. NOAA. CLASS. Available online: https://www.avl.class.noaa.gov/saa/products/welcome (accessed on 18 August 2021).
73. Amazon. Earth on AWS. Available online: https://aws.amazon.com/earth/?ncl=h_ls (accessed on 18 August 2021).
74. Google Earth Engine. Available online: https://explorer.earthengine.google.com/#workspace (accessed on 29 July 2021).
75. ZoomEarth. Available online: https://zoom.earth/ (accessed on 18 August 2021).
76. MAXAR. Open Data. Available online: https://www.maxar.com/open-data/ (accessed on 18 August 2021).

77. JAXA. ALOS 3D WORLD. Available online: https://www.eorc.jaxa.jp/ALOS/aw3d30/l_map_v2003.htm (accessed on 18 August 2021).
78. ESRI. Esri ArcGIS Online Image Services. Available online: http://www.arcgis.com/home/gallery.html (accessed on 18 August 2021).
79. NERC. CEDA Archive (Part of NERC Environmental Data Service). Available online: http://archive.ceda.ac.uk/ (accessed on 18 August 2021).
80. NASA. Alaska Satellite Facility. Available online: https://asf.alaska.edu/ (accessed on 18 August 2021).
81. Probst, L.; Frideres, L.; Cambier, B.; Duval, J.P.; Roth, M.; Lu-Dac, C. *Space Tech and Services Applications Related to Earth Observation*; European Union Business Innovation Observatory: Luxembourg, 2016; Available online: https://ec.europa.eu/docsroom/documents/16591/attachments/1/translations/en/renditions/native (accessed on 10 January 2022).
82. ESA. Thematic Exploitation Platforms Overview. Available online: https://eo4society.esa.int/thematic-exploitation-platforms-overview/ (accessed on 17 August 2021).
83. FAO. WaPOR Database Methodology: Level 1 Data. Remote Sensing for Water Productivity Technical Report: Methodology Series. Available online: http://www.fao.org/fileadmin/user_upload/faoweb/RS-WP/pdf_files/Web_WaPOR-beta_Methodology_document_Level1.pdf (accessed on 18 August 2021).
84. Gonzalez-Roglich, M.; Zvoleff, A.; Noon, M.; Liniger, H.; Fleiner, R.; Harari, N.; Garcia, C. Synergizing global tools to monitor progress towards land degradation neutrality: Trends.Earth and the World Overview of Conservation Approaches and Technologies sustainable land management database. *Environ. Sci. Policy* **2019**, *93*, 34–42. [CrossRef]
85. Hansen, M.C.; Potapov, P.V.; Moore, R.; Hancher, M.; Turubanova, S.A.; Tyukavina, A.; Thau, D.; Stehman, S.V.; Goetz, S.J.; Loveland, T.R.; et al. High-resolution global maps of 21st-century forest cover change. *Science* **2013**, *342*, 850–853. [CrossRef]
86. Li, W.; MacBean, N.; Ciais, P.; Defourny, P.; Lamarche, C.; Bontemps, S.; Houghton, R.A.; Peng, S. Gross and net land cover changes in the main plant functional types derived from the annual ESA CCI land cover maps (1992–2015). *Earth Syst. Sci. Data* **2018**, *10*, 219–234. [CrossRef]
87. Marselis, S.M.; Abernethy, K.; Alonso, A.; Armston, J.; Baker, T.R.; Bastin, J.F.; Bogaert, J.; Boyd, D.S.; Boeckx, P.; Burslem, D.F.R.P.; et al. Evaluating the potential of full-waveform lidar for mapping pan-tropical tree species richness. *Glob. Ecol. Biogeogr.* **2020**, *29*, 1799–1816. [CrossRef]
88. Showstack, R. Global Forest Watch Initiative Provides Opportunity for Worldwide Monitoring. *Eos Trans. Am. Geophys. Union* **2014**, *95*, 77–79. [CrossRef]
89. UNESCAP. *Innovative Big Data Approaches for Capturing and Analyzing Data to Monitor and Achieve the SDGs*; United Nations: Bangkok, Thailand, 2017; p. 137. Available online: https://hdl.handle.net/20.500.12870/2862 (accessed on 15 October 2021).
90. Ferreira, B.; Iten, M.; Silva, R.G. Monitoring sustainable development by means of earth observation data and machine learning: A review. *Environ. Sci. Eur.* **2020**, *32*, 1–17. [CrossRef]
91. Lewis, A.; Oliver, S.; Lymburner, L.; Evans, B.; Wyborn, L.; Mueller, N.; Raevksi, G.; Hooke, J.; Woodcock, R.; Sixsmith, J.; et al. The Australian Geoscience Data Cube—Foundations and lessons learned. *Remote Sens. Environ.* **2017**, *202*, 276–292. [CrossRef]
92. CEOS. Available online: https://ceos.org/ard/ (accessed on 15 October 2021).
93. Butler, D. Earth observation enters next phase. *Nature* **2014**, *508*, 160–161. [CrossRef]
94. Guo, H.-D.; Zhang, L.; Zhu, L.-W. Earth observation big data for climate change research. *Adv. Clim. Change Res.* **2015**, *6*, 108–117. [CrossRef]
95. Garcia-Pineda, O.; Staples, G.; Jones, C.E.; Hu, C.; Holt, B.; Kourafalou, V.; Graettinger, G.; DiPinto, L.; Ramirez, E.; Streett, D.; et al. Classification of oil spill by thicknesses using multiple remote sensors. *Remote Sens. Environ.* **2020**, *236*, 1–15. [CrossRef]
96. Badarinath, K.V.S.; Sharma, A.R.; Kharol, S.K. Forest fire monitoring and burnt area mapping using satellite data: A study over the forest region of Kerala State, India. *Int. J. Remote Sens.* **2011**, *32*, 85–102. [CrossRef]
97. Muckenhuber, S. High resolution sea ice monitoring using space borne Synthetic Aperture Radar. Ph.D. Thesis, the University of Bergen, Bergen, Norway, 2017.
98. Simonneaux, V.; Duchemin, B.; Helson, D.; Er-Raki, S.; Olioso, A.; Chehbouni, A.G. The use of high-resolution image time series for crop classification and evapotranspiration estimate over an irrigated area in central Morocco. *Int. J. Remote Sens.* **2010**, *29*, 95–116. [CrossRef]
99. Jeong, S.J.; Ho, C.H.; Choi, S.D.; Kim, J.; Lee, E.J.; Gim, H.J. Satellite data-based phenological evaluation of the nationwide reforestation of South Korea. *PLoS ONE* **2013**, *8*, e58900. [CrossRef] [PubMed]
100. Kaplan, G.; Avdan, U. Mapping and Monitoring Wetlands Using Sentinel-2 Satellite Imagery. *ISPRS Ann. Photogramm. Remote Sens. Spat. Inf. Sci.* **2017**, *IV*, 271–277. [CrossRef]
101. Sallustio, L.; De Toni, A.; Strollo, A.; Di Febbraro, M.; Gissi, E.; Casella, L.; Geneletti, D.; Munafo, M.; Vizzarri, M.; Marchetti, M. Assessing habitat quality in relation to the spatial distribution of protected areas in Italy. *J. Env. Manag.* **2017**, *201*, 129–137. [CrossRef] [PubMed]
102. Andries, A.; Murphy, R.J.; Morse, S.; Lynch, J. Earth Observation for Monitoring, Reporting, and Verification within Environmental Land Management Policy. *Sustainability* **2021**, *13*, 9105. [CrossRef]
103. Sawaya, K.E.; Olmanson, L.G.; Heinert, N.J.; Brezonik, P.L.; Bauer, M.E. Extending satellite remote sensing to local scales: Land and water resource monitoring using high-resolution imagery. *Remote Sens. Environ.* **2003**, *88*, 144–156. [CrossRef]

104. Liu, L.; Tang, H.; Caccetta, P.; Lehmann, E.A.; Hu, Y.; Wu, X. Mapping afforestation and deforestation from 1974 to 2012 using Landsat time-series stacks in Yulin District, a key region of the Three-North Shelter region, China. *Environ. Monit. Assess* **2013**, *185*, 9949–9965. [CrossRef] [PubMed]

105. Hislop, S.; Haywood, A.; Jones, S.; Soto-Berelov, M.; Skidmore, A.; Nguyen, T.H. A satellite data driven approach to monitoring and reporting fire disturbance and recovery across boreal and temperate forests. *Int. J. Appl. Earth Obs. Geoinf.* **2020**, *87*, 102034. [CrossRef]

106. Sheffield, J.; Wood, E.F.; Pan, M.; Beck, H.; Coccia, G.; Serrat-Capdevila, A.; Verbist, K. Satellite Remote Sensing for Water Resources Management: Potential for Supporting Sustainable Development in Data-Poor Regions. *Water Resour. Res.* **2018**, *54*, 9724–9758. [CrossRef]

107. Hede, A.N.H.; Koike, K.; Kashiwaya, K.; Sakurai, S.; Yamada, R.; Singer, D.A. How can satellite imagery be used for mineral exploration in thick vegetation areas? *Geochem. Geophys. Geosystems* **2017**, *18*, 584–596. [CrossRef]

108. Zhang, X.; Friedl, M.A.; Schaaf, C.B.; Strahler, A.H.; Hodges, J.C.F.; Gao, F.; Reed, B.C.; Huete, A. Monitoring vegetation phenology using MODIS. *Remote Sens. Environ.* **2003**, *84*, 471–475. [CrossRef]

109. Liu, Y.; Zhou, Y.; Lu, N.; Tang, R.; Liu, N.; Li, Y.; Yang, J.; Jing, W.; Zhou, C. Comprehensive assessment of Fengyun-3 satellites derived soil moisture with in-situ measurements across the globe. *J. Hydrol.* **2021**, *594*, 125949. [CrossRef]

110. Zhou, T.; Geng, Y.; Chen, J.; Pan, J.; Haase, D.; Lausch, A. High-resolution digital mapping of soil organic carbon and soil total nitrogen using DEM derivatives, Sentinel-1 and Sentinel-2 data based on machine learning algorithms. *Sci. Total Env.* **2020**, *729*, 138244. [CrossRef] [PubMed]

111. Bolch, T. Climate change and glacier retreat in northern Tien Shan (Kazakhstan/Kyrgyzstan) using remote sensing data. *Glob. Planet. Change* **2007**, *56*, 1–12. [CrossRef]

112. Fretwell, P.T.; Trathan, P.N.; Scales, K.; Bouchet, P. Discovery of new colonies by Sentinel2 reveals good and bad news for emperor penguins. *Remote Sens. Ecol. Conserv.* **2020**, *7*, 139–153. [CrossRef]

113. Merchant, C.J.; Embury, O.; Bulgin, C.E.; Block, T.; Corlett, G.K.; Fiedler, E.; Good, S.A.; Mittaz, J.; Rayner, N.A.; Berry, D.; et al. Satellite-based time-series of sea-surface temperature since 1981 for climate applications. *Sci Data* **2019**, *6*, 223. [CrossRef]

114. Groom, S.; Sathyendranath, S.; Ban, Y.; Bernard, S.; Brewin, R.; Brotas, V.; Brockmann, C.; Chauhan, P.; Choi, J.-k.; Chuprin, A.; et al. Satellite Ocean Colour: Current Status and Future Perspective. *Front. Mar. Sci.* **2019**, *6*. [CrossRef]

115. Nerem, R.S.; Beckley, B.D.; Fasullo, J.T.; Hamlington, B.D.; Masters, D.; Mitchum, G.T. Climate-change–driven accelerated sea-level rise detected in the altimeter era. *Proc. Natl. Acad. Sci. USA* **2018**, *115*, 2022–2025. [CrossRef]

116. Xing, Q.; An, D.; Zheng, X.; Wei, Z.; Wang, X.; Li, L.; Tian, L.; Chen, J. Monitoring seaweed aquaculture in the Yellow Sea with multiple sensors for managing the disaster of macroalgal blooms. *Remote Sens. Environ.* **2019**, *231*, 111279. [CrossRef]

117. Biermann, L.; Clewley, D.; Martinez-Vicente, V.; Topouzelis, K. Finding Plastic Patches in Coastal Waters using Optical Satellite Data. *Sci Rep* **2020**, *10*, 5364. [CrossRef] [PubMed]

118. Melet, A.; Teatini, P.; Le Cozannet, G.; Jamet, C.; Conversi, A.; Benveniste, J.; Almar, R. Earth Observations for Monitoring Marine Coastal Hazards and Their Drivers. *Surv. Geophys.* **2020**, *41*, 1489–1534. [CrossRef]

119. Fretwell, P.T.; Staniland, I.J.; Forcada, J. Whales from space: Counting southern right whales by satellite. *PLoS ONE* **2014**, *9*, e88655. [CrossRef] [PubMed]

120. Transforming Satellite Data into Weather Forecasts. Available online: https://eos.org/science-updates/transforming-satellite-data-into-weather-forecasts (accessed on 17 August 2021).

121. Liou, Y.-A.; Kar, S. Evapotranspiration Estimation with Remote Sensing and Various Surface Energy Balance Algorithms—A Review. *Energies* **2014**, *7*, 2821–2849. [CrossRef]

122. Brown, A.; Hayward, T.; Timmis, R.; Wade, R.; Pope, R.; Trent, T.; Boesch, H.; Guillo, R. *Satellite Measurements of Air Quality and Greenhouse Gases: Application to Regulatory Activities Report*; Environment Agency: Bristol, UK, 2021; p. 156.

123. Chahine, M.; Pagano, T.; Aumann, H.; Atlas, R.; Barnet, C.; Blaisdell, J.; Chen, L.; Divakarla, M.; Fetzer, E.; Goldberg, M.; et al. AIRS: Improving Weather Forecasting and Providing New Data on Greenhouse Gases. *Bull. Am. Meteorol. Soc. BULL. AMER. METEOROL. SOC.* **2006**, *87*, 911–926. [CrossRef]

124. Ialongo, I.; Virta, H.; Eskes, H.; Hovila, J.; Douros, J. Comparison of TROPOMI/Sentinel-5 Precursor NO_2 observations with ground-based measurements in Helsinki. *Atmos. Meas. Tech.* **2020**, *13*, 205–218. [CrossRef]

125. Boersma, K.F.; Eskes, H.J.; Veefkind, J.P.; Brinksma, E.J.; Sneep, M.; van den Oord, G.H.J.; Levelt, P.F.; Stammes, P.; Gleason, J.F.; Bucsela, E.J.; et al. Near-real time retrieval of tropospheric NO_2 from OMI. *Atmos. Chem. Phys.* **2007**, *7*, 2103–2118. [CrossRef]

126. Gauthier, J.-F.; Germain, S. From Data to Actionable Insight: Monitoring Fugitive Methane Emissions at Oil and Gas Facilities Using Satellites. In Proceedings of the Abu Dhabi International Petroleum Exhibition & Conference, Abu Dhabi, United Arab Emirates, 11–14 November 2019.

127. Bucsela, E.J.; Celarier, E.A.; Wenig, M.O.; Gleason, J.F.; Veefkind, J.P.; Boersma, K.F.; Brinksma, E.J. Algorithm for NO/sub 2/ vertical column retrieval from the ozone monitoring instrument. *IEEE Trans. Geosci. Remote Sens.* **2006**, *44*, 1245–1258. [CrossRef]

128. Carboni, E.; Mather, T.A.; Schmidt, A.; Grainger, R.G.; Pfeffer, M.A.; Ialongo, I.; Theys, N. Satellite-derived sulfur dioxide (SO 2) emissions from the 2014–2015 Holuhraun eruption (Iceland). *Atmos. Chem. Phys.* **2019**, *19*, 4851–4862. [CrossRef]

129. Le Cozannet, G.; Kervyn, M.; Russo, S.; Ifejika Speranza, C.; Ferrier, P.; Foumelis, M.; Lopez, T.; Modaressi, H. Space-Based Earth Observations for Disaster Risk Management. *Surv. Geophys.* **2020**, *41*, 1209–1235. [CrossRef]

130. Joyce, K.E.; Samsonov, S.V.; Levick, S.R.; Engelbrecht, J.; Belliss, S. Mapping and monitoring geological hazards using optical, LiDAR, and synthetic aperture RADAR image data. *Nat. Hazards* **2014**, *73*, 137–163. [CrossRef]
131. Mateo-Garcia, G.; Veitch-Michaelis, J.; Smith, L.; Oprea, S.V.; Schumann, G.; Gal, Y.; Baydin, A.G.; Backes, D. Towards global flood mapping onboard low cost satellites with machine learning. *Sci Rep* **2021**, *11*, 7249. [CrossRef] [PubMed]
132. Chen, B.; Peng, K.; Parkinson, C.; Bertozzi, A.L.; Slough, T.L.; Urpelainen, J. Modeling illegal logging in Brazil. *Res. Math. Sci.* **2021**, *8*, 1–21. [CrossRef]
133. Hodler, R.; Raschky, P.A. Regional Favoritism. *Q. J. Econ.* **2014**, *129*, 995–1033. [CrossRef]
134. Elvidge, C.; Zhizhin, M.; Baugh, K.; Hsu, F.-C. Automatic Boat Identification System for VIIRS Low Light Imaging Data. *Remote Sens.* **2015**, *7*, 3020–3036. [CrossRef]
135. Martinez, L. How Much Should We Trust the Dictator's GDP Growth Estimates? *Univ. Chic. Harris Sch. Public Policy* **2019**. [CrossRef]
136. Straka, W.; Seaman, C.; Baugh, K.; Cole, K.; Stevens, E.; Miller, S. Utilization of the Suomi National Polar-Orbiting Partnership (NPP) Visible Infrared Imaging Radiometer Suite (VIIRS) Day/Night Band for Arctic Ship Tracking and Fisheries Management. *Remote Sens.* **2015**, *7*, 971–989. [CrossRef]
137. Hoppe, E.; Bruckno, B.; Campbell, E.; Acton, S.; Vaccari, A.; Stuecheli, M.; Bohane, A.; Falorni, G.; Morgan, J. Transportation Infrastructure Monitoring Using Satellite Remote Sensing. In *Materials and Infrastructures 1*; Torrenti, J.-M., Torre, F.L., Eds.; John Wiley & Sons: Paris, France, 2016; pp. 185–198.
138. ESA. The Impact of Space Data on Smart Transport and Logistics. Available online: https://business.esa.int/news/impact-space-data-smart-transport-logistics (accessed on 17 August 2021).
139. Sutton, P.; Roberts, D.; Elvidge, C.; Baugh, K. Census from Heaven: An estimate of the global human population using night-time satellite imagery. *Int.,J. Remote Sens.* **2010**, *22*, 3061–3076. [CrossRef]
140. Ghosh, T.; Anderson, S.; Elvidge, C.; Sutton, P. Using Nighttime Satellite Imagery as a Proxy Measure of Human Well-Being. *Sustainability* **2013**, *5*, 4988–5019. [CrossRef]
141. Zhao, F.; Song, L.; Peng, Z.; Yang, J.; Luan, G.; Chu, C.; Ding, J.; Feng, S.; Jing, Y.; Xie, Z. Night-Time Light Remote Sensing Mapping: Construction and Analysis of Ethnic Minority Development Index. *Remote Sens.* **2021**, *13*, 2129. [CrossRef]
142. Engstrom, R.N.; Hersh, J.; Newhouse, D. *Poverty from Space: Using High-Resolution Satellite Imagery for Estimating Economic Well-Being*; Policy Research Working Paper No. 8284; World Bank: Washington, DC, USA, 2017; Available online: https://openknowledge.worldbank.org/handle/10986/29075 (accessed on 18 January 2022).
143. Henderson, J.V.; Storeygard, A.; Weil, D.N. A Bright Idea for Measuring Economic Growth. *Am. Econ. Rev.* **2011**, *101*, 194–199. [CrossRef]
144. Al-Bilbisi, H. Spatial Monitoring of Urban Expansion Using Satellite Remote Sensing Images: A Case Study of Amman City, Jordan. *Sustainability* **2019**, *11*, 2260. [CrossRef]
145. Singhal, A.; Sahu, S.; Chattopadhyay, S.; Mukherjee, A.; Bhanja, S.N. Using night time lights to find regional inequality in India and its relationship with economic development. *PLoS ONE* **2020**, *15*, e0241907. [CrossRef]
146. Warth, G.; Braun, A.; Assmann, O.; Fleckenstein, K.; Hochschild, V. Prediction of Socio-Economic Indicators for Urban Planning Using VHR Satellite Imagery and Spatial Analysis. *Remote Sens.* **2020**, *12*, 1730. [CrossRef]
147. Shi, K.; Yu, B.; Huang, Y.; Hu, Y.; Yin, B.; Chen, Z.; Chen, L.; Wu, J. Evaluating the Ability of NPP-VIIRS Nighttime Light Data to Estimate the Gross Domestic Product and the Electric Power Consumption of China at Multiple Scales: A Comparison with DMSP-OLS Data. *Remote Sens.* **2014**, *6*, 1705–1724. [CrossRef]
148. Doll, C.N.H.; Pachauri, S. Estimating rural populations without access to electricity in developing countries through night-time light satellite imagery. *Energy Policy* **2010**, *38*, 5661–5670. [CrossRef]
149. Froehlich, A.; Tăiatu, C.M. Practical Use of Satellite Data in Support of Human Rights. In *Space in Support of Human Rights*; Springer: Berlin/Heidelberg, Germany, 2020; pp. 49–124.
150. Quinn, J.A.; Nyhan, M.M.; Navarro, C.; Coluccia, D.; Bromley, L.; Luengo-Oroz, M. Humanitarian applications of machine learning with remote-sensing data: Review and case study in refugee settlement mapping. *Philos. Trans. R. Soc. A Math. Phys. Eng. Sci.* **2018**, *376*, 20170363. [CrossRef]
151. McDonald, G.G.; Costello, C.; Bone, J.; Cabral, R.B.; Farabee, V.; Hochberg, T.; Kroodsma, D.; Mangin, T.; Meng, K.C.; Zahn, O. Satellites can reveal global extent of forced labor in the world's fishing fleet. *Proc. Natl. Acad. Sci. USA* **2021**, *118*, e2016238117. [CrossRef]
152. Lai, K.Y.; Sarkar, C.; Ni, M.Y.; Cheung, L.W.T.; Gallacher, J.; Webster, C. Exposure to light at night (LAN) and risk of breast cancer: A systematic review and meta-analysis. *Sci. Total Environ.* **2021**, *762*, 143159. [CrossRef] [PubMed]
153. Malone, J.B.; Bergquist, R.; Martins, M.; Luvall, J.C. Use of Geospatial Surveillance and Response Systems for Vector-Borne Diseases in the Elimination Phase. *Trop. Med. Infect. Dis.* **2019**, *4*, 15. [CrossRef]
154. Anderson, K.; Ryan, B.; Sonntag, W.; Kavvada, A.; Friedl, L. Earth observation in service of the 2030 Agenda for Sustainable Development. *Geo-Spat. Inf. Sci.* **2017**, *20*, 77–96. [CrossRef]
155. GEO. *EO4SDG Earth Observations in Service of the 2030 Agenda for Sustainable Development*; Group on Earth Observations: Geneva, Switzerland, 2019; p. 32. Available online: https://earthobservations.org/documents/gwp20_22/eo_for_sustainable_development_goals_ip.pdf (accessed on 10 January 2022).

156. O'Connor, B.; Moul, K.; Pollini, B.; DeLamo, X.; Simonson, W. EO4SDG: Compendium of Earth Observation Contributions to the SDG Targets and Indicators. 2020, p. 165. Available online: https://eo4society.esa.int/2021/01/15/compendium-of-eo-contributions-to-the-sdgs-just-released/ (accessed on 18 January 2022).

157. Estoque, R. A Review of the Sustainability Concept and the State of SDG Monitoring Using Remote Sensing. *Remote Sens.* **2020**, *12*, 1770. [CrossRef]

158. DigitalGlobe. *Satellite Data's Role in Supporting Sustainable Development Goals Empowering Organizations with Earth Observation Geospatial Information & Big Data*; DigitalGlobe: Denver, CO, USA, 2020; p. 22.

159. Curzi, G.; Modenini, D.; Tortora, P. Large Constellations of Small Satellites: A Survey of Near Future Challenges and Missions. *Aerospace* **2020**, *7*, 133. [CrossRef]

160. PlanetLab. Available online: https://www.planet.com (accessed on 17 August 2021).

161. Engstrom, R.; Newhouse, D.; Soundararajan, V. Estimating small-area population density in Sri Lanka using surveys and Geo-spatial data. *PLoS ONE* **2020**, *15*, e0237063. [CrossRef]

162. DigitalGlobe. Available online: https://blog.maxar.com/earth-intelligence/2017/eliminating-malaria-with-the-power-of-the-crowd (accessed on 17 August 2021).

163. Gender and Urban Mobility in Chile. Available online: https://thegovlab.org/project/project-gender-and-urban-mobility-in-chile (accessed on 18 August 2021).

164. The Living Lib. Available online: https://thelivinglib.org/selected-readings-on-data-gender-and-mobility (accessed on 20 August 2021).

165. Gauvin, L.; Tizzoni, M.; Piaggesi, S.; Young, A.; Adler, N.; Verhulst, S.; Ferres, L.; Cattuto, C. Gender gaps in urban mobility. *Humanit. Soc. Sci. Commun.* **2020**, *7*, 11. [CrossRef]

166. Purdy, R.; Harris, R.; Carver, J.; Slater, D. Action B8: Remote sensing–Phase 1 (Research) Final Report. 2017. Available online: https://www.sepa.org.uk/media/311269/lsw-b8-phase-1-final-report-v10.pdf (accessed on 18 January 2022).

167. The Conversation. Available online: https://theconversation.com/the-eye-in-the-sky-that-can-spot-illegal-rubbish-dumps-from-space-98395 (accessed on 17 August 2021).

168. Paoli, A.D.; Addeo, F. Assessing SDGs: A methodology to measure sustainability. *Athens, J. Soc. Sci.* **2019**, *6*, 229–250. [CrossRef]

169. Tremblay, D.; Fortier, F.; Boucher, J.F.; Riffon, O.; Villeneuve, C. Sustainable development goal interactions: An analysis based on the five pillars of the 2030 agenda. *Sustain. Dev.* **2020**, *28*, 1584–1596. [CrossRef]

170. IRP; UNEP. *Resource Efficiency for Sustainable Development: Key Messages for the Group of 20*; United Nations Environment Programme: Nairobi, Kenya, 2018; p. 49. Available online: https://www.resourcepanel.org/reports/resource-efficiency-sustainable-development (accessed on 18 January 2022).

171. Geniaux, G. *Sustainable Development Indicator Frameworks and Initiatives*; SEAMLESS: Avignon, France, 2009; p. 150.

172. Azevedo, G.; Costa, H.; Farias Filho, J. Measuring Well-Being Through OECD Better Life Index: Mapping the Gaps. In Proceedings of the International Joint Conference on Industrial Engineering and Operations Management, Rio de Janeiro, Brazil, 8–11 July 2020.

173. Schwartz, S.H. An Overview of the Schwartz Theory of Basic Values. *Online Read. Psychol. Cult.* **2012**, *2*. [CrossRef]

174. Rutstein, S.O.; Rojas, G. *Guide to DHS statistics- Demographic and Health Surveys*; ICF: Rockville, MD, USA, 2006; p. 171.

175. Hicks, C.C.; Levine, A.; Agrawal, A.; Basurto, X.; Breslow, S.J.; Carothers, C.; Charnley, S.; Coulthard, S.; Dolsak, N.; Donatuto, J. Engage key social concepts for sustainability. *Science* **2016**, *352*, 38–40. [CrossRef]

176. King, C. *Partnership Effectiveness Continuum: A Research-Based Tool for Use in Developing, Assessing, and Improving Partnerships*; Education Development Center: Waltham, MA, USA, 2014; p. 21.

177. Pham, T.-T.-H.; Apparicio, P.; Gomez, C.; Weber, C.; Mathon, D. Towards a rapid automatic detection of building damage using remote sensing for disaster management. *Disaster Prev. Manag.* **2014**, *23*, 53–66. [CrossRef]

178. Patino, J.E.; Duque, J.C.; Pardo-Pascual, J.E.; Ruiz, L.A. Using remote sensing to assess the relationship between crime and the urban layout. *Appl. Geogr.* **2014**, *55*, 48–60. [CrossRef]

179. Ghosh, T.; Anderson, S.; Powell, R.L.; Sutton, P.C.; Elvidge, C.D. Estimation of Mexico's Informal Economy and Remittances Using Nighttime Imagery. *Remote Sens.* **2009**, *1*, 418–444. [CrossRef]

180. Yazdani, M.; Nguyen, M.H.; Block, J.; Crawl, D.; Zurutuza, N.; Kim, D.; Hanson, G.; Altintas, I. Scalable Detection of Rural Schools in Africa Using Convolutional Neural Networks and Satellite Imagery. In Proceedings of the 2018 IEEE/ACM International Conference on Utility and Cloud Computing Companion (UCC Companion), Zurich, Switzerland, 17–20 December 2018; pp. 142–147.

181. UNICEF. Available online: https://projectconnect.unicef.org/ (accessed on 18 August 2021).

182. Min, B.; Gaba, K.M.; Sarr, O.F.; Agalassou, A. Detection of rural electrification in Africa using DMSP-OLS night lights imagery. *Int. J. Remote Sens.* **2013**, *34*, 8118–8141. [CrossRef]

183. Dugoua, E.; Kennedy, R.; Urpelainen, J. Satellite data for the social sciences: Measuring rural electrification with night-time lights. *Int. J. Remote Sens.* **2018**, *39*, 2690–2701. [CrossRef]

184. Pekel, J.F.; Cottam, A.; Gorelick, N.; Belward, A.S. High-resolution mapping of global surface water and its long-term changes. *Nature* **2016**, *540*, 418–422. [CrossRef]

185. Andries, A.; Morse, S.; Murphy, R.; Woolliams, E.; Jim, L. Assessing Education from Space: Using satellite Earth Observation to assess pupil density in primary schools in rural areas of Nigeria. In Proceedings of the Sustainability in Transforming Societies of the 26th Annual Conference of the International Sustainable Development Research Society, Budapest, Hungary, 15–17 July 2020; p. 1008.
186. Andries, A.; Morse, S.; Murphy, R.J.; Lynch, J.; Woolliams, E.R. Assessing education from space: Using satellite Earth Observation to quantify overcrowding in primary schools in rural areas of Nigeria. *Sustainability* 2021, in press.
187. Bhatt, S.; Weiss, D.; Cameron, E.; Bisanzio, D.; Mappin, B.; Dalrymple, U.; Battle, K.; Moyes, C.; Henry, A.; Eckhoff, P. The effect of malaria control on Plasmodium falciparum in Africa between 2000 and 2015. *Nature* **2015**, *526*, 207–211. [CrossRef] [PubMed]
188. Evans, J.; van Donkelaar, A.; Martin, R.V.; Burnett, R.; Rainham, D.G.; Birkett, N.J.; Krewski, D. Estimates of global mortality attributable to particulate air pollution using satellite imagery. *Env. Res* **2013**, *120*, 33–42. [CrossRef]
189. Mohamed, M.F. Satellite data and real time stations to improve water quality of Lake Manzalah. *Water Sci.* **2015**, *29*, 68–76. [CrossRef]
190. SDG Indicator Metadata 13.2.2. Available online: https://unstats.un.org/sdgs/metadata/files/Metadata-13-02-02.pdf (accessed on 20 October 2021).
191. Copernicus Climate Change Service. Available online: https://www.iup.uni-bremen.de/carbon_ghg/docs/C3S/CDR3_2003-2018/PQAR/C3S_D312b_Lot2.2.3.2-v1.0_PQAR-GHG_MAIN_v3.1.pdf (accessed on 17 August 2021).
192. Climate Data Store. Available online: https://cds.climate.copernicus.eu/#!/home (accessed on 15 September 2021).
193. Sentinel-5P Air Quality Data Look Promising. Available online: https://www.ecmwf.int/en/newsletter/157/news/sentinel-5p-air-quality-data-look-promising (accessed on 15 October 2021).
194. Magro, C.; Nunes, L.; Gonçalves, O.C.; Neng, N.R.; Nogueira, J.M.; Rego, F.C.; Vieira, P. Atmospheric Trends of CO and CH$_4$ from Extreme Wildfires in Portugal Using Sentinel-5P TROPOMI Level-2 Data. *Fire* **2021**, *4*, 25. [CrossRef]
195. UNEP; CCAC. *Global Methane Assessment: Benefits and Costs of Mitigating Methane Emissions*; United Nations Environment Programme: Nairobi, Kenya, 2021; p. 173.
196. Wu, D.; Lin, J.C.; Oda, T.; Kort, E.A. Space-based quantification of per capita CO$_2$ emissions from cities. *Environ. Res. Lett.* **2020**, *15*, 035004. [CrossRef]
197. Pandey, S.; Gautam, R.; Houweling, S.; van der Gon, H.D.; Sadavarte, P.; Borsdorff, T.; Hasekamp, O.; Landgraf, J.; Tol, P.; van Kempen, T.; et al. Satellite observations reveal extreme methane leakage from a natural gas well blowout. *Proc. Natl. Acad. Sci. USA* **2019**, *116*, 26376–26381. [CrossRef]
198. Bamber, J.; Bates, P.; Brindley, H.; Evans, B.; Jackson, T.; Merchant, C.; Davey, M.P.; Palmer, P.; Scott, M.; Spencer, T. Space-based Earth Observations for Climate Security. COP 26 Universityies Network Briefing. 2021. Available online: https://www.gla.ac.uk/media/Media_792662_smxx.pdf (accessed on 18 January 2022).
199. Seeds, D. India Lights. Available online: http://india.nightlights.io/#/nation/2006/12 (accessed on 17 August 2021).
200. Medium-Voltage Distribution (Predictive). Available online: https://energydata.info/dataset/medium-voltage-distribution-predictive (accessed on 15 September 2021).
201. Model Documentation. Available online: https://github.com/facebookresearch/many-to-many-dijkstra (accessed on 17 August 2021).
202. UNESC. Exploring Space Technologies for Sustainable Development and the Benefits of International Research Collaboration in this Context. Available online: https://unctad.org/system/files/official-document/ecn162020d3_en.pdf (accessed on 17 August 2021).
203. FAO. Challenges and Opportunities for Reporting on SDG Indicators-Information Seminar for Permanent Representatives. 2020. Available online: https://www.fao.org/3/ca9190en/ca9190en.pdf (accessed on 18 January 2022).
204. Sudmanns, M.; Tiede, D.; Lang, S.; Bergstedt, H.; Trost, G.; Augustin, H.; Baraldi, A.; Blaschke, T. Big Earth data: Disruptive changes in Earth observation data management and analysis? *Int J Digit Earth* **2019**, *13*, 832–850. [CrossRef]
205. MacFeely, S. The Big (data) Bang: Opportunities and Challenges for CompilingSDGIndicators. *Glob. Policy* **2019**, *10*, 121–133. [CrossRef]
206. Perez, A.; Yeh, C.; Azzari, G.; Burke, M.; Lobell, D.; Ermon, S. Poverty prediction with public landsat 7 satellite imagery and machine learning. *arXiv* **2017**, arXiv:1711.03654.
207. Malarvizhi, K.; Kumar, S.V.; Porchelvan, P. Use of high resolution Google Earth satellite imagery in landuse map preparation for urban related applications. *Procedia Technol.* **2016**, *24*, 1835–1842. [CrossRef]
208. Bernknopf, R.; Kuwayama, Y.; Gibson, R.; Blakely, J.; Mabee, B.; Clifford, T.J.; Quayle, B.; Epting, J.; Hardy, T.; Goodrich, D. *The Cost-Effectiveness of Satellite Earth Observations to Inform a Post-Wildfire Response*; Valuables an RFF/NASA CONSORTIUM; RFF: Washington, DC, USA, 2019; Available online: https://media.rff.org/documents/Valuables_Wildfires.pdf (accessed on 10 January 2022).
209. Vuolo, F.; Essl, L.; Atzberger, C. Costs and benefits of satellite-based tools for irrigation management. *Front. Environ. Sci.* **2015**, *3*, 1–12. [CrossRef]
210. Cerbaro, M.; Morse, S.; Murphy, R.; Lynch, J.; Griffiths, G. Challenges in Using Earth Observation (EO) Data to Support Environmental Management in Brazil. *Sustainability* **2020**, *12*, 10411. [CrossRef]
211. UN. Integrating Statistical Geospatial and Other Big Data to Leave no one Behind 5 March 2018. Available online: https://media.un.org/en/asset/k1w/k1wkg1785y (accessed on 24 August 2021).

212. UN. Virtual High-Level Forum Events on United Nations Global Geospatial. Available online: https://ggim.un.org/meetings/2020/Virtual_HLF (accessed on 25 August 2021).
213. Tscherning, K.; Helming, K.; Krippner, B.; Sieber, S.; y Paloma, S.G. Does research applying the DPSIR framework support decision making? *Land Use Policy* **2012**, *29*, 102–110. [CrossRef]
214. Pissourios, I.A. An interdisciplinary study on indicators: A comparative review of quality-of-life, macroeconomic, environmental, welfare and sustainability indicators. *Ecol. Indic.* **2013**, *34*, 420–427. [CrossRef]
215. Hák, T.; Moldan, B.; Dahl, A.L. *Sustainability Indicators: A Scientific Assessment*; Island Press: Washington, DC, USA, 2007.
216. Andries, A.M.S.; Lynch, J.M.; Woolliams, E.R.; Fonweban, J.; Murphy, R.J. Translation of Earth Observation data into sustainable development indicators: An analytical framework. In Proceedings of the 24th International Sustainable Development Research Society Conference—Actions for a Sustainable World: From Theory to Practice, Messina, Italy, 13–15 June 2018; p. 17.
217. Masó, J.; Serral, I.; Domingo-Marimon, C.; Zabala, A. Earth observations for sustainable development goals monitoring based on essential variables and driver-pressure-state-impact-response indicators. *Int. J. Digit. Earth* **2019**, *13*, 217–235. [CrossRef]
218. Cozzolino, E.; Lasta, C.A. Use of VIIRS DNB satellite images to detect jigger ships involved in the Illex argentinus fishery. *Remote Sens. Appl. : Soc. Environ.* **2016**, *4*, 167–178. [CrossRef]
219. Park, J.; Lee, J.; Seto, K.; Hochberg, T.; Wong, B.A.; Miller, N.A.; Takasaki, K.; Kubota, H.; Oozeki, Y.; Doshi, S.; et al. Illuminating dark fishing fleets in North Korea. *Sci. Adv.* **2020**, *6*, eabb1197. [CrossRef] [PubMed]
220. Bell, S.; Morse, S. Participatory approaches for the development and evaluation of sustainability indicators. In *Routledge HandBook of Sustainability Indicators*; Routledge: Oxford, UK, 2018; pp. 188–203.

Article

Earth Observation for Monitoring, Reporting, and Verification within Environmental Land Management Policy

Ana Andries *, Richard J. Murphy, Stephen Morse and Jim Lynch

Centre for Environment and Sustainability, Faculty of Engineering and Physical Sciences, University of Surrey, Guildford GU2 7XH, UK; rj.murphy@surrey.ac.uk (R.J.M.); s.morse@surrey.ac.uk (S.M.); j.lynch@surrey.ac.uk (J.L.)
* Correspondence: a.andries@surrey.ac.uk

Abstract: The main aim of the new agricultural scheme, Environmental Land Management, in England is to reward landowners based on their provision of 'public goods' while achieving the goals of the 25 Year Environment Plan and commitment to net zero emission by 2050. Earth Observation (EO) satellites appear to offer an unprecedented opportunity in the process of monitoring, reporting, and verification (MRV) of this scheme. In this study, we worked with ecologists to determine the habitat–species relationships for five wildlife species in the Surrey Hills 'Area of Outstanding Natural Beauty' (AONB), and this information was used to examine the extent to which EO satellite imagery, particularly very high resolution (VHR) imagery, could be used for habitat assessment, via visual interpretation and automated methods. We show that EO satellite products at 10 m resolution and other geospatial datasets enabled the identification and location of broadly suitable habitat for these species and the use of VHR imagery (at 1–4 m spatial resolution) allowed valuable insights for remote assessment of habitat qualities and quantity. Hence, at a fine scale, we obtained additional habitats such as scrub, hedges, field margins, woodland and tree characteristics, and agricultural practices that offer an effective source of information for sustainable land management. The opportunities and limitations of this study are discussed, and we conclude that there is considerable scope for it to offer valuable information for land management decision-making and as support and evidence for MRV for incentive schemes.

Keywords: very high resolution satellite data; earth observation; habitat suitability; habitat assessment; Environmental Land Management

Citation: Andries, A.; Murphy, R.J.; Morse, S.; Lynch, J. Earth Observation for Monitoring, Reporting, and Verification within Environmental Land Management Policy. *Sustainability* **2021**, *13*, 9105. https://doi.org/10.3390/su13169105

Academic Editor: Andrew Kirby

Received: 1 July 2021
Accepted: 9 August 2021
Published: 14 August 2021

Publisher's Note: MDPI stays neutral with regard to jurisdictional claims in published maps and institutional affiliations.

1. Introduction

The sustainability of land-use and -cover systems are determined by the interaction between natural resources (including soils, water, plants, and wildlife), climate, and human activities while ensuring a long-term productivity of these resources and maintaining a balance within the environmental functions [1]. Hence, sustainable land management (SLM) has been debated extensively in the literature [2–4] and can be defined as "a knowledge-based procedure that helps to integrate land, water, biodiversity, and environmental management to meet rising food and fibre demands while sustaining ecosystem services and livelihoods" [5]. At the global level, the United Nations Food and Agriculture Organisation (FAO) supports integrated land resource planning strategies through a wide range of comprehensive approaches, tools, and measures adapted to different biophysical and socio-economic contexts when aiming for sustainable land resilience and avoid degradation [6].

Land-use policy impacts land management and therefore affects the extent and state of ecosystem services [7]. The importance of decision support for mainstreaming and scaling up of SLM formulated by FAO led to each country shaping its own set of policies, programmes, and schemes that determine the extent to which SLM is taken into account in resource-management decisions [8]. Hence, at the national level, the UK Environment

Bill [9] and Agriculture Bill [10] set the foundations for implementing the 25 Year Environment Plan and led to the establishment of the Environmental Land Management (ELM) scheme.

The ELM scheme is a new programme of intensive policy development over the period 2019–2024 with implementation post-2025 by the Department for Environment, Food and Rural Affairs (DEFRA) replacing the previous EU Common Agricultural Policy. The ELMs is based on three schemes, Sustainable Farming Incentive (SFI), Local Nature Recovery (LNR), and Landscape Recovery (LR). The overall aim of these is to use public money to pay farmers and land managers in England to deliver a set of 'public goods' that cover clean air, clean and plentiful water, thriving plants and wildlife, protection from and mitigation of environmental hazards, beauty, heritage and engagement, and mitigation of and adaptation to climate change [11]. Nevertheless, land-use management and decision making that prioritises only one type of ecosystem service without considering those occurring within and between ecosystems results in policy failure [12].

While 'public goods' and 'ecosystem services' have different theoretical backgrounds [13,14] from an environmental perspective, DEFRA's range of public goods are often identified with some of the typical ecosystem services [15,16]. Therefore, ecosystem services are known as the benefits obtained directly or indirectly from ecosystems that support human life and enhance social welfare [12]. They are typically categorised as supporting services, provisioning services, regulating services, and services that support cultural needs.

In terms of measuring biodiversity conservation and ecosystem services, there are freely available integrated modelling tools such as Integrated Valuation of Ecosystem Services and Trade-offs (InVEST) [17], Co$tingNature [18], Multiscale Integrated Model of Ecosystem Services [19], Protected Area Benefits Assessment Tool (PA-BAT) [20], Ecosystem Services Toolkit (EST) [21], Toolkit for Ecosystem Services Site-based Assessment (TESSA) [22], Multiscale Integrated Model of Ecosystem Services (MIMES) [23], and Artificial Intelligence for Ecosystem Services (ARIES) [24]. All these can help with measuring, modelling, and valuing ecosystem services and thus help support decision-making about resources management and planning [25].

Land-use management activities have profoundly altered many ecosystems and therefore SLM approaches, and nature-based solutions are needed across policies [26,27]. Integrating nature-based solutions has been well debated in the literature within a range of applications on land-use management [28], land and soil degradation [28], urban planning [26,29], and polluted land [30]. Likewise, delivering nature-based interventions can maximise ecosystem services and lead towards net zero greenhouse gas emissions [28,31].

As part of the ELM policy development, DEFRA established 57 Test and Trails with stakeholders to understand how ELMs could be shaped in order to work in a real-life environment across a wide range of circumstances and to help understand the barriers and opportunities of new approaches [32]. The Surrey Hills Area of Outstanding Natural Beauty (SH-AONB) Test and Trial, entitled 'Making Space for Nature', is one of these 57 Test and Trails and is one of the 12 AONB-based ones that together make up the largest Defra Test and Trial project group for the ELM scheme. The main aim of this test and trial is to understand what local nature recovery and enhanced access represent for a 'designated landscape' and how this can be delivered through ELMs [33].

Currently, DEFRA [16,32] is exploring a range of ways for monitoring, reporting, and verification (MRV) of ELM scheme agreements. They are aware that ground-based surveys and physical inspections can be prohibitively expensive, and they want to consider the use of satellite and geospatial data, alongside other technologies that can offer cost-effective processes for the MRV of ELM and similar activities [16,34,35]. The spectral, spatial, and temporal resolution capabilities of modern Earth Observation (EO) satellites certainly appear to offer an unparalleled opportunity relevant to such ELM and MRV application through their capacities to assess:

- biodiversity and habitat type [36–40]
- habitat quality [41,42] and model habitat suitability [43,44]
- species identity and diversity [45–50]
- species behaviour [49,51].

Habitat suitability models seek to predict the optimal habitat for a species based on the assessment of habitat attributes such as habitat structure, habitat type, and spatial arrangements between habitat features. The models use established correlations between the presence of particular species and known ecological niche elements to analyse geographical space (e.g., from earth observation at various scales) to locate those same niche elements. This can then be used to create predictive location maps and other representations of the distribution and quantities of relevant habitat for a given species or group of species [44,52].

According to Turner et al. [53], there are two general approaches to quantify and model biodiversity using satellite and airborne remote sensing: direct mapping of species composition (based on tree canopy) and land use and cover [54–56] and indirect sensing by using environmental variables proxies such as: primary productivity (chlorophyll and ocean colour), climate (rainfall, soil moisture, and phenology), habitat structure (topography and vertical canopy structure) [44]. These approaches can be performed using visual photo/image interpretation [57] and computing/automated methods [57,58]. Visual interpretation is a 'manual' method based on the visual detection, identification, and spatial localisation of different objects and terrain shapes [59], whilst the automated methods typically use supervised and unsupervised image classification approaches (object and pixel image analysis) via classification algorithms (e.g., maximum-likelihood, support vector machine, random forest, and neural networks) [55,60–62].

Spatial resolution is an important attribute for the remote assessment of opportunities for biodiversity as it describes the level of spatial detail which can be seen in the image. Generally, higher spatial detail requires a sensor with a narrower field of view, hence less spatial coverage per image scene. Satellite sensors with a smaller field of view (e.g., Worldview and Superview have a 16 km swath) are generally constrained by low revisit times, unless the satellite is combined with phased orbit of the satellite constellation, which can offer a two day revisit for any point on Earth. Coarser spatial resolution sensors can image larger areas (e.g., Sentinel-2 can image strips 290 km wide) in one overpass of the satellite sensor with more regular repeat cycles [63]. There are also several important biodiversity trade-offs when considering the spatial resolution of a satellite sensor. For example, low-resolution data are perfectly adequate for monitoring status of broad land cover (LC)/use (LU) [64] and land change at regional level [65], while higher-resolution data are often desirable for monitoring at a local scale, individual protected areas [66,67] and agricultural practices [68], habitat loss, fragmentation and degradation, and ecosystem services [69,70] and for substituting for physical inspection [34,68]. Moreover, Mairota et al. [47] demonstrated that VHR satellite data can help mapping biodiversity and modelling habitat across different scales (landscape, plot, and patch) which can be used for identifying specific taxonomic groups. One other benefit of using VHR satellite data (0.5–5 m, depending on the optical sensor) is to discriminate some aspects of the habitat characteristics which normally cannot be seen at the coarser resolution [37]. For instance, case studies have been carried out using VHR optical satellite or aerial imagery for the mapping and monitoring of moorland to map vegetation patterns, evidence of ditching, and degree of peat exposure or burn histories [71,72].

The potential for assessing biodiversity and ecosystem status can be obtained through EO-derived land-cover and land-use datasets, spatial information, and in situ data [37]. In terms of the land-cover classification system at the global scale (refers to scales smaller than 1:250,000 and using NOAA Advanced Very High Resolution Radiometer (AVHRR) satellites), the most representative examples are USGS product [73], EarthSat GeoCover Land Cover Legend [74], the UN/FAO Land Cover Classification System [75], Global Observation of Forest Cover [76], and Global Observation of Land Cover Dynamics (GOFC/GOLD) [77]. Furthermore, land-cover classification at a regional scale (between 1:250,000 and 1:100,000) often use a medium spatial resolution from satellites such as Landsat TM, SPOTHRV/XS, IRS-1C/LISS, Landsat ETM+, and MODIS, offering classification systems such as CORINE [78] and AFRICOVER [79]. At the national scale, land-cover classification systems are available at smaller scales (between 1:5000–1:10,000) and are commonly based on remote sensing data and in situ surveys. Some products of land-cover classification systems at the national scale are National Land Cover Data Classification System [80], National Land Survey Classification System, and Centre for Ecology and Hydrology (CEH) Land Cover Map [81].

The Centre for Ecology and Hydrology (CEH) has developed its Land Cover Map spatial framework, which uses satellite data from 2015–2019 from the European Space Agency's Copernicus Sentinel-1 and Sentinel-2 satellites. These have a high resolution (10 m) enabling the distinction between 21 land-cover classes on the basis of Sentinel imagery and linked to Biodiversity Action Plan (BAP) Broad Habitats [82]. However, it has been noted [83] that discrepancies can exist in the habitat terminology, revealing the need for additional ecological expert knowledge and potentially VHR data to help resolve such ambiguities (e.g., in plant height estimations). VHR data have a wide application across many sectors, including assessing biodiversity and habitat quality [47]. While UK CEH LC and Crops are products that have a dynamic role for identifying main habitats, land cover/use and change and connectivity, and to support land parcel boundary mapping and principal crop types, VHR imagery can capture field boundary features (e.g., hedgerows and field margins), seasonal aspects related to agricultural management practices, vegetation patterns, evidence of ditches, forest canopy, etc. [72,83,84].

Insufficient collaboration with biodiversity experts is frequently cited as a limitation in using remote sensing for monitoring biodiversity and developing relevant indicators [85]. A deep understanding of habitat–species relationships must be available to use remotely sensed information effectively. Gaps still exist in our knowledge regarding the potential of EO imagery at the higher resolutions for projects such as the SH-AONB 'Making Space for Nature' Defra Test and Trial:

1. VHR satellite data use in identifying the availability of suitable and connected habitat, and
2. how effectively do EO approaches at the highest resolutions add new insights to overall habitat assessments for various species of importance for biodiversity conservation and recovery.

Our aim in the present research was, therefore, to work with ecologists to establish habitat criteria for five example species/species groups (henceforth referred to as the 'species') and then to use this information to examine the potential role of EO in habitat evaluation for these species at particular sites and more widely in the SH-AONB. The research was conducted in three linked steps (see Figure 1): Step 1, knowledge acquisition of species–habitat requirements; Step 2, use of readily-available EO and similar resources to evaluate broad habitat suitability and connectivity for the species; and Step 3, exploration of the additional contribution of the increasingly available Very High Resolution (VHR) EO data (0.7–4 m in this study) for habitat assessment.

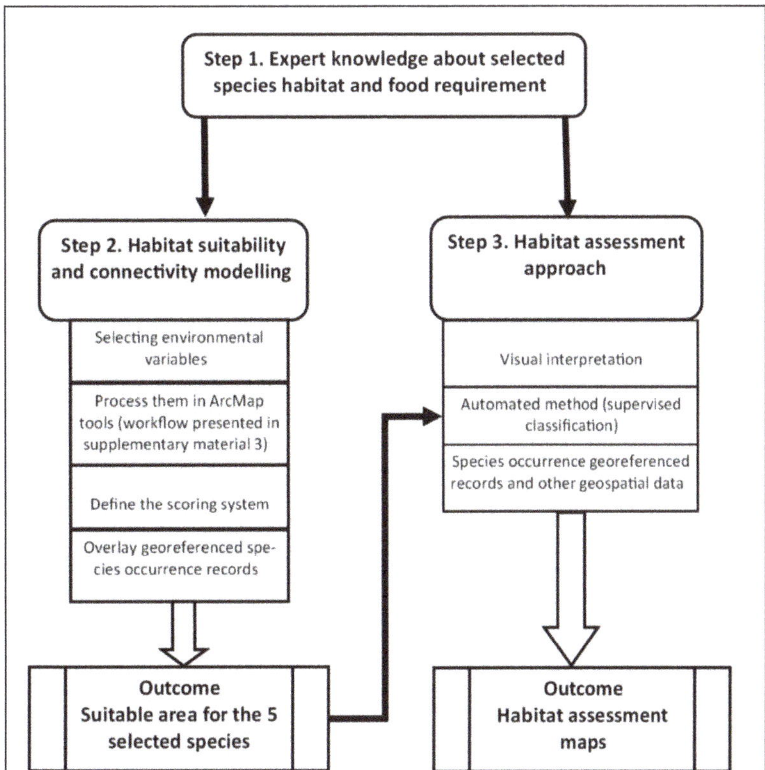

Figure 1. Workflow, analysis, and outcomes in this study.

2. Materials and Methods

Figure 1 summarises the overall research workflow, methods, and outcomes of the three steps applied in this study.

2.1. Study Area and Sites Selected

The Surrey Hills was one of the first landscapes in the UK to be 'designated' as an 'Area of Outstanding Natural Beauty' (AONB) in 1958. It should be noted here that AONB is not a subjective term of the authors but a formal one created by the UK Government. In June 2000, it was confirmed that AONBs have the same level of landscape quality and share the same level of protection as National Parks, and the primary purpose of an AONB is stated as *'to conserve and enhance the natural beauty of the area'*. There are 34 AONBs in England, and together they cover 15% of the total land area. Notably, AONBs and National Parks are found in Category V—landscapes managed mainly for conservation and recreation—of the International Union for the Conservation of Nature and Natural Resources' global network of protected landscapes [86].

The Surrey Hills AONB covers 422 km^2 with myriad landowners and community nature spaces. These include a wide diversity of broad priority habitats in the UK Biodiversity Action Plan (BAP) such as ancient woodland, broadleaved mixed and yew woodland, coniferous woodland, arable and horticulture, improved grassland, acid grassland, calcareous grassland, heather, freshwater, built-up areas, and traditional orchards. These provide habitat for BAP priority species, 85 specially protected species, and at least 300 species recognised as being a priority for conservation [87].

We selected for intensive study three farm sites within the SH-AONB area that comprise different habitats and biodiversity opportunities (Figure 2):

- Hampton Estate, located in the Greensand Plateau: Shackleford (~739 ha), 51.2116° N, 0.7034° W (Site 1) Sondes Place Farm, near Dorking, between Greensand Valley (Pippbrook and Tillingbourne) and The North Downs (~78.6 ha), 51.2304° N, 0.3425° W, (Site 2)
- Landbarn Farm also located near Dorking, between Greensand Valley (Pippbrook and Tillingbourne) and The North Downs (~100 ha), 51.2309° N, 0.3740° W, (Site 3).

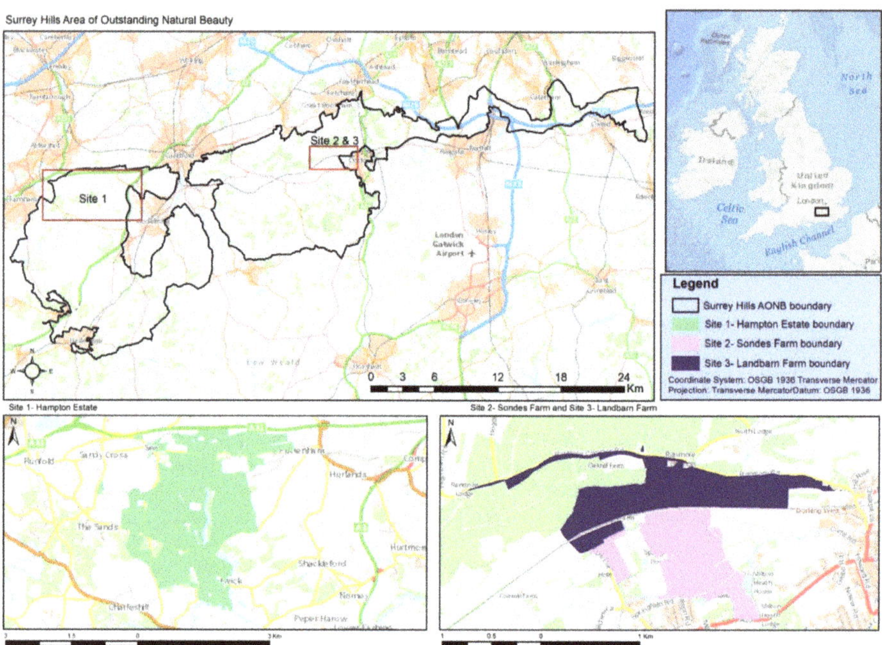

Figure 2. Locations of SH-AONB and the three intensive study sites.

2.2. Resources for Readily Available EO and Other Data (Research for Step 2)

The following datasets were used for Step 2 and to complement the use of VHR imagery for the habitat suitability assessment carried out in Step 3 of this research:

2.2.1. UK Centre for Ecology and Hydrology Land Cover Map 2019 (LCM2019) (20 m Classified Pixels) and Crop 2019

The Centre for Ecology and Hydrology (CEH) has recently made available the UK Land Cover Map 2019 (LCM2019) [56,88], a product based on satellite images (mainly Sentinel-2), digital cartography, and machine-learning techniques (e.g., bootstrap training combined with random forest classifier). LCM2019 provides information about physical materials on the Earth's surface; these can be natural such as vegetation, freshwater, inland rocks, etc., and non-natural materials such as buildings or urbanised areas. LCM2019 has 21 classes using similar UK BAP Broad Habitats categories except for a couple of inconsistencies. Mapping relationships and discrepancies in definitions between the CEH LC 2019 and UK BAP Broad Habitats categories are outlined in [56,88].

The SH-AONB is covered by 11 of the CEH LC classes illustrated and listed in Figure 3 for the whole SH-AONB and for our three intensive study sites. Further details of the SH-AONB LC and UK BAP Broad Habitats classes are given in Supplementary Materials (Table S1) [56,82]. UK CEH Crops 2019 product was obtained using Sentinel-2. The crops from 2019 in SH-AONB were identified as winter wheat, winter oats, spring wheat, winter barley, spring barley, oilseed rape, field beans, potatoes, maize, peas, and other crops (which might include other cereals, root crops, early potatoes, and vegetables) (Figure 4).

The UK CEH LCM2019 and Crops2019 products were obtained free of charge in raster and vector format from the Digimap Edina platform [89] under University of Surrey's license.

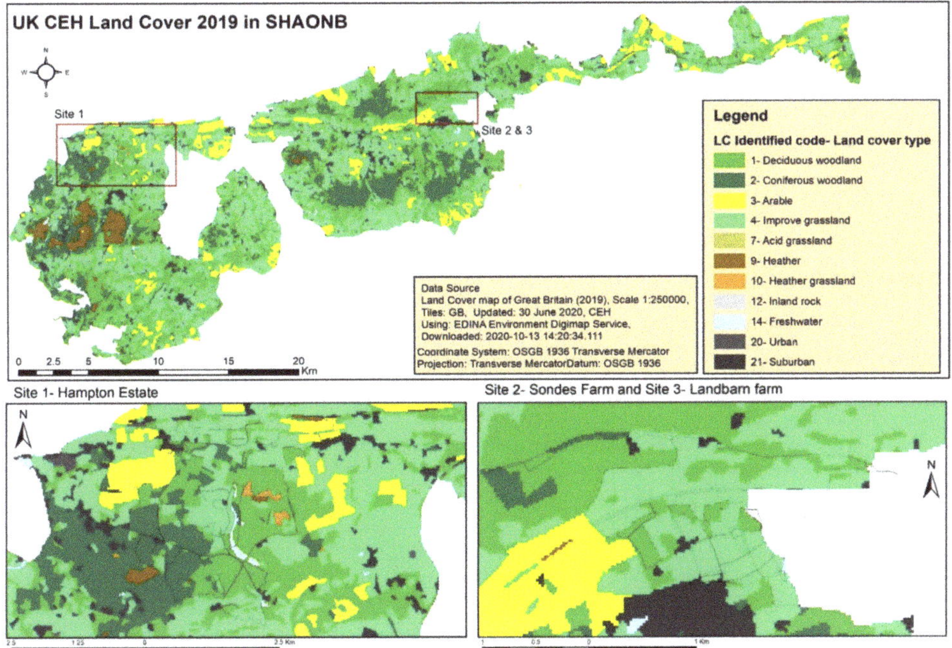

Figure 3. UK CEH Land Cover classes in the SH-AONB (locations of the three intensive study sites shown in red boxes).

2.2.2. Priority Habitat Inventory (England)

This is a spatial dataset that describes the geographic extent and location of 41 habitats, which is a separate classification from BAP Broad Habitats relevant to the Natural Environment and Rural Communities Act (2006). Among these 41 classes, we extracted only the calcareous grassland (chalk grassland) that helped for modelling the habitat suitability for the Small Blue butterfly. The data are free open access provided by Natural England [90].

2.2.3. OS Open Rivers and Open Street Map GIS Shapefiles

We extracted the watercourses from OS Open Rivers dataset provided by DigiMap Edina [89] and small roads such as footpaths and bridleways from Open Street Map [91], which has been used mainly for assessing the habitat suitability for dragonflies and damselflies and the silver-washed fritillary butterfly.

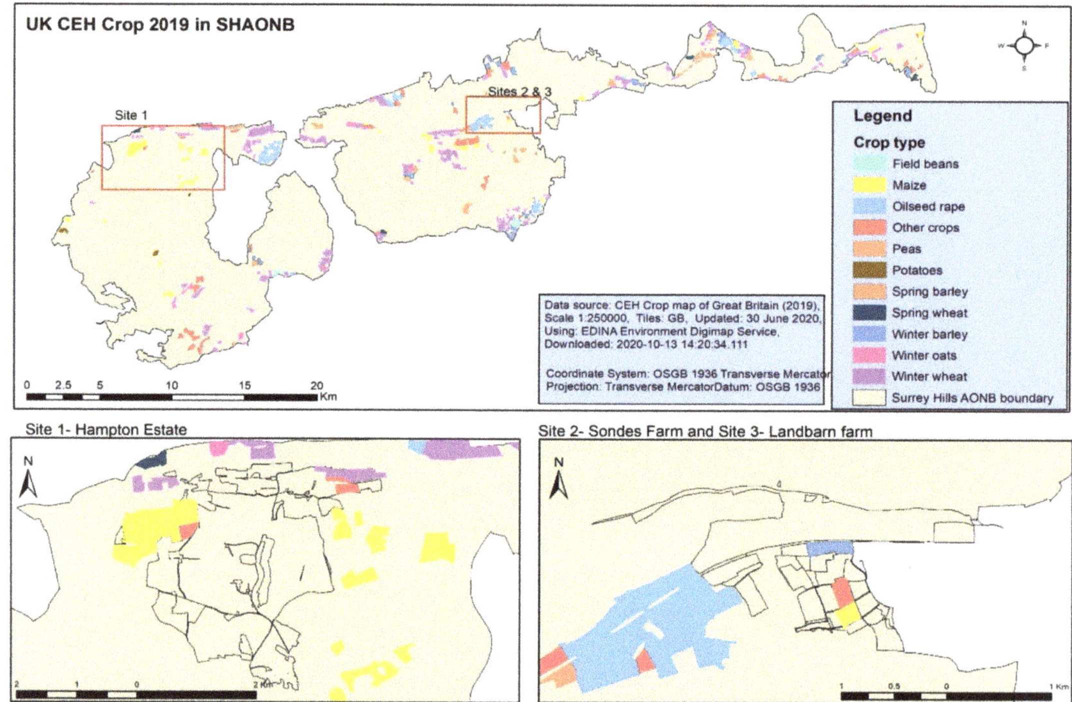

Figure 4. UK CEH Crops 2019 in SH-AONB and the three intensive study sites (red boxes).

2.2.4. Digital Elevation Model (DEM)

The DEM generally refers to a representation of a bare terrain surface or a subset of it, excluding features such as vegetation, buildings, or bridges. DEM is often useful for flood disaster evaluations or water-flow estimation models, land-use studies, and geological applications, aspect, slope, etc. DEM can be obtained from different data types (e.g., satellites and Lidar), but in this study, we used Google Earth Engine (GEE) [92], which provides free open access to the NASA Shuttle Radar Topography Mission (SRTM) Digital Elevation 30 m. SRTM digital elevation data are based on an international research effort that obtained digital elevation models on a near-global scale [93].

2.2.5. Species Occurrence Records

We obtained occurrence data for all five species from the past 5–12 years. These datasets were requested from a variety of organisations (Table 1). The original records requested were collected from crowd-sourcing biodiversity data via the NBN Atlas and iRecord (dragonflies and damselflies) and conducted surveys (dormouse, skylark, and butterflies). For instance, the UK British Monitoring Scheme (UKBMS), which is run by CEH, Butterfly Conservation, British Trust for Ornithology, and Joint Nature Conservation Committee (JNCC), uses a structured methodology to monitor at least 56 butterfly species across over 4000 UK sites.

Table 1. Selected species occurrence records and their providers.

Species	Species Scientific Name	Main Habitat Requirement	Provider of the Species Occurrence Records	Collection Period
Silver-washed fritillary butterfly	*Argynnis paphia*	Woodland/scrub	Butterfly Conservation (BC)	2015–2019
Small blue butterfly	*Cupido minimus*	Chalk grassland	Butterfly Conservation (BC)	2015–2019
Skylark	*Alauda arvensis*	Pasture/Arable	Surrey Bird Club (SBC)	2010–2019
Hazel dormouse	*Muscardinus avellanarius*	Hedgerow	National Dormouse Database (NDD)	2008–2019
Dragonflies Damselflies	*Anisoptera* spp. *Zygoptera* spp.	Inland water	British Dragonfly Society (BDS)	2010–2019

2.3. Resources for Very High Resolution (VHR) Satellite Imagery (Research for Step 3)

We used VHR Multispectral (MS) and Panchromatic (PAN) data from 3 different satellite sources, DMC3, Superview, and Komsat 3/3A, purchased from Earth-i, for performing the habitat suitability assessments and the associated automated classification maps. The description of the satellites' camera mode, spectral and spatial resolutions, and the acquisition date are given in Table 2. These satellite images were requested to be between spring and late summer when most tree crowns were leafy and crops had just been harvested; thus, we could discern different patterns, shapes, and textures to understand the landscape and habitat requirements of the selected species. As time-series images of the same location from 2016 until 2020 were available, we could also analyse for habitat change.

Table 2. VHR satellite data characteristics.

Satellite Name	Camera Modes	Spatial Resolution and Bands	Date of Acquisition	Sites Covered
DMC3	MS and PAN	MS = 4 m (blue, green, red, NIR) PAN = 1 m	6 May 2016	Site 2 and 3
			12 August 2016	Site 1
			6 May 2018	Site 2 and 3
			30 June 2018	Site 1
			20 April 2019	Site 2 and 3
			29 August 2019	Site 1
			25 June 2020	Site 1
Superview-1	MS and PAN	MS = 2 m (blue, green, red, NIR) PAN = 0.5 m	5 July 2017	Site 2 and 3
			19 May 2018	Site 1
			15 July 2018	Site 1
Komsat-3	MS and PAN	MS = 2.8 m (blue, green, red, NIR) PAN = 0.7 m	20 April 2018	Site 1
			6 May 2020	Site 2 and 3

2.4. Analysis

The following presents the analytical methods applied in each of the three research Steps summarised in Figure 1.

2.4.1. Step 1—Expert Knowledge Elicitation on the Habitat and Food Requirements of the Species

Expert knowledge was obtained from ecologists from the Surrey Wildlife Trust and Butterfly Conservation and from literature review to understand the relationship between the selected species and their habitat and food requirements in the SH-AONB. Likewise, the expert consultations guided us also in the selection of the five species, which play a role as indicators of an appropriate and healthy habitat both for themselves and in supporting the wider ecosystem and maintaining its balance [94]. In consequence, the presence of such key species indicates a healthy habitat, its good management providing food and shelter for these species and suggesting thriving biodiversity. These represent an indirect contribution to other 'public goods' such as improvements in carbon storage, improving clean air and water quality, enhancing the landscape, mitigating climate change, and promoting heritage engagement with the environment, well-being, and protection from environmental hazards.

The five species and their requirements are presented in Supplementary Materials Table S2 [95–103].

2.4.2. Step 2—Habitat Suitability and Connectivity Modelling Using Existing EO and Other Resources

The ecological information on the basic needs of these species for food, habitat type, dispersal movement, and breeding (Table S3 [98,102,103]) was integrated with variables such as land cover, crop type, footpaths, topographical parameters, watercourses, settlements, etc., that could be observed in sufficient detail from the EO. These EO-observable variables played an equal role in evaluating the habitat suitability and its connectivity for the studied species. This habitat suitability evaluation process was applied both at the level of the three intensive study sites and also at the whole SH-AONB area level.

The following workflow was performed for the EO-based habitat suitability and connectivity modelling:

(a) Choose the environmental variables (factors) based on species habitat and food requirement

We used 6 different environmental variables, which can be placed into 3 categories including land cover [104], topography [105], and anthropogenic [106] (Table 3). In our modelling, we used land cover (UK CEH LCM2019), crop type (UK CEH Crop Type 2019— only for Skylark), and chalk grassland (priority habitat inventory—only for modelling the Small Blue butterfly suitable habitat) as a means of habitat type, vegetation structure, and food availability. Land cover plays a role as a key variable in this analysis, and it was used to model all habitat suitability of the studied species. The anthropogenic variables selected in this study are footpaths and settlements (urban and suburban areas). These variables (depending on the species–habitat relationship) are likely to affect the connectivity within the habitats. For instance, the settlements vector layer was used as a pressure factor for Skylark nesting habitat; thus, we created a 100 m buffer around the settlement polygons. However, well-managed footpaths and bridleways that cross the broadleaved woodland can be a positive aspect in maintaining a Silver-washed fritillary butterfly population. Moreover, the topographic variable extracted from DEM was needed to ascertain the slope degree, which is an important factor in understanding the small blue butterfly's favoured habitat.

(b) Georeferenced species occurrence records

In this study, the occurrence records of the species were combined with (overlaid on) the results of habitat suitability analysis. This was considered as a further indicator for the presence of healthy and appropriate habitat, thus reinforcing the potential of a location as suitable for the species.

Table 3. Environmental variables used in developing habitat suitability model.

Species	Environment Variables	Data Source	Variable Type
Silver-washed fritillary butterfly	Habitat type	UK CEH LCM2019	Land cover
	Broadleaved woodland edge	UK CEH LCM2019	Land cover
	Footpaths through broadleaved woodland	Open Street Map	Anthropogenic
Small blue butterfly	Habitat type	UK CEH LCM2019	Land cover
	Chalk grassland	Priority habitat inventory (England)	Land cover
	Slope degree	DEM	Topographic
Skylark	Habitat type	UK CEH LCM2019	Land cover
	Crop type	UK CEH Crops2019	Land cover
	Settlements	UK CEH LCM2019	Anthropogenic
Hazel dormouse	Habitat type	UK CEH LCM2019	Land cover
Dragonflies and Damselflies	Habitat type	UK CEH LCM2019	Land cover
	Slow-flowing watercourse (ditches, brooks, stream, rivulets, and rills)	OS Open Rivers	Land cover

(a)   ArcMap tools

The datasets presented in Section 3, UK CEH LCM2019, UK CEH Crop type 2019, species occurrence georeferenced records, OS Open Rivers, Open Street map footpaths, and slope derived from DEM, were processed through a wide variety of tools in ArcGIS 10.6.1 (e.g., Extract, Buffer, Reclassify, etc.) to design the habitat suitability and connectivity of the monitored species. A detailed workflow is presented in Supplementary Materials Table S4.

(b) Scoring system (from 1 to 5) based on species habitat and food requirement

Table 4 explains the rationale behind the scoring system. Scores were assigned from 1 to 5, where 1 is attributed to the least suitable habitat and 5 represents the most suitable habitat of a species.

Table 4. Habitat scoring system and rationale.

Score	Habitat Suitability	Rationale
1	Least suitable/unsuitable	Species cannot survive due to the lack of food
2	Low suitability	Species cannot cross through that area (depending on species and the area size) due to the lack of food.
3	Moderately suitable habitat	Species may travel through to reach other more suitable areas, but it is unlikely to find food sources.
4	Suitable habitat	It may be used occasionally but dependant on other factors such as food availability or ideal requirements to sustain a breeding population.
5	Highly suitable habitat	Provides the best habitat for nesting, breeding, or food (depending on the species).

2.4.3. Step 3—Habitat Assessment Approach and Contribution of VHR EO Data

This step explored the use of VHR EO data, via visual interpretation (manual) and supervised image classification (automated). This VHR-based habitat assessment was performed at the three intensive study sites by combining:

-   prior expert knowledge about the studied species
-   either or both visual interpretation and automated classification of features of the VHR imagery
-   species occurrence georeferenced records and OS Open Rivers (only for Dragonfly and Damselfly species).

The advantage of very high spatial resolution (between 0.7–4 m in this study) was to unlock rich information for the species by distinguishing pattern, shape, texture, shadows, and colours (see Figure S1) and to reveal observable detail about the habitat(s) such as field boundary characteristics (e.g., hedgerows and field margins), agricultural management practices, ditches, woodland canopy structure, type of woodlands, continuity of the habitat over time, the interconnectedness of different types of habitats, etc. Patterns, shapes, texture, shadows, and colours were discriminated in "true colour" bands and translated into key features (see Table S5 [107] and Figure S1).

3. Results

The outcomes of the Step 2 (habitat suitability and connectivity) and Step 3 (habitat assessment via visual interpretation and supervised classification methods) for the five species are presented in this section. The information from Step 1, expert knowledge elicitation (complemented with appropriate records), is not presented because it is embedded as underpinning knowledge for the other two steps (see Table S2 [95–103] and Table S3 [98,102,103]).

3.1. Step 2—Habitat Suitability and Connectivity Analysis

We designed our representation and mapping of the habitat suitability and connectivity for our chosen species in the SH-AONB and our three intensive study sites using a 1–5 scoring system (1 = least suitable, 5 = highly suitable, see Table 4). This representation for each species was derived from the Step 1 expert knowledge elicitation of the habitat requirements (e.g., type, structure, and food requirements, see Table S1 [56,82] and Table S3 [98,102,103]) integrated with the readily available datasets and resources (e.g., UK CEH LCM2019, UK CEH Crop, DEM, Priority Habitat Inventory, OS Open Rivers, and Open Street Map).

3.1.1. Silver-Washed Fritillary (SWF) Butterfly

Figure 5 shows the suitable habitats of SWF butterflies in SH-AONB and the three intensive study sites. As the most suitable habitat (score 5) that can sustain an SWF population is placed in mixed deciduous woodland with sunny, flowery rides, and glades, and its edges, heather, and scrub, we included a 20 m buffer alongside footpaths that cross broadleaved woodlands and a 20 m buffer edge of the broadleaved forests. In addition, a score of 5 was assigned to the heather and heather grassland land cover classes. Scores of 4 and 3 were assigned to broadleaved woodland and grassland respectively, as the SWF butterfly is a powerful flier, albeit restricted to areas of Viola flower presence which grows at the margins of broadleaved, open, and sunny grassland and heathland. Scores of 2 and 1 were assigned to Coniferous woodland and Arable, Freshwater, and Inland rock, respectively, as these habitats are poor for this species due to the lack of food or host for their eggs. In terms of habitat connectivity, a positive linking of the available favourable habitat can be observed at our three intensive study sites. The potential benefits of higher resolution at the local scale than what is available in this Step 2 evaluation would be to enable improved understanding of the canopy openness and the effects of management at woodland edges and in hedgerows (as a key habitat for understanding connectivity). This is explored further in Section 3.2.1. In addition, at the SH-AONB scale, we can observe the availability of many areas of suitable habitats that can support this species. However, the best areas are somewhat 'patchy' in distribution, with some isolation of one area from another.

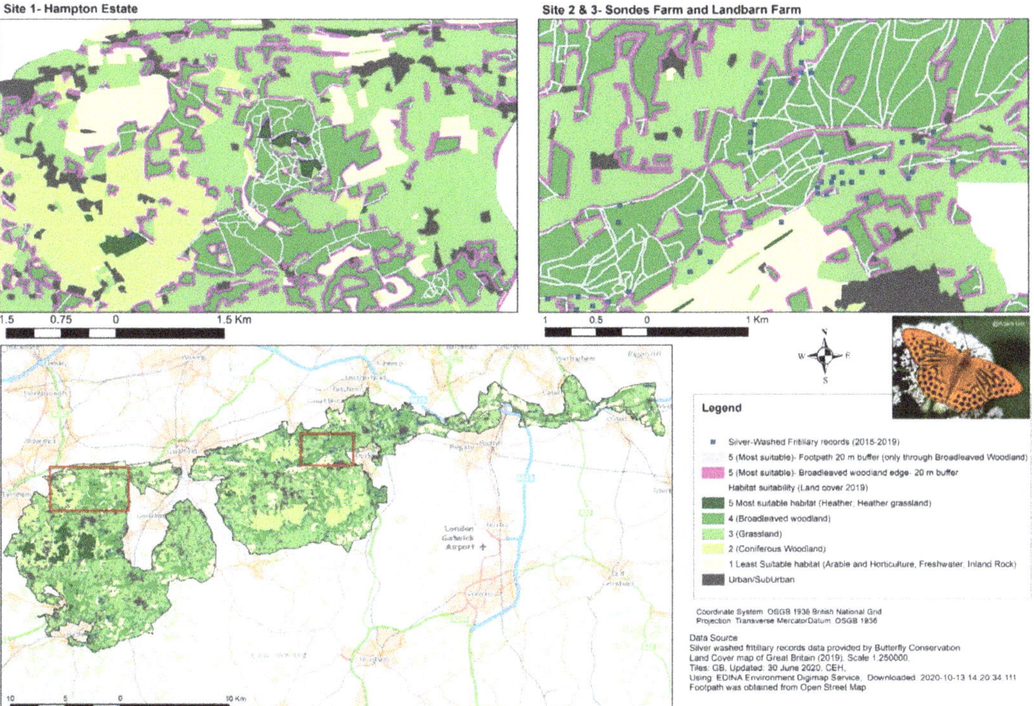

Figure 5. Silver-washed fritillary butterfly—Step 2 habitat suitability assessment in the SH-AONB and the three intensive study sites (red boxes).

3.1.2. Small Blue (SB) Butterfly

Figure 6 presents the suitable habitats for the SB butterfly, based on two environment variables: slope degree (Maps 1 and 2) and land-cover classes (Map 3). In addition, it overlays the records of SB butterfly from 2015 to 2019 which can indirectly confirm the presence of the Kidney Vetch plant (as this is only source of food for the caterpillars). It can be observed in both approaches (Maps 1 and Map 3) that most of the SB butterfly occurrence has been recorded across areas of chalk grassland (with a high slope degree), heather, and improved grassland.

In Map 3, the land-cover classes are classified based on our scoring system and the species habitat requirements. Therefore, at this Step 2 level, it is possible to observe several areas of potentially highly suitable habitat (score 4 and 5) that may offer the best opportunities for additional habitat creation for the SB butterfly, e.g., by seeding of Kidney Vetch in autumn or winter. However, further detailed examination of these broad potential habitat locations to find fully optimal sites (e.g., scrub or sparse eroding vegetation with bare ground, calcareous sites such as old quarries, gravel pits, and disused railways) would be advantageous and could be accomplished by site visits or, potentially, via a higher spatial resolution of EO imagery. Likewise, Map 3 illustrates the existence of many scattered areas of suitable habitat which can lead to the isolation of SB populations. Hence, the map from this Step 2 level can be useful for designing the introduction or restoration of connections between colonies (e.g., by even quite small habitat creation interventions) as a vital aspect of the long-term conservation of this species. Due to its good dispersal movement potential (efficient and light wings), maintaining the existing and new habitats and continuity among them depends on the farming activities and agricultural practices such as avoiding overgrazing by the livestock and fertilizer spray drift (see Table S3 [98,102,103]).

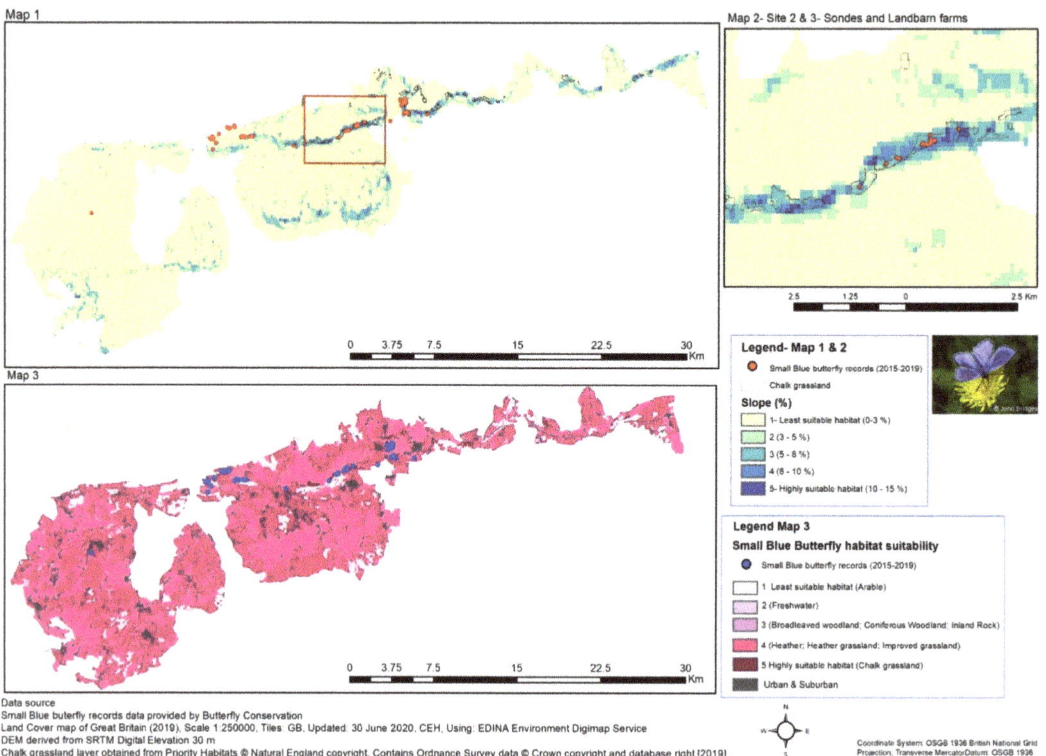

Figure 6. Small Blue butterfly—Step 2 habitat suitability assessment in the SH-AONB and the two intensive study sites.

3.1.3. Skylark

Skylarks select nesting sites at ground level with vegetation height about 20–50 cm to give easy access to the nest but also near to field margins that can provide food (insects). Certain crops such as spring and winter barley, spring wheat, and late-cut hay meadows are also ideal for nesting; however, nests are only successful if the field is not cut or grazed between early April and the end of May (see Table S3 [98,102,103]). Figure 7 shows suitable habitat for Skylark nesting in the SH-AONB and the three intensive study sites based on the CEH LCM2019 and CEH Crops 2019 classification and settlements. In addition, species record points between 2010 and 2019 can be observed across most of the suitable areas. However, the food requirement for this species cannot be assessed at this Step 2 resolution, and higher detail maps are required. Moreover, in terms of connectivity, Figure 7 illustrates that many sites ideal for nesting show quite high degrees of separation. Interpretation of the significance of this connectivity feature should include awareness that this species has a short-distance dispersal movement, and its home range has been noted as 1.08 ha [68].

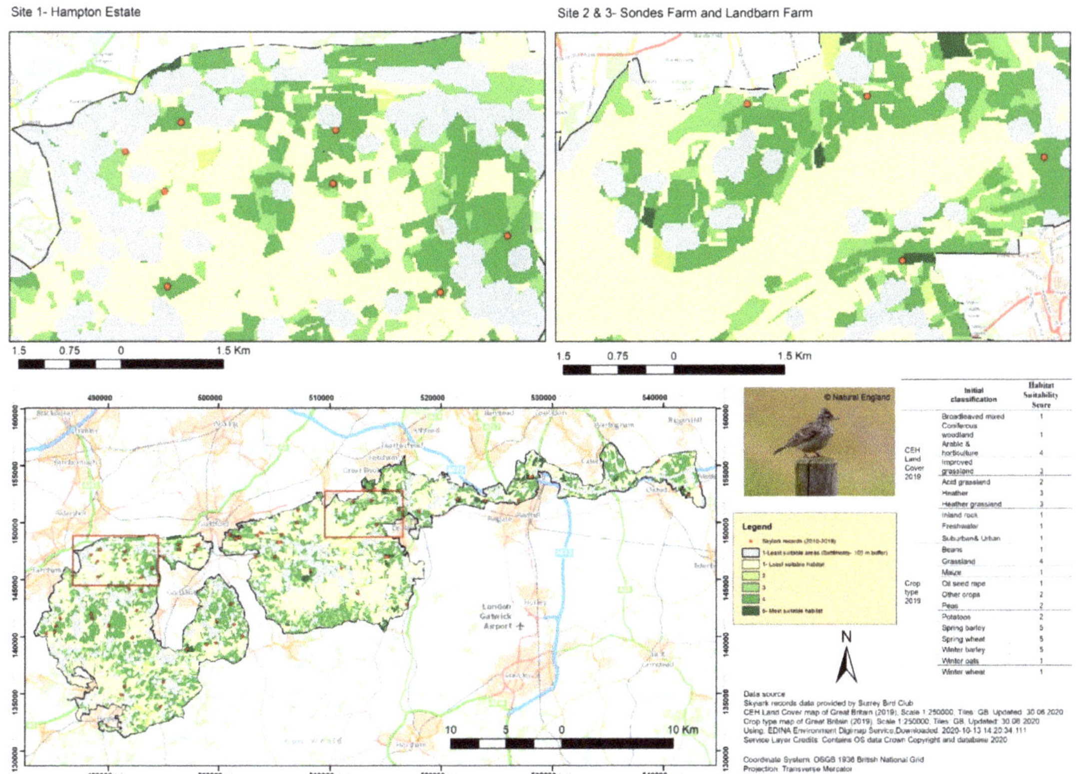

Figure 7. Skylark nesting—Step 2 habitat suitability assessment in the SH-AONB and the three intensive study sites.

3.1.4. Hazel/Common Dormouse

Habitat suitable for the Dormouse requires medium height (5–10 m) broadleaved woodland, scrub, heather, and tall hedgerows, which were the main driver for the five suitability classes identified based on the LCM2019 classes, as shown in Figure 8. A large amount of suitable habitat can be identified in the SH-AONB and the intensive study sites, but in some places, these are interrupted or fragmented by unfavourable habitats. It is notable that most of the dormouse occurrence records have been registered in those habitats that were assigned a score of 5 and 4; however, there are other suitable habitat areas in which dormouse occurrence has not been registered (e.g., across Site1).

3.1.5. Dragonflies and Damselflies

Figure 9 presents habitat suitability analysis for dragonflies and damselflies in Step 2. Despite a wide variety of these species (around 40) in the SH-AONB integrated into this approach, all rely mainly on clear and shallow waters, which can be part of a lake, pond, river, stream, ditch, or canal. Highly suitable habitats occur in Thursley National Reserve and its proximities, where most of the dragonflies and damselflies sightings have been recorded. As these species have preference for slow-flow watercourses, we created a 20 m buffer alongside all ditches, brooks, etc., as highly suitable habitat. Suburban and urban areas have not been included in the classification, although nature lovers may well have created small ponds in their gardens, and ditches and ponds can be present in municipal areas; hence, some occurrences have been recorded in suburban/urban sites.

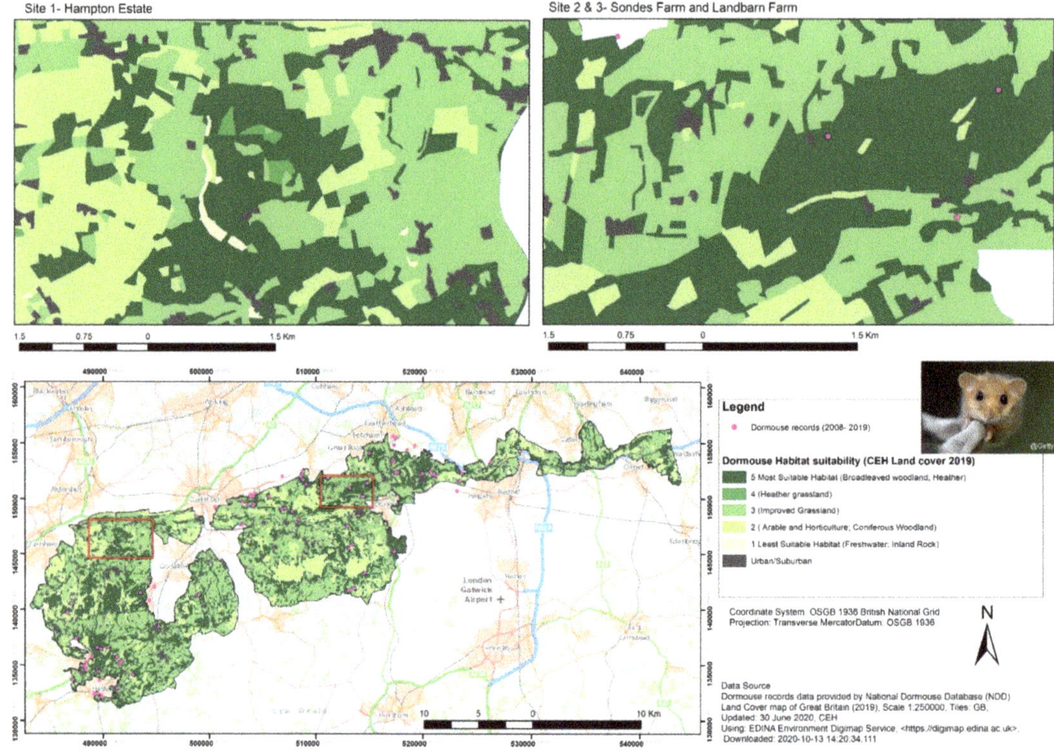

Figure 8. Dormouse—Step 2 habitat suitability assessment in the SH-AONB and the three intensive study sites.

3.2. Step 3—Habitat Assessment Analysis with VHR Imagery

The aim of the Step 3 approach was to investigate any additional value that VHR EO data can add to the understanding of available habitat for the five species from the Step 2 analysis. The analysis below is for the intensive study sites via the visual interpretation and the automated supervised image classification approaches. Interpretation of the VHR images by either approach applied the direct and indirect associations established between spectral bands (visible and PAN), textural, structural features, and plant diversity and habitat opportunities. The analysis also included examples of assessment over a period of time, opportunities to create/improve/extend suitable habitat, pressure factors, and dispersal movements for the species (see Table S3 [98,102,103]).

The VHR imagery-based analysis enabled observation of a larger range of species-associated habitat land-cover classes than in the LCM2019 set. These additional classes were canopy openness, scrub, hedgerows, water plants, trees shade, individual trees, and artificial features (tracks, pavement, etc). Visual interpretation of habitat assessment is time consuming, and particularly for larger areas, automated supervised image classification can play a very useful role in supporting efficient habitat quality and quantity assessment.

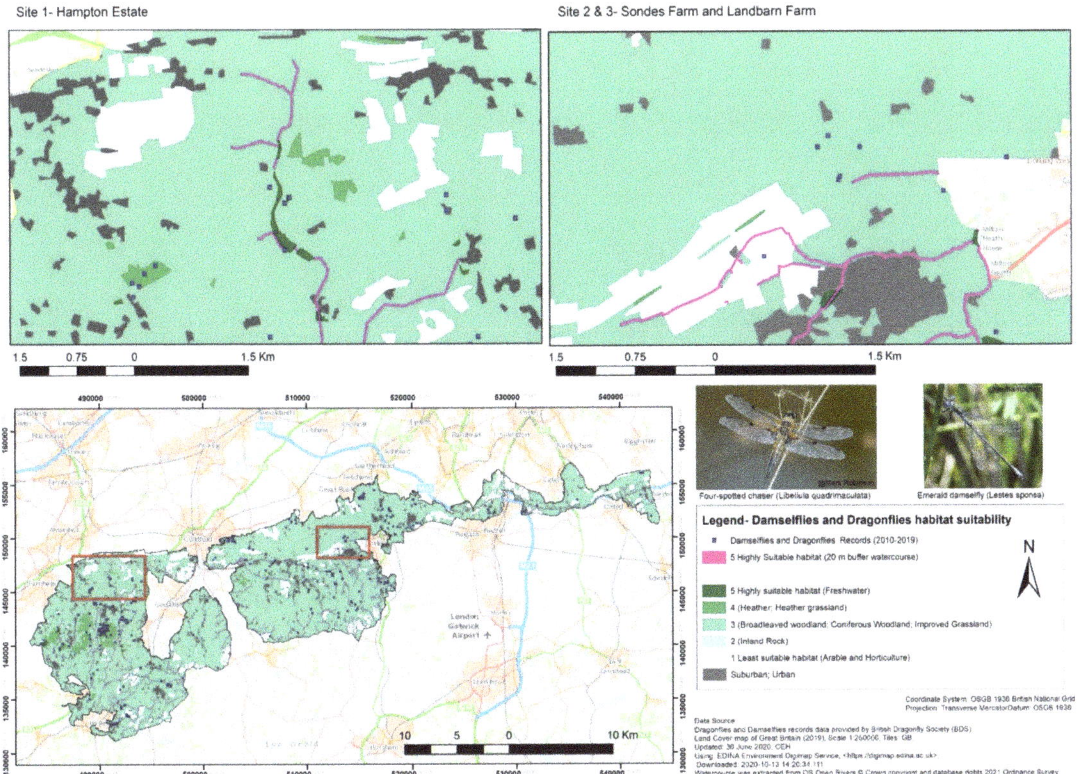

Figure 9. Dragonflies and damselflies—Step 2 habitat suitability assessment in the SH-AONB and the three intensive study sites.

The results of the VHR use are presented below in the same sequence of species as the results for Step 2. However, in this case, while noting habitat features relevant to the selected species our primary focus was on evaluating any 'additionality' that the VHR analysis can offer over and above that available from Step 2.

3.2.1. Silver-Washed Fritillary Butterfly

Figure 10 shows a well-vegetated area (Site 3) consisting of broadleaved and coniferous woodland, grazed grassland, maintained scrub patches, arable fields, ditches, hedges, etc. The broadleaved woodland exhibits several patches of open areas, glades and open rides that allow sunlight to come through, sunny and open woodland edges, and scattered scrub. These are all positive habitat features for this species and are revealed in finer detail than in the Step 2 results. The records of SWF sightings between 2015 and 2019 emphasise the presence of the species in such areas and also that such sightings coincide with walking pathways. The VHR images shows additional features such as other vegetation (providing additional shade) in places along such pathways across open grassland that also coincide with recorded sightings. In terms of assessment between July 2017 and May 2020, the detailed habitat features are mainly consistent between the two, indicating stability in the provision of appropriate habitat over this time series. Interestingly, there has been the removal of a small block of coniferous woodland over this time (see red circled area in Figure 10; shown as present in LCM2019 Figure 5 of Step 2) and the now open area observable in the 2020 imagery can provide additional habitat for this species.

Figure 10. Silver-washed fritillary butterfly—Step 3 VHR habitat assessment, intensive study Site 3, visual interpretation (red circles indicate area of coniferous woodland removed between 2017 and 2020).

Figure 11 (Map 1) provides specific visual detail indicating highly desirable habitat for the SWF within the broadleaved woodland (canopy openness, glades and other spaces, open rides, etc.) that are absent from the Step 2 representation (Map 3). Similar detail is available for the grassland areas in which highly desirable scattered tree cover and associated shade is present. Map 2 shows that these highly desirable habitat features are also available from the supervised image classification indicating that they are tractable to automatic recognition and mapping. In this mapping, one conspicuous open ride is readily identified in the broadleaved woodland (red linear pattern in north-south orientation in Map 2) together with other areas of canopy openness and scattered trees, making location and calibration of these habitat features readily assessed.

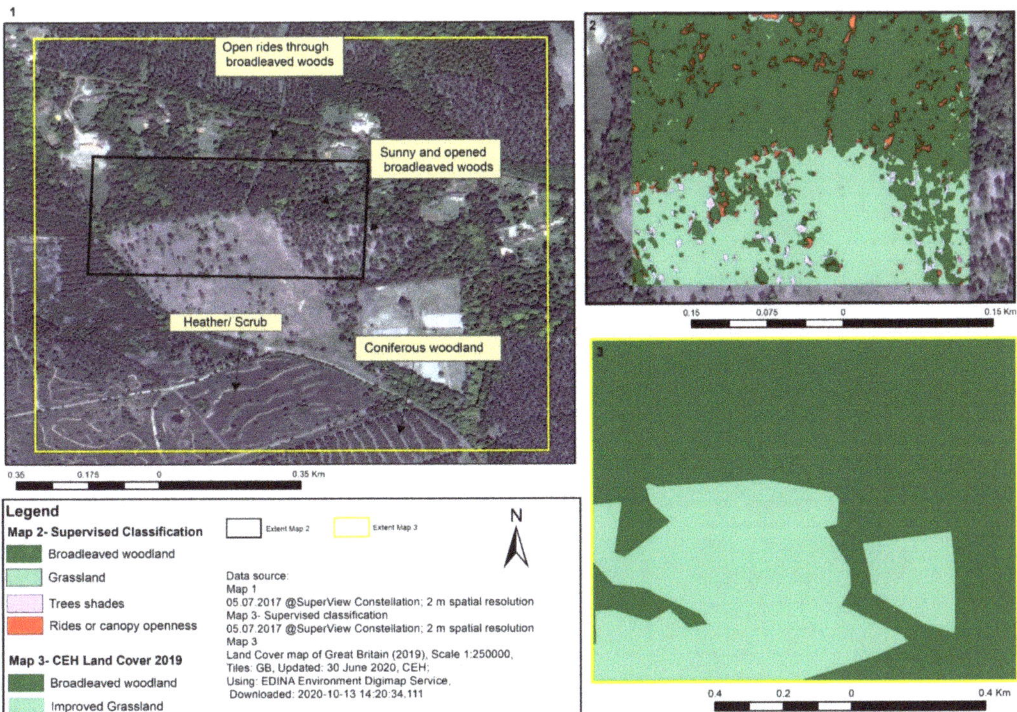

Figure 11. Silver-washed fritillary butterfly—Step 3 VHR habitat assessment, intensive study Site 1, visual interpretation (Map 1), supervised image classification (Map 2). Map 1 contains visual interpretation of 2 m resolution Superview satellite image; black box indicates the same area of view in Map 2; yellow box indicates the same area of view in Map 3; Map 3 shows the yellow box area of Map 1 overlaid with CEH LCM2019 broadleaved woodland and improved grassland land-cover classes from Step 2 (scored 4 and 5, respectively, as highly suitable habitat for the species).

3.2.2. Small Blue Butterfly

Figure 12 presents a time series of VHR (ranging from 0.7 to 2 m) DMC3 and Superview images in PAN and MS (visible bands) from 2016 to 2020 to assess the consistency of the SB butterfly habitat provision over this time period. As the SB butterfly presence is strongly linked to the condition of Kidney Vetch flower, even though this cannot be identified directly by satellite images, patches of scrub, sunny open space, and some areas of ungrazed grassland are readily discerned (the quality of grassland can be differentiated—the light grey (PAN) and light green (MS) are those areas with grazed grassland: the dark grey and green is tall grass and scrub), and all these types of areas are positive for the Kidney Vetch. In addition, detailed assessment of the habitat observable from the VHR imagery over the 2016–2020 period provides evidence that the habitat has been stable and well maintained over this 5-year period, thus providing support for the SB population.

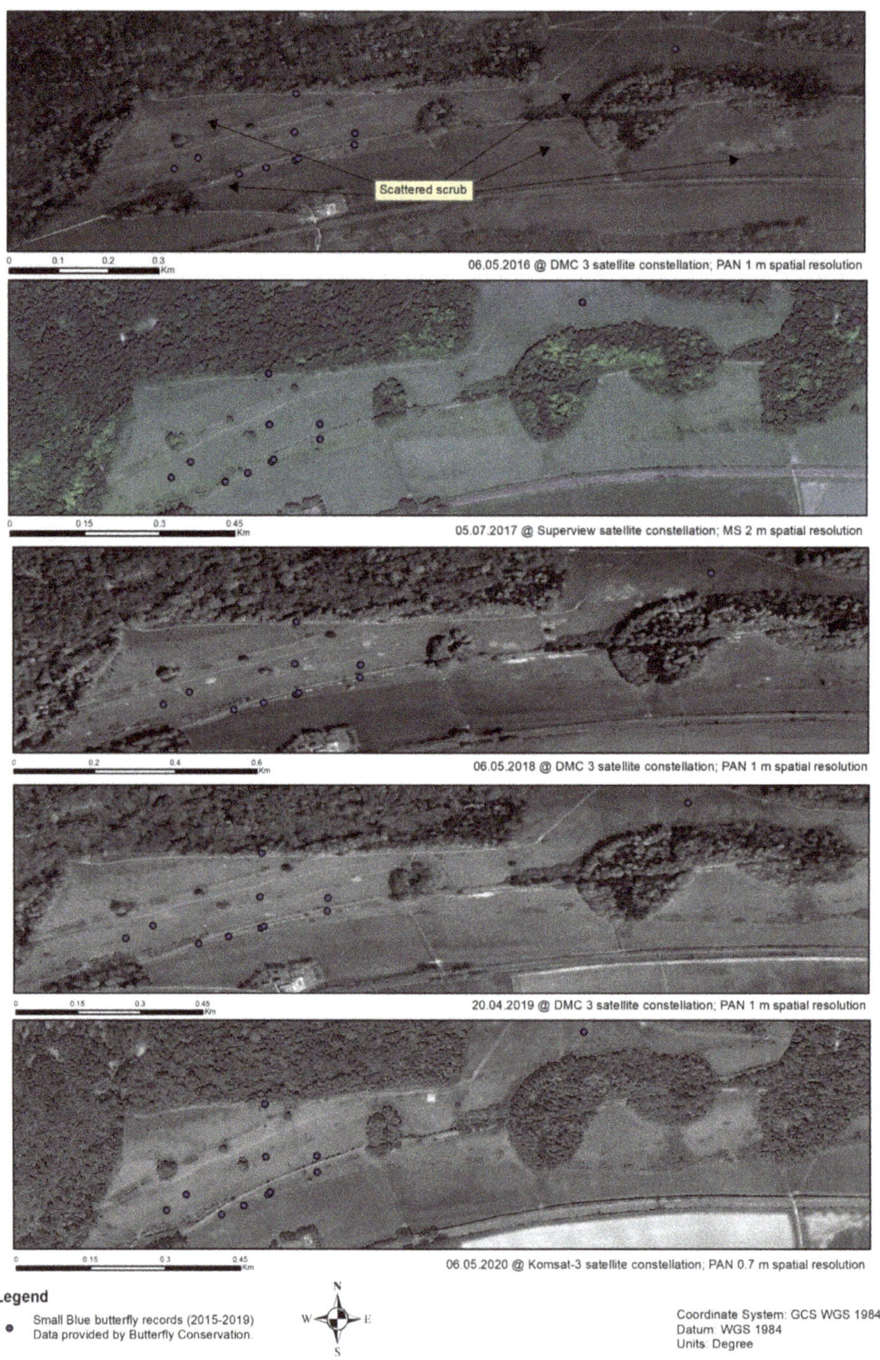

Figure 12. Small blue butterfly—Step 3 VHR habitat assessment, intensive study Site 3, visual interpretation over time series.

Figure 13 illustrates the use of VHR imagery for automated quantification of the habitat relevant for the SB butterfly in a 0.5 km^2 area within Site 3. The histogram shows the percentage of different land-cover classes using supervised image classification in which the VHR imagery has provided much finer detail, thus obtaining the areas of scrub, unimproved grassland, and footpaths, classes that are not available via LCM2019 data. For example, the automated quantification shows that 13% of the selected area is scrub associated with 46% of open grassland, which together offer highly favoured habitat for the SB.

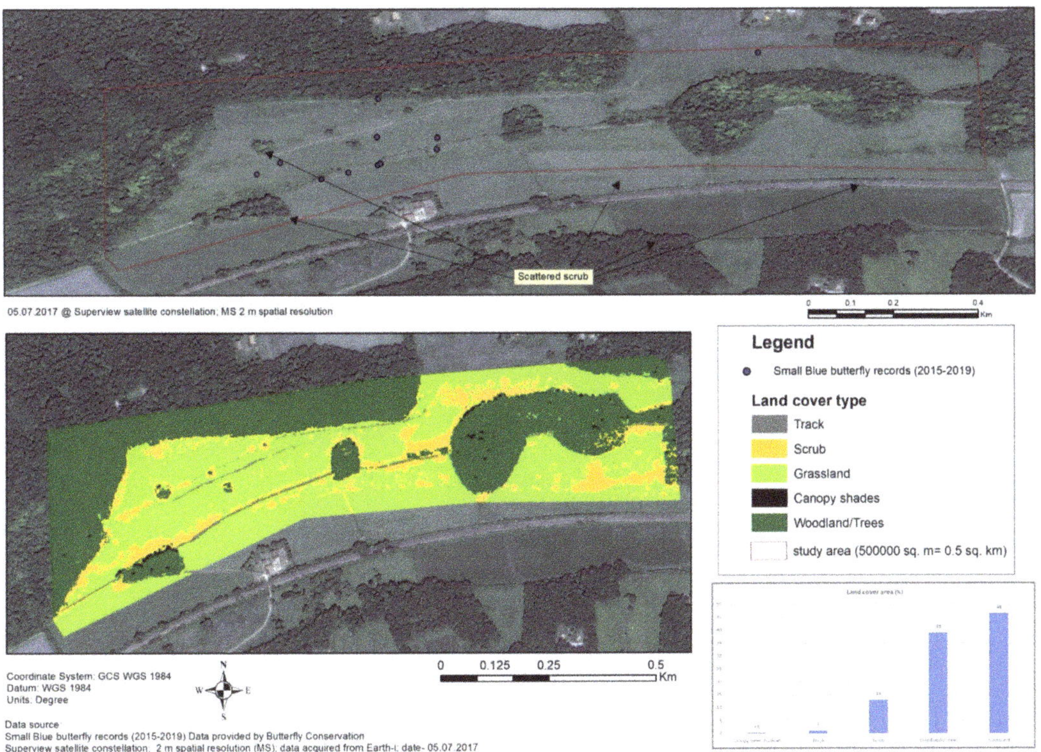

Figure 13. Small Blue butterfly—Step 3 VHR habitat assessment, intensive study Site 2, visual interpretation, supervised image classification with quantification.

3.2.3. Skylark

This bird can nest in silage fields (grazed pasture) if the field is not grazed between early April and the end of May. Figure 14 presents an example of monitoring such habitat suitable for Skylark nesting in four grass/silage fields at intensive study Site 1, by using VHR satellite images in mid-May and again in late-June 2018 and species sighting recorded in the same year. The fields are clearly uncut in May 2018, thus maintaining this habitat for Skylark nesting and rearing, and they are showing as cut (for silage) at the end of June 2018, after this important period for the Skylark. It is clear that the detail available from the VHR imagery is sufficient to assess the status of these fields (cut/uncut), and therefore, with appropriate overflies and image collection frequencies, VHR can be used readily to monitor the grass management in support of Skylark nesting habitat. We believe that the Skylark seen in 2018 was successfully maintained by the agricultural practices in place at that time.

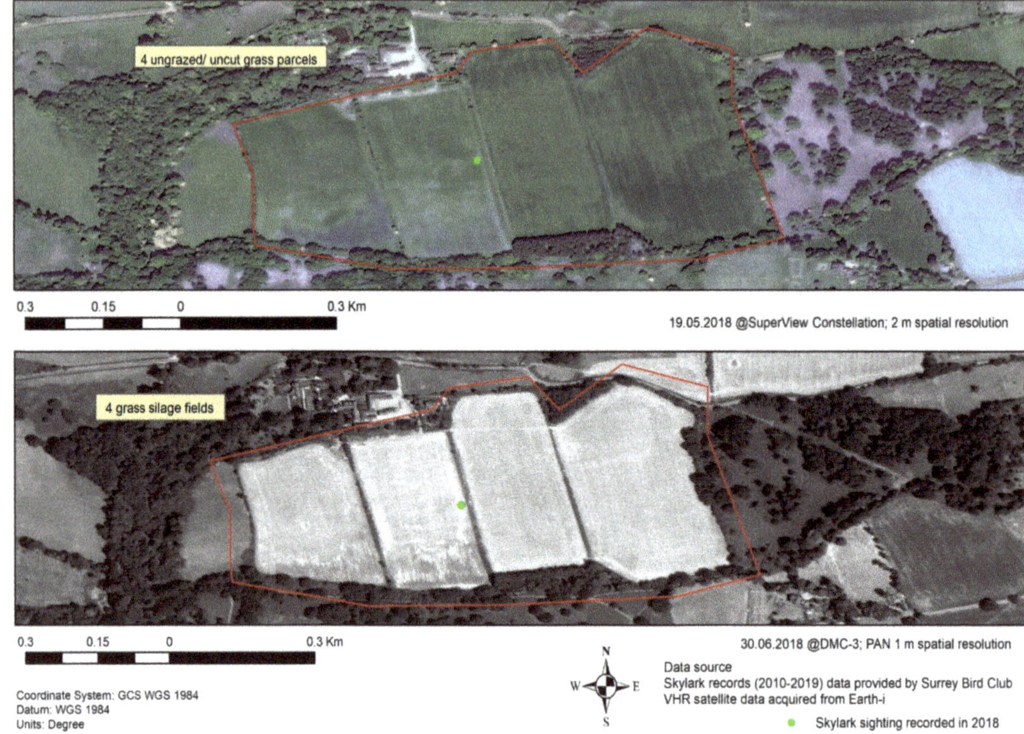

Figure 14. Skylark Nesting—Step 3 VHR habitat assessment, intensive study Site 1, visual interpretation.

Figure 15 shows a grass field at intensive study Site 2 in 2017 and 2020. This location is a suitable nesting habitat for Skylark (skylark recorded there in 2017). The VHR imagery (Map 1 Figure 15) shows the field with full grass cover in early-July 2017, indicating that it had not been cut during the Skylark breeding season. Furthermore, the grass margins alongside hedgerows (lighter green (Map 1) and lighter grey (Map 2) Figure 15) are clearly present with a width of ~10 m in both 2017 and 2020. These two practices, grass cutting time and field margins, plus the presence of hedgerows and neighbouring woodland (supporting insect and seed diversity), play a crucial role for the Skylark by providing a good physical location, source of food, and materials for nesting.

3.2.4. Dormouse

Favoured habitat for this species is deciduous woodland (preferring yew, rowan, and hazel during ranging), hedgerows, and dense scrub. Figure 16 illustrates a well-connected network of favourable dormouse habitat. In terms of the habitat quality, the dormouse population recorded in 2017 had a suitable habitat in both 2017 and 2020, noted as (+) in Figure 16, due to the dense and wider hedges and young woodland (bright green or light grey). However, there are some gaps in hedgerows, noted as (−) in Figure 16, which could have a negative impact on the species dispersal. Farmers and land managers can restore and/or add to hedgerows and fill such gaps to better link existing habitat areas for improved foraging and dispersal for this species.

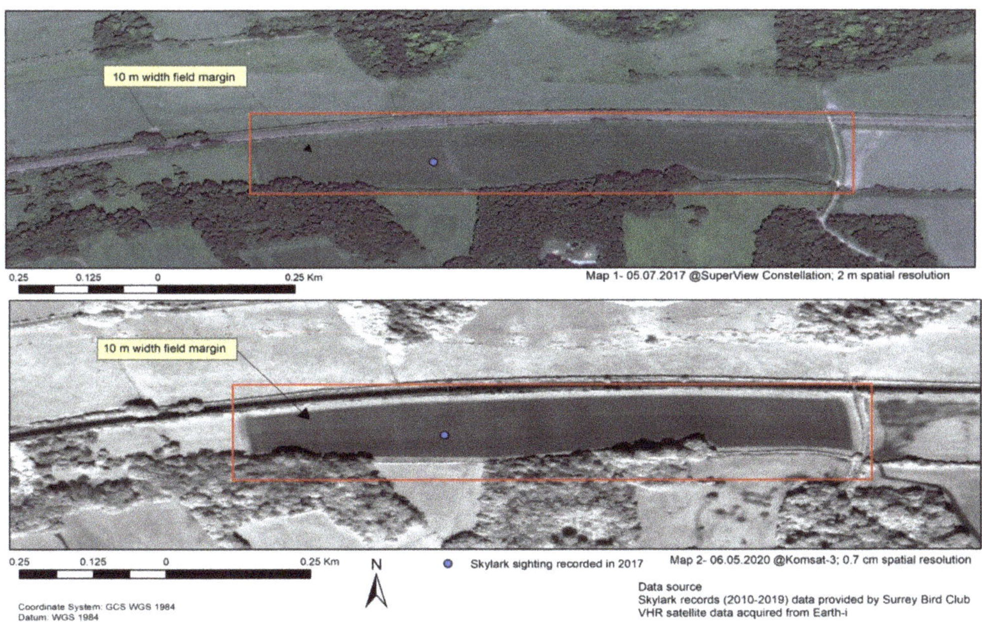

Figure 15. Skylark—Step 3 VHR habitat assessment, intensive study Site 1, visual interpretation.

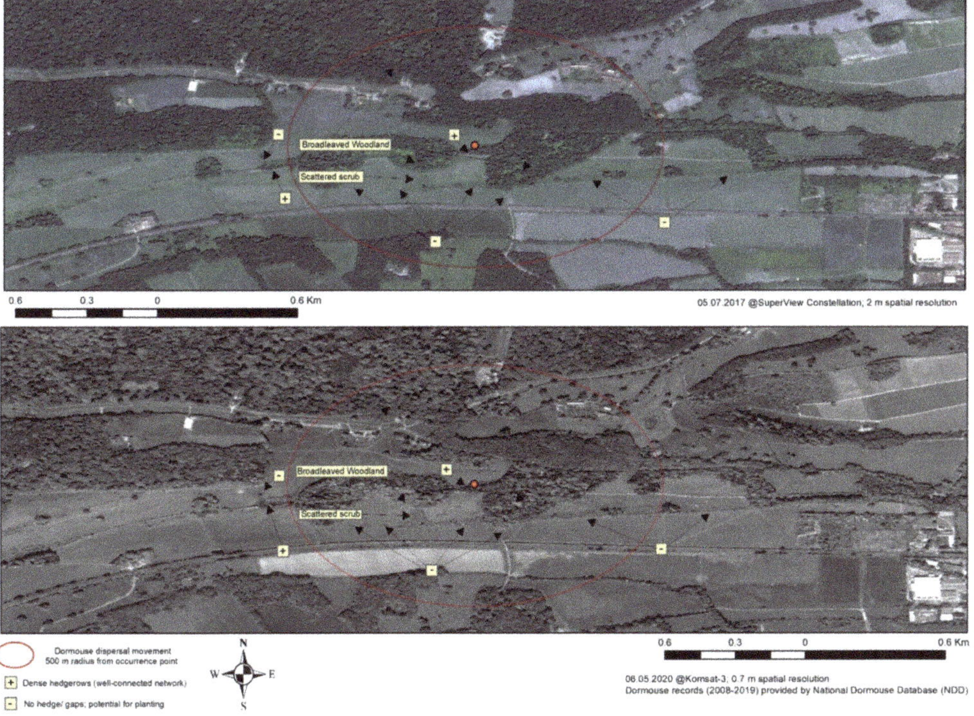

Figure 16. Dormouse—Step 3 VHR habitat assessment, intensive study Sites 2 and 3, visual interpretation.

Figure 17 shows a comparison of VHR imagery at 2 m resolution from SuperView (Maps 1 and 3, Figure 17) with the same area at 10 m resolution from of Sentinel 2 (as per LCM2019) (Map 2, Figure 17). As for the other species, the VHR imagery permits recognition and location of a much more granular classification and range of land-cover types for dormouse habitat assessment, e.g., scrub, patchy tree or tall shrub cover and shading as strong positive features. This is also tractable to automatic representation by the supervised image classification approach (Map 3, Figure 17).

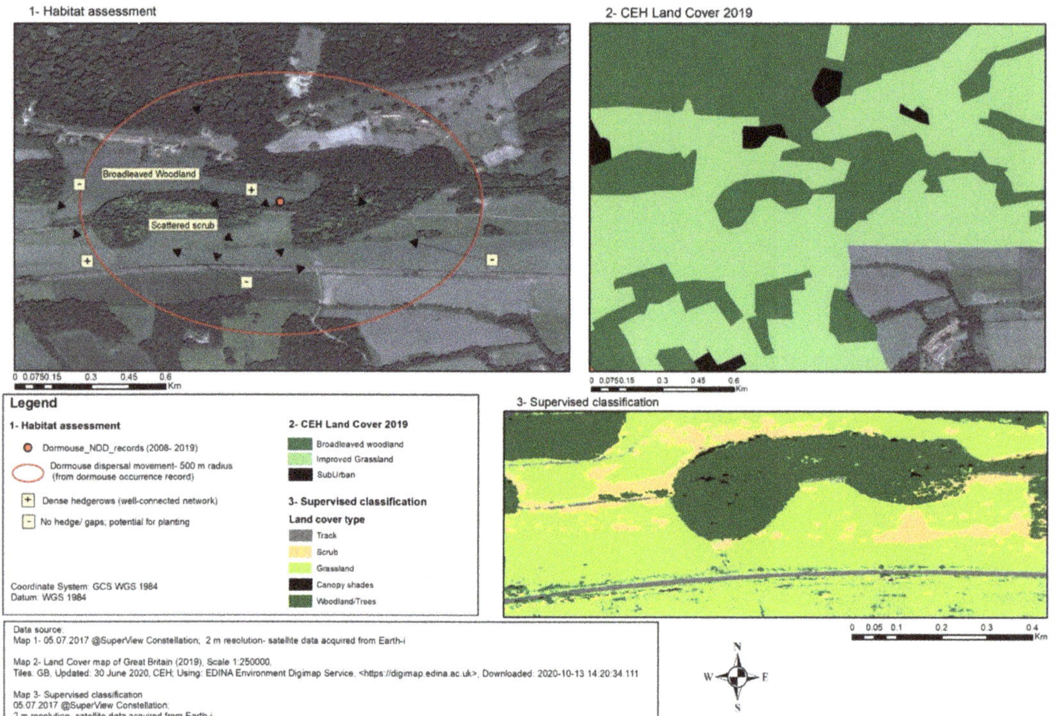

Figure 17. Dormouse—Step 3 VHR habitat assessment, intensive study Sites 2 and 3, visual interpretation and supervised image classification.

3.2.5. Dragonflies and Damselflies

Two of the main pressures on dragonflies and damselflies are (i) poor water quality from leaf litter, extensive growths of filamentous algae, water coloured green by planktonic algae, and nutrient inputs from agricultural practices, and (ii) livestock access to bankside vegetation. Therefore, Figure 18 shows that several variations in the habitat qualities for dragonflies and damselflies that can be distinguished through the use of VHR imagery for habitat assessment. The imagery from July 2017 and May 2020 permits recognition of positive habitat features such as thick bankside vegetation alongside ditches and a buffer zone of at least 10 m arounds the ponds that have not been affected by livestock both of which also provides reasonable protection against nutrients inputs from fertilisers, and from herbicides and pesticides. Negative features are also readily observed: for example, in 2017, the pond situated in the centre of the image (marked as red hexagon in Figure 18) was mostly covered by vegetation with extensive growth of aquatic plants. This is less apparent in 2020, which could be due to the low rainfall and higher temperature in the 2017 season.

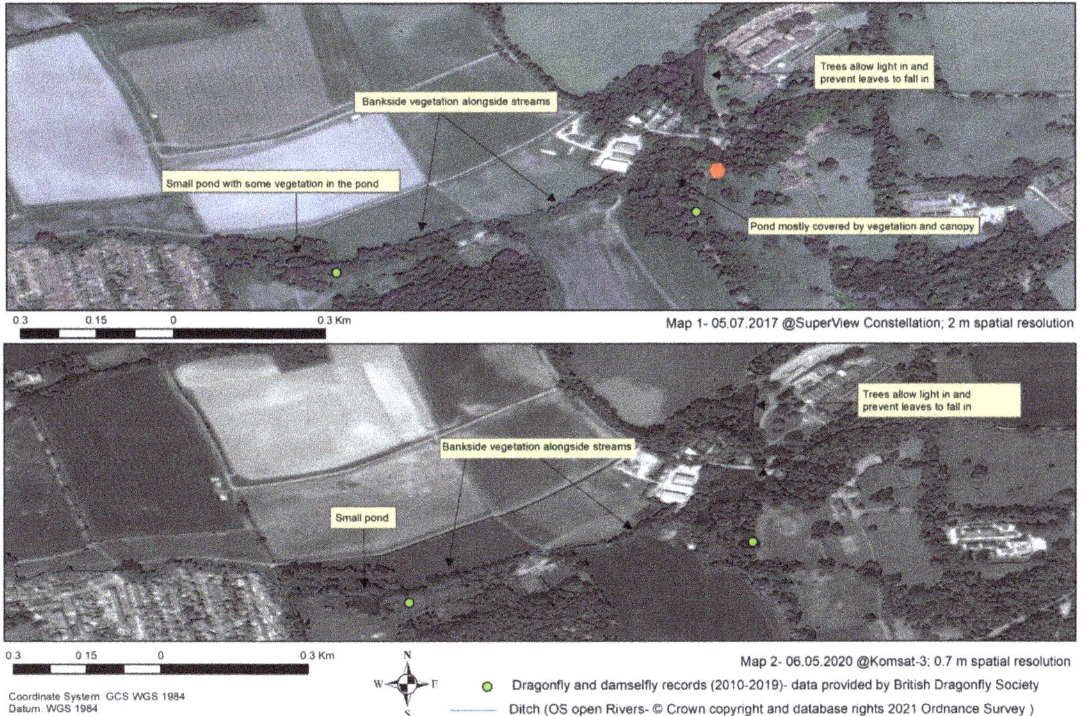

Figure 18. Dragonflies and damselflies—Step 3 VHR habitat assessment, intensive study Site 2, visual interpretation.

In Figure 19, the two VHR images from May 2018 and June 2020 present an array of open ponds in which habitat assessment features are easily recognised. The tree canopies are not overhanging, but in some areas, the light brown colouration of the water indicates the presence of sediments. This cloudiness could occur due to relatively shallow waters, if the pond is located on clay, or it can be caused by fish or ducks stirring up bottom sediments. Overall, it indicates a favourable and healthy habitat for dragonflies and damselflies, which concurs with the sighting records.

Many of the dragonfly and damselfly species in the SH-AONB have habitat preferences for clear ponds with open water surfaces and slow water flows (e.g., ditches). The upper two VHR images in Figure 20 shows a substantial amount of aquatic vegetation within the pond boundary with substantial tree cover surrounding. These conditions are not ideal habitat for these species. In comparison, the lower pair of images in Figure 20 show a high degree of openness of the water surface and an absence of aquatic plants or extensive shading and overhanging trees, indicating the bulk of the pond area provides high quality habitat for the species. The supervised image classification images in Figure 20 show that these main habitat features are tractable to automated assessment.

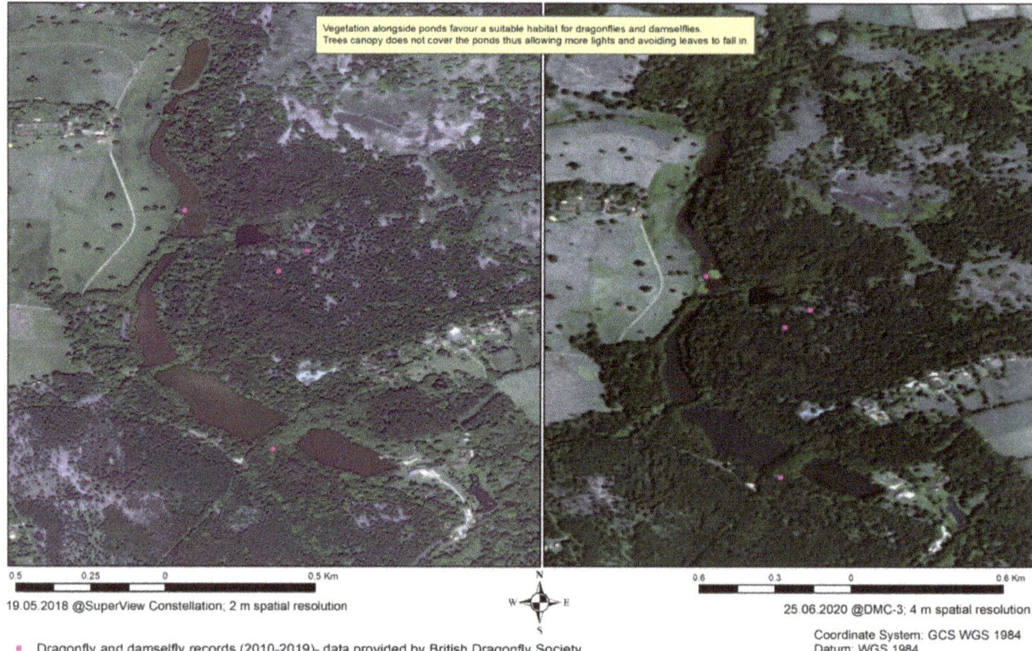

Figure 19. Dragonflies and damselflies—Step 3 VHR habitat assessment, intensive study Site 1, visual interpretation.

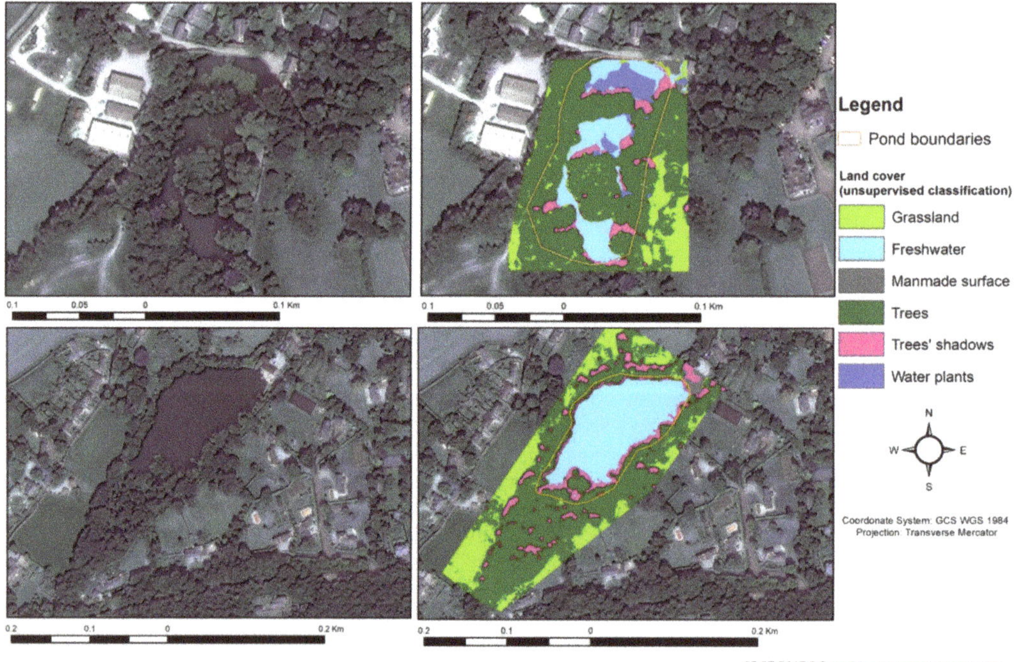

Figure 20. Dragonflies and damselflies—Step 3 VHR habitat assessment, intensive study Site 2 and surroundings, visual interpretation and supervised image classification.

Detailed visual interpretation of the lower VHR imagery in Figure 20 indicates, however, some limitation in the precision of the automated supervised image classification in that the tapered, southwest part of the pond is classified as trees but is actually an overgrown part of the pond area. Close visual assessment shows that, while the main part of the pond, especially its more northern part, is very good habitat for these species, the lower southwest part exhibits a mix of freshwater (little open surface), trees, and other vegetation. A time series of imagery of his pond would enable interpretation of whether this condition of the pond's southwest part was due to the encroachment of vegetation into a previously open area of the pond or, conversely, if the cause was inundation of a previously vegetated area. This detail, readily observable in the VHR imagery here, provides a good demonstration of the value that VHR EO can bring, for example by indicating where management interventions can best be targeted to maintain or enhance available, high-quality habitat for particular species such as the dragonflies and damselflies in this case.

4. Discussion

This present study has examined the use of EO and geospatial datasets with a particular focus on the additional insight that VHR imagery (0.7–4 m resolution) can bring to the understanding of the availability, connectivity, and assessment at various scales of habitat suitable for five key species in the SH-AONB. Efforts to use the capabilities of EO for the purposes examined here depend crucially on combining knowledge on the habitat needs of the species in question (acquired from expert consultations and published sources) with that on EO technology and image interpretation. There are clearly important synergies here between these groups and use of the EO-derived data needed to have the input of experts who have in-depth knowledge of the ecology of the species. Similar image classifications have been accomplished in the literature [37]. The present research has provided several examples of the ways in which such a combination can be applied to habitat assessment.

EO-derived land-cover and land-use products have been used successfully for mapping and monitoring biodiversity and modelling species habitat suitability [36,37,43,44,64,108] and for assessing ecosystem services [108,109]. Hence, at the scale of the whole SH-AONB, as well as for our three intensive study sites, the readily-available EO and geospatial datasets provide valuable resources enabling identification of the areas of broadly appropriate habitat for these species, albeit at a fairly 'coarse' level of resolution. In general terms, this provides a semiquantitative snapshot of potential habitat availability for the various species which may be particularly useful as a broad guide to the prevalence and location of the habitats across the whole SH-AONB. At the level of individual farms and similar landholdings, the readily available EO and geospatial datasets again provide the capability to broadly map suitable habitat at the individual property scale, e.g., several hectares. However, as is apparent from the habitat needs of these species outlined in Table S3 [98,102,103] and in the detailed analyses, the high-level land classifications in the very useful LCM2019 data and similar/complementary resources only permit a quite general assessment of the availability and distribution of suitable habitat. A key question driving our research was whether information potentially available from the VHR satellite imagery could provide the level of detail that would enable a much more confident assessment of the quantity and quality of the habitat present for the given species. Further questions about such VHR imagery and its interpretation were related to whether such information would be relevant for habitat management and, potentially, for MRV systems for habitat provision under policies such as ELMs. These aspects form the basis of the discussion below.

VHR satellite data have proven efficient and accurate in assessing habitat quality [41,42]. The VHR data (0.7–4 m) used in Step 3 of this research allowed detailed analysis of image texture, patterns, colours, meanings, and identification and quantification of areas of scrub, hedges, field margins, different tree species types, etc. This was a considerable extension to the 'granularity' of habitat assessment over that available from the readily available resources analysed in Step 2 (which relied upon resolutions of 10 m or more and other geospatial datasets). Specifically, the benefits of using the VHR imagery as demonstrated in our results were a considerably extended capability to:

- Identify a wider range of landcover types relevant to the habitat preferences of the species than the range of classes in LCM2019. Notable amongst these was scrub (an aspect of preferred habitat of small blue butterfly, Silver-washed fritillary butterfly, and Hazel dormouse), 'patchiness' small open spaces in broadleaved woodland (Silver-washed fritillary butterfly), and recognition of small areas of single trees (Hazel dormouse, Silver-washed fritillary butterfly, and small blue butterfly).
- Assess qualitative differences within a given land class/habitat type—examples include the ability to resolve uncut and cut grassland as shown in the habitat assessment for the Skylark and, by inference, whether grassland areas are grazed or are tall/scrubby (the Kidney Vetch flower is associated with the latter, and the SB butterfly is dependent on this plant).
- Automate quantitative assessment of the habitat/land cover and automate the representation of the extended range of land-cover/habitat types via supervised image classification. This raises the potential to efficiently quantify at a finely resolved level (e.g., 1 to a few m), the provision and location of suitable habitats for these, and most likely, many other species. Given the availability of suitable VHR image coverage, even quite large areas (up to 23 km^2) are tractable to this finely resolved habitat assessment approach.
- Evaluate negative aspects of habitat provision such as discontinuities in valuable 'corridors' such as hedgerows (important avenues for migration and movement for wildlife and therefore a highly distinctive habitat) and diversity of 'qualities' within a given habitat type (such as ponds where features such as the quality of adjacent vegetation are also important).
- Evaluate habitat provision and quality efficiently and readily over time, offering valuable information for management decision making, e.g., urgency of intervention or responses and for efficient MRV for incentive schemes and policies.
- Assess habitat in 'non-traditional' urban and suburban areas by using the more finely resolved observation offered by VHR imagery to analyse the distribution and connectedness of relevant habitat types in such settings, and to support the nature recovery potential of private gardens, municipal and public spaces, and the built environment.

These benefits from the use of VHR imagery within this research are compelling. There are also a number of improvements and limitations that can be recognised. For example, species–habitat knowledge is a critical factor in the VHR interpretation as noted at the beginning of this discussion. It would have been advantageous to have had more extensive data, particularly for the occurrence records of the given species (and for those that they rely on for food). For instance, the SWF butterfly relies on the presence of Viola flowers and SB butterfly on the Kidney Vetch and having the records of the distribution and detailed preferences of these two plants would likely have allowed the creation of a more accurate habitat suitability analysis from VHR observations. Regarding access to VHR imagery, a more comprehensive time sequence for the imagery would have allowed unequivocal demonstration of land-use management directly supporting habitat provision. A good example of this was the full habitat assessment benefit that would have been possible if VHR imagery had been available for the precise harvesting time of silage grass cutting in relation to the Skylark. It is apparent from the levels of VHR resolution examined that the results presented here reflect habitat availability and its characteristics. The observations do not confirm if the species in question is actually present or not. The debate about whether the key metric for biodiversity conservation and enhancement should be (i) the availability of appropriate habitat (and its 'connectedness') or, (ii) whether it should be based strictly on species occurrence, lies outside the scope of this research. Therefore, it should be recognised clearly that the scope of the VHR imagery used here is at the level of habitat recognition and availability, rather than as a proof of species presence (although occurrence of the species was recorded where available in the image presentations).

The precision in time and resolution for assessing areas of land, as demonstrated in this research, clearly offer huge potential as primary evidence for the MRV of habitat and other benefits that are needed for the correct distribution of the incentives associated with ELM policy. Capitalising on this potential requires improved coverage in both time and area at these levels of resolution and at acceptable cost. EO data are becoming more and more widely used in governance and planning at all scales, and the demonstration of its cost-effectiveness in similar applications [34,110] provides good reason to be optimistic that ongoing R&D and applications development enable the near-term use of VHR imagery in MRV for ELM and related policy.

Finally, we wish to reiterate the usefulness of EO data to inform biodiversity conservation and environmental management as highlighted here and by remote sensing specialists and ecologists in recent years [37,47,64,66]. When EO is used in this field, it requires ongoing collaboration between ecologists and EO experts to develop a common, shared understanding about the relationship between the species selected and their habitats, terminology (e.g., regarding land-cover classes), guidance for understanding the opportunities to create/improve/extend suitable habitat for the species, and recognition of ecological and environmental pressures on the species. Such collaboration is essential to realising the outstanding potential for EO data to effectively support (i) ecologically informed environmental management decisions and (ii) the provision of habitat as a much-needed 'public good'.

5. Conclusions

We have drawn the following main conclusions from this research:

- Satellite spatial resolution is decisive in terms of assessing biodiversity and habitats. VHR data (at approximately 1–4 m) offers great potential for habitat suitability and connectivity assessment for the five wildlife species in this research and, most likely, for many more.
- Automated habitat suitability assessment using VHR imagery is feasible and provides valuable, ecologically meaningful information
- The expert insights of ecologists on the species–habitat relationships examined here provide key underpinning knowledge to enable use to be made of the potential of VHR satellite data for habitat assessment.
- VHR data and imagery offer great potential for use in habitat management at the scale of individual properties (farms, etc.) and at a whole-landscape scale. It provides an effective source of information of value for land management and environmental decision making and as potential evidence for the MRV relevant to ELM and similar policies.

Supplementary Materials: The following are available online at https://www.mdpi.com/article/10.3390/su13169105/s1, Table S1. UK CEH Land Cover 2019 classes description; Table S2. Habitat requirements of 5 studied species; Table S3. Species habitat requirement and opportunities to create/improve/extend suitable habitat for selected species; Table S4. Workflow of modelling the habitat suitability and connectivity; Table S5. Habitat assessment approach and contribution of VHR satellite imagery; and Figure S1. Examples of VHR satellite data.

Author Contributions: Conceptualization, A.A., R.J.M., S.M. and J.L.; methodology, A.A., R.J.M., S.M. and J.L.; software, A.A.; validation, A.A., R.J.M., S.M. and J.L.; formal analysis, A.A., investigation, A.A.; resources, A.A., S.M. and R.J.M.; data curation, A.A.; writing—original draft preparation, A.A.; writing—review and editing, A.A., S.M., R.J.M. and J.L., visualization, A.A.; supervision, S.M., R.J.M. and J.L., project administration, R.J.M.; funding acquisition, S.M., R.J.M. and J.L. All authors have read and agreed to the published version of the manuscript.

Funding: This research was part-funded by Natural Environment Research Council (NERC) SCENARIO Doctoral Training Partnership, Grant/Award NE/L002566/1, CASE award partner the

National Physical Laboratory (NPL) and by UKRI/Research England SPF funding via the University of Surrey. The APC was funded by the University of Surrey.

Institutional Review Board Statement: Not applicable.

Informed Consent Statement: Not applicable.

Data Availability Statement: Some dataset presented in this study are openly available in [90–93]. Restrictions apply to the availability of the UK CEH LCM2019 and Crops2019 products. Data were obtained from Digimap Edina, available from the https://digimap.edina.ac.uk/ with the permission of University of Surrey.

Acknowledgments: This research was supported by the SCENARIO Doctoral Training Partnership of the UK Natural Environment Research Council (NERC) with PhD funding for the first author. We thank the National Physical Laboratory (NPL) for co-funding as the CASE award partner and, in particular, for co-supervision by Emma Woolliams of NPL for AA's PhD research. We also would like to thank University of Surrey SPF UKRI/Research England support for AA for her secondment *"Earth Observation for Advanced Monitoring, Reporting and Verification for Environmental Land Management (ELM) policy in the Surrey Hills Area of Outstanding Natural Beauty (SH-AONB)"* to conduct this research with the SH-AONB Defra Test and Trial. We are grateful to the ecological experts for their generous consultations, including Surrey Wildlife Trust, in particular, Mike Waite, and Butterfly Conservation, SH-AONB and its personnel, and the team delivering SH-AONB Nature Recovery Defra Test and Trial. Likewise, Earth-i are acknowledged for provision of the pre-processed VHR satellite imagery and for associated support.

Conflicts of Interest: The authors declare no conflict of interest.

References

1. Motavalli, P.; Nelson, K.; Udawatta, R.; Jose, S.; Bardhan, S. Global achievements in sustainable land management. *Int. Soil Water Conserv. Res.* **2013**, *1*, 1–10. [CrossRef]
2. Schwilch, G.; Bestelmeyer, B.; Bunning, S.; Critchley, W.; Herrick, J.; Kellner, K.; Liniger, H.; Nachtergaele, F.; Ritsema, C.; Schuster, B.; et al. Experiences in monitoring and assessment of sustainable land management. *Land Degrad. Dev.* **2011**, *22*, 214–225. [CrossRef]
3. Hans, H. Concepts of sustainable land management. *ITC J.* **1997**, *3/4*, 210–215.
4. Zscheischler, J.; Rogga, S. Innovations for Sustainable Land Management—A Comparative Case Study. In *Sustainable Land Management in a European Context: A Co-Design Approach*; Weith, T., Barkmann, T., Gaasch, N., Rogga, S., Strauß, C., Zscheischler, J., Eds.; Springer International Publishing: New York, NY, USA, 2021; pp. 145–164.
5. World Bank. *Sustainable Land Management*; The World Bank: Washington, DC, USA, 2006; p. 108.
6. Sustainable Soil and Land Management and Climate Change. Available online: http://www.fao.org/climate-smart-agriculture-sourcebook/production-resources/module-b7-soil/chapter-b7-1/en/ (accessed on 29 July 2021).
7. Bryan, B. Incentives, land use, and ecosystem services: Synthesizing complex linkages. *Environ. Sci. Policy* **2013**, *27*, 124–134. [CrossRef]
8. Bastidas Fegan, S. *The DS-SLM Sustainable Land Management Mainstreaming Tool-Decision Support for Mainstreaming and Scaling up Sustainable Land Management*; FAO: Rome, Italy, 2019; p. 44.
9. Environment Bill. Available online: https://services.parliament.uk/bills/2019-21/environment.html (accessed on 29 July 2021).
10. Agriculture Bill. Available online: https://services.parliament.uk/bills/2019-21/agriculture.htm (accessed on 29 July 2021).
11. DEFRA. Available online: https://www.gov.uk/government/publications/environmental-land-management-schemes-overview/environmental-land-management-scheme-overview (accessed on 29 July 2021).
12. Deng, X.; Li, Z.; Gibson, J. A review on trade-off analysis of ecosystem services for sustainable land-use management. *J. Geogr. Sci.* **2016**, *26*, 953–968. [CrossRef]
13. Dwyer, J.; Short, C.; Berriet-Solliec, M.; Gael-Lataste, F.; Pham, H.-V.; Affleck, M.; Courtney, P.; Déprès, C. *Public Goods and Ecosystem Services from Agriculture and Forestry–Towards a Holistic Approach: Review of Theories and Concepts*; European Commission: Brussels, Belgium, 2015; pp. 1–37.
14. Hejnowicz, A.P.; Hartley, S.E. *New Directions: A Public Goods Approach to Agricultural Policy Post-Brexit*; CECAN: Guildford, UK, 2018; pp. 1–42.
15. Rodgers, C. Delivering a better natural environment? The Agriculture Bill and future agri-environment policy. *Environ. Law Rev.* **2019**, *21*, 38–48. [CrossRef]
16. DEFRA. Environmental Land Management: Policy Discussion, ELMS Consulation Document. Available online: https://consult.defra.gov.uk/elm/elmpolicyconsultation/supporting_documents/ELM%20Policy%20Discussion%20Document%20230620.pdf (accessed on 29 July 2021).

17. Nelson, E.; Mendoza, G.; Regetz, J.; Polasky, S.; Tallis, H.; Cameron, D.; Chan, K.; Daily, G.C.; Goldstein, J.; Kareiva, P.M.; et al. Modeling multiple ecosystem services, biodiversity conservation, commodity production, and tradeoffs at landscape scales. *Front. Ecol. Environ.* **2009**, *7*, 4–11. [CrossRef]
18. Mulligan, M.; van Soesbergen, A.; Hole, D.G.; Brooks, T.M.; Burke, S.; Hutton, J. Mapping nature's contribution to SDG 6 and implications for other SDGs at policy relevant scales. *Remote Sens. Environ.* **2020**, *239*, 111671. [CrossRef]
19. Boumans, R.; Roman, J.; Altman, I.; Kaufman, L. The Multiscale Integrated Model of Ecosystem Services (MIMES): Simulating the interactions of coupled human and natural systems. *Ecosyst. Serv.* **2015**, *12*, 30–41. [CrossRef]
20. Ivanic, K.-Z.; Stolton, S.; Arango, C.F.; Dudley, N. *Protected Areas Benefits Assessment Tool + (PA-BAT+): A Tool to Assess Local Stakeholder Perceptions of the Flow of Benefits from Protected Areas*; IUCN, International Union for Conservation of Nature: Gland, Switzerland, 2020; p. 84.
21. Preston, S.; Raudsepp-Hearne, C. *Ecosystem Services Toolkit. Completing and Using Ecosystem Service Assessment for Decision-Making: An Interdisciplinary Toolkit for Managers*; European Commission: Brussels, Belgium, 2017.
22. TESSA (Toolkit for Ecosystem Service Site-Based Assessment). Available online: https://ecosystemsknowledge.net/tessa-toolkit-ecosystem-service-site-based-assessment (accessed on 29 July 2021).
23. MIMES. Available online: https://ipbes.net/ar/node/29397?page=15 (accessed on 29 July 2021).
24. ARIES. Available online: https://aries.integratedmodelling.org/ (accessed on 29 July 2021).
25. Neugarten, R.; Langhammer, P.F.; Osipova, E.; Bagstad, K.J.; Bhagabati, N.; Butchart, S.H.; Dudley, N.; Elliott, V.; Gerber, L.R.; Gutiérrez-Arellano, C.; et al. *Tools for Measuring, Modelling, and Valuing Ecosystem Services: Guidance for Key Biodiversity Areas, Natural World Heritage Sites, and Protected Areas*; IUCN, International Union for Conservation of Nature: Gland, Switzerland, 2018; p. 70.
26. Pan, H.; Zhang, L.; Cong, C.; Deal, B.; Wang, Y. A dynamic and spatially explicit modeling approach to identify the ecosystem service implications of complex urban systems interactions. *Ecol. Indic.* **2019**, *102*, 426–436. [CrossRef]
27. Cohen-Shacham, E.; Andrade, A.; Dalton, J.; Dudley, N.; Jones, M.; Kumar, C.; Maginnis, S.; Maynard, S.; Nelson, C.R.; Renaud, F.G.; et al. Core principles for successfully implementing and upscaling Nature-based Solutions. *Environ. Sci. Policy* **2019**, *98*, 20–29. [CrossRef]
28. Keesstra, S.; Nunes, J.; Novara, A.; Finger, D.; Avelar, D.; Kalantari, Z.; Cerdà, A. The superior effect of nature based solutions in land management for enhancing ecosystem services. *Sci. Total Environ.* **2018**, *610–611*, 997–1009. [CrossRef]
29. Pan, H.; Page, J.; Cong, C.; Barthel, S.; Kalantari, Z. How ecosystems services drive urban growth: Integrating nature-based solutions. *Anthropocene* **2021**, *35*, 100297. [CrossRef]
30. Song, Y.; Kirkwood, N.; Maksimović, Č.; Zheng, X.; O'Connor, D.; Jin, Y.; Hou, D. Nature based solutions for contaminated land remediation and brownfield redevelopment in cities: A review. *Sci. Total Environ.* **2019**, *663*, 568–579. [CrossRef]
31. Natural Capital Committee. *Advice on Using Nature Based Interventions to Reach Net Zero Greenhouse Gas. Emission*; UK Department for Environment, Food\Rural Affairs: Westminster, UK, 2020.
32. DEFRA. Environmental Land Management Tests and Trials Quarterly Evidence Report. Available online: https://www.gov.uk/government/publications/environmental-land-management-tests-and-trials (accessed on 9 June 2021).
33. Surrey Hills AONB. Making Space for Nature. 2020. Available online: https://mk0surreyhillsnfif4k.kinstacdn.com/wp-content/uploads/2020/02/Item-5-Making-Space-for-Nature.pdf (accessed on 29 July 2021).
34. Sadlier, G.; Flytkjær, R.; Sabri, S.; Robin, N. *Value of Satellite-Derived Earth Observation Capabilities to the UK Government Today and by 2020*; London Economics: Boston, MA, USA, 2018.
35. Geospatial Commission. Unlocking the Power of Location: The UK's Geospatial Strategy. Available online: https://www.gov.uk/government/publications/unlocking-the-power-of-locationthe-uks-geospatial-strategy/unlocking-the-power-of-location-the-uks-geospatial-strategy-2020-to-2025 (accessed on 29 July 2021).
36. Kuenzer, C.; Ottinger, M.; Wegmann, M.; Guo, H.; Wang, C.; Zhang, J.; Dech, S.; Wikelski, M. Earth observation satellite sensors for biodiversity monitoring: Potentials and bottlenecks. *Int. J. Remote Sens.* **2014**, *35*, 6599–6647. [CrossRef]
37. Lucas, R.; Blonda, P.; Bunting, P.; Jones, G.; Inglada, J.; Arias, M.; Kosmidou, V.; Petrou, Z.I.; Manakos, I.; Adamo, M.; et al. The Earth Observation Data for Habitat Monitoring (EODHaM) system. *Int. J. Appl. Earth Obs. Geoinf.* **2015**, *37*, 17–28. [CrossRef]
38. Vihervaara, P.; Auvinen, A.-P.; Mononen, L.; Törmä, M.; Ahlroth, P.; Anttila, S.; Böttcher, K.; Forsius, M.; Heino, J.; Heliölä, J.; et al. How Essential Biodiversity Variables and remote sensing can help national biodiversity monitoring. *Glob. Ecol. Conserv.* **2017**, *10*, 43–59. [CrossRef]
39. Rocchini, D.; Marcantonio, M.; Da Re, D.; Chirici, G.; Galluzzi, M.; Lenoir, J.; Ricotta, C.; Torresani, M.; Ziv, G. Time-lapsing biodiversity: An open source method for measuring diversity changes by remote sensing. *Remote Sens. Environ.* **2019**, *231*, 111192. [CrossRef]
40. Randin, C.F.; Ashcroft, M.B.; Bolliger, J.; Cavender-Bares, J.; Coops, N.C.; Dullinger, S.; Dirnböck, T.; Eckert, S.; Ellis, E.; Fernández, N.; et al. Monitoring biodiversity in the Anthropocene using remote sensing in species distribution models. *Remote Sens. Environ.* **2020**, *239*, 111626. [CrossRef]
41. Townsend, P.A.; Lookingbill, T.R.; Kingdon, C.; Gardner, R.H. Spatial pattern analysis for monitoring protected areas. *Remote Sens. Environ.* **2009**, *113*, 1410–1420. [CrossRef]

42. Sallustio, L.; De Toni, A.; Strollo, A.; Di Febbraro, M.; Gissi, E.; Casella, L.; Geneletti, D.; Munafò, M.; Vizzarri, M.; Marchetti, M. Assessing habitat quality in relation to the spatial distribution of protected areas in Italy. *J. Environ. Manag.* **2017**, *201*, 129–137. [CrossRef]
43. Crawford, B.A.; Maerz, J.C.; Moore, C.T. Expert-Informed Habitat Suitability Analysis for At-Risk Species Assessment and Conservation Planning. *J. Fish. Wildl. Manag.* **2020**, *11*, 130–150. [CrossRef]
44. Ahmadipari, M.; Yavari, A.; Ghobadi, M. Ecological monitoring and assessment of habitat suitability for brown bear species in the Oshtorankooh protected area, Iran. *Ecol. Indic.* **2021**, *126*, 107606. [CrossRef]
45. McMahon, C.R.; Howe, H.; Hoff, J.V.D.; Alderman, R.; Brolsma, H.; Hindell, M. Satellites, the All-Seeing Eyes in the Sky: Counting Elephant Seals from Space. *PLoS ONE* **2014**, *9*, e92613. [CrossRef]
46. Fretwell, P.T.; Staniland, I.; Forcada, J. Whales from Space: Counting Southern Right Whales by Satellite. *PLoS ONE* **2014**, *9*, e88655. [CrossRef] [PubMed]
47. Mairota, P.; Cafarelli, B.; Labadessa, R.; Lovergine, F.P.; Tarantino, C.; Lucas, R.; Nagendra, H.; Didham, R.K. Very high resolution Earth observation features for monitoring plant and animal community structure across multiple spatial scales in protected areas. *Int. J. Appl. Earth Obs. Geoinf.* **2015**, *37*, 100–105. [CrossRef]
48. Rocchini, D.; Hernandez-Stefanoni, J.L.; He, K.S. Advancing species diversity estimate by remotely sensed proxies: A conceptual review. *Ecol. Inform.* **2015**, *25*, 22–28. [CrossRef]
49. Fretwell, P.T.; Trathan, P.N. Discovery of new colonies by Sentinel2 reveals good and bad news for emperor penguins. *Remote Sens. Ecol. Conserv.* **2021**, *7*, 139–153. [CrossRef]
50. Räsänen, A.; Aurela, M.; Juutinen, S.; Kumpula, T.; Lohila, A.; Penttilä, T.; Virtanen, T. Detecting northern peatland vegetation patterns at ultra-high spatial resolution. *Remote Sens. Ecol. Conserv.* **2020**, *6*, 457–471. [CrossRef]
51. Klimetzek, D.; Stăncioiu, P.T.; Paraschiv, M.; Niţă, M.D. Ecological Monitoring with Spy Satellite Images—The case of Red Wood Ants in Romania. *Remote Sens.* **2021**, *13*, 520. [CrossRef]
52. Rotenberry, J.T.; Preston, K.L.; Knick, S.T. Gis-Based Niche Modeling for Mapping Species' Habitat. *Ecology* **2006**, *87*, 1458–1464. [CrossRef]
53. Turner, W.; Spector, S.; Gardiner, N.; Fladeland, M.; Sterling, E.; Steininger, M. Remote sensing for biodiversity science and conservation. *Trends Ecol. Evol.* **2003**, *18*, 306–314. [CrossRef]
54. Haines-Young, R.H.; Potschin, M.B.; Deane, R.; Porter, K. Policy Impact and Future Options for Countryside Survey. Final Report. Available online: https://www.nottingham.ac.uk/cem/pdf/FOFCS_FinalReport_Revised_August2014.pdf (accessed on 29 July 2021).
55. Morton, R.D.; Rowland, C.S. Developing and Evaluating an Earth Observation-enabled ecological land cover time series system. *JNCC Rep.* **2015**, *563*, 2–7.
56. Morton, R.D.; Marston, C.G.; O'Neil, A.W.; Rowland, C.S. Land Cover Map 2019 (land parcels, GB). *NERC Environ. Inf. Data Cent.* **2020**. [CrossRef]
57. Feranec, J.M.G.; Hazeu, G. *European Landscape Dynamics (Chapter 5. Interpretation of Satellite Images)*; Taylor & Francis: Oxfordshire, UK, 2016; pp. 33–41.
58. UN. *Earth Observations for Official Statistics Satellite Imagery and Geospatial Data Task Team Report*; United Nations Satellite Imagery and Geo-spatial Data Task Team: Rio de Janeiro, Brazil, 2017; p. 170.
59. Svatonova, H. Analysis of Visual Interpretation of Satellite Data. *ISPRS-Int. Arch. Photogramm. Remote Sens. Spat. Inf. Sci.* **2016**, *XLI-B2*, 675–681. [CrossRef]
60. Campbell, J.B.; Wynne, R.H. *Introduction to Remote Sensing*, 5th ed.; The Guilford Press: New York, NY, USA, 2011; pp. 130–158.
61. Thomas, N.P.; Benning, I.L.; Ching, V.M. *Classification of Remotely Sensed Images*; Taylor\Francis: Oxfordshire, UK, 1987; p. 267.
62. Ibarrola-Ulzurrun, E.; Marcello, J.; Gonzalo-Martin, C. Assessment of Component Selection Strategies in Hyperspectral Imagery. *Entropy* **2017**, *19*, 666. [CrossRef]
63. Kumar, P.S.J.; Huan, T.L. *Earth Science and Remote Sensing Applications*; Springer: Berlin/Heidelberg, Germany, 2018.
64. Anderson, C.B. Biodiversity monitoring, earth observations and the ecology of scale. *Ecol. Lett.* **2018**, *21*, 1572–1585. [CrossRef]
65. Ernst, C.; Mayaux, P.; Verhegghen, A.; Bodart, C.; Christophe, M.; Defourny, P. National forest cover change in Congo Basin: Deforestation, reforestation, degradation and regeneration for the years 1990, 2000 and 2005. *Glob. Chang. Biol.* **2013**, *19*, 1173–1187. [CrossRef] [PubMed]
66. Nagendra, H.; Mairota, P.; Marangi, C.; Lucas, R.; Dimopoulos, P.; Honrado, J.; Niphadkar, M.; Mucher, S.; Tomaselli, V.; Panitsa, M.; et al. Satellite Earth observation data to identify anthropogenic pressures in selected protected areas. *Int. J. Appl. Earth Obs. Geoinf.* **2015**, *37*, 124–132. [CrossRef]
67. Adamo, M.; Tomaselli, V.; Tarantino, C.; Vicario, S.; Veronico, G.; Lucas, R.; Blonda, P. Knowledge-Based Classification of Grassland Ecosystem Based on Multi-Temporal WorldView-2 Data and FAO-LCCS Taxonomy. *Remote Sens.* **2020**, *12*, 1447. [CrossRef]
68. DEFRA. Roadmap for the Use of Earth Observation across Defra 2015–2020 Report. Available online: https://assets.publishing.service.gov.uk/government/uploads/system/uploads/attachment_data/file/488133/defra-earth-obs-roadmap-2015.pdf (accessed on 29 July 2021).
69. Secades, C.; O'Connor, B.; Brown, C.; Walpole, M. *Review of the Use of Remotely-Sensed Data for Monitoring Biodiversity Change and Tracking Progress towards the Aichi Biodiversity Targets*; UNEP-WCMC: Cambridge, UK, 2014; p. 183.

70. Barbosa, C.D.A.; Atkinson, P.; Dearing, J. Remote sensing of ecosystem services: A systematic review. *Ecol. Indic.* **2015**, *52*, 430–443. [CrossRef]

71. SIRS. Feasibility Study about the Mapping and Monitoring of Green Linear Features Based on VHR Satellites Imagery. Available online: https://land.copernicus.eu/user-corner/technical-library/study-lead-by-sirs (accessed on 29 July 2021).

72. Making Earth Observation Work (MEOW) for UK Biodiversity Monitoring and Surveillance, Phase 4: Testing Applications in Habitat Condition Assessment A Report to the Department for Environment, Food and Rural Affairs, Prepared by Environment Systems. Available online: http://randd.defra.gov.uk/Document.aspx?Document=13900_BE0119_MEOW4_Report_Final.pdf (accessed on 29 July 2021).

73. Anderson, J.R.; Hardy, E.E.; Roach, J.T.; Witmer, R.E. *A Land Use and Land Cover Classification System for Use with Remote Sensor Data*; US Government Printing Office: Washington, DC, USA, 1976.

74. Yang, H.; Li, S.; Chen, J.; Zhang, X.; Xu, S. The Standardization and Harmonization of Land Cover Classification Systems towards Harmonized Datasets: A Review. *ISPRS Int. J. Geo-Inf.* **2017**, *6*, 154. [CrossRef]

75. Jansen, L.J.; Di Gregorio, A. Land-use data collection using the "land cover classification system": Results from a case study in Kenya. *Land Use Policy* **2003**, *20*, 131–148. [CrossRef]

76. Headquarters, C.; Skole, D.; Salas, W.; Taylor, V. *Global Observation of Forest Cover: Fine Resolution Data and Product Design Strategy, Report of a Workshop*; GOFC-GOLD: Washington, DC, USA, 1998.

77. Global Observation for Forest Cover and Land Dynamics (GOFC/GOLD). Available online: http://www.gofcgold.wur.nl (accessed on 19 July 2021).

78. Venter, Z.; Sydenham, M. Continental-Scale Land Cover Mapping at 10 m Resolution over Europe (ELC10). *Remote Sens.* **2021**, *13*, 2301. [CrossRef]

79. Africover. Available online: http://www.fao.org/3/bd854e/bd854e.pdf (accessed on 29 July 2021).

80. National Land Cover Database 2019 (NLCD2019) Legend. Available online: https://www.mrlc.gov/data/legends/national-land-cover-database-2019-nlcd2019-legend (accessed on 29 July 2021).

81. National Land Use Database: Land Use and Land Cover Classification. Available online: https://assets.publishing.service.gov.uk/government/uploads/system/uploads/attachment_data/file/11493/144275.pdf (accessed on 29 July 2021).

82. Jackson, D.L. Guidance on the Interpretation of the Biodiversity Broad Habitat Classification (Terrestrial and Freshwater Types): Definitions and the Relationship with Other Classifications. JNCC Report No. 307, JNCC, Peterborough. Available online: https://data.jncc.gov.uk/data/0b7943ea-2eee-47a9-bd13-76d1d66d471f/JNCC-Report-307-SCAN-WEB.pdf (accessed on 29 July 2021).

83. Society, T.R. Observing the Earth: Expert Views on Environmental Observation for the UK. Available online: https://royalsociety.org/~/media/policy/projects/environmental-observation/environmental-observations-report.pdf (accessed on 29 July 2021).

84. Tansey, K.; Chambers, I.; Anstee, A.; Denniss, A.; Lamb, A. Object-oriented classification of very high resolution airborne imagery for the extraction of hedgerows and field margin cover in agricultural areas. *Appl. Geogr.* **2009**, *29*, 145–157. [CrossRef]

85. Pettorelli, N.; Safi, K.; Turner, W. Satellite remote sensing, biodiversity research and conservation of the future. *Philos. Trans. R. Soc. B Biol. Sci.* **2014**, *369*, 20130190. [CrossRef]

86. SH AONB Surrey Hills Management Plan 2020–2025. Available online: https://mk0surreyhillsnfif4k.kinstacdn.com/wp-content/uploads/2019/12/Surrey-Hills-Management-Plan-Web-72-SP-1.pdf (accessed on 29 July 2021).

87. Trust, S.W. Biodiversity and Planning in Surrey. Available online: https://surreynaturepartnership.files.wordpress.com/2019/10/biodiversity-planning-in-surrey-revised_post-revision-nppf_mar-2019.pdf (accessed on 29 July 2021).

88. Morton, R.D.; Marston, C.G.; O'Neil, A.W.; Rowland, C.S. *Land Cover Map 2017, 2018, 2019 (25 m Rasterised Land Parcels, GB)*; NERC Environmental Information Data Centre: Lancaster, UK, 2020. [CrossRef]

89. DigiMap EDINA. Available online: https://digimap.edina.ac.uk (accessed on 29 July 2021).

90. Open Government Licence, Priority Habitat Inventory (England). Available online: https://data.gov.uk/dataset/4b6ddab7-6c0f-4407-946e-d6499f19fcde/priority-habitat-inventory-england (accessed on 29 July 2021).

91. Export Open Street Map. Available online: https://www.openstreetmap.org/export#map=6/42.088/12.564 (accessed on 4 September 2020).

92. Google Earth Engine. Available online: https://explorer.earthengine.google.com/#workspace (accessed on 29 July 2021).

93. Farr, T.G.; Rosen, P.A.; Caro, E.; Crippen, R.; Duren, R.; Hensley, S.; Kobrick, M.; Paller, M.; Rodriguez, E.; Roth, L.; et al. The Shuttle Radar Topography Mission. *Rev. Geophys.* **2007**, *45*, 2. [CrossRef]

94. Garibaldi, A.; Turner, N. Cultural Keystone Species: Implications for Ecological Conservation and Restoration. *Ecol. Soc.* **2004**, *9*, 3. [CrossRef]

95. Hötker, H.; Oppermann, R.; Jahn, T.; Bleil, R. Protection of biodiversity of free living birds and mammals in respect of the effects of pesticides. *Jul.-Kühn-Arch.* **2013**, *442*, 91–92.

96. Donald, P.; Buckingham, D.; Moorcroft, D.; Muirhead, L.; Evans, A.; Kirby, W. Habitat use and diet of skylarks Alauda arvensis wintering on lowland farmland in southern Britain. *J. Appl. Ecol.* **2001**, *38*, 536–547. [CrossRef]

97. Chamberlain, D.E.; Crick, H.Q. Population declines and reproductive performance of Skylarks Alauda arvensis in different regions and habitats of the United Kingdom. *IBIS* **2008**, *141*, 38–51. [CrossRef]

98. Wilson, J.D.; Evans, J.; Browne, S.J.; King, J.R. Territory Distribution and Breeding Success of Skylarks Alauda arvensis on Organic and Intensive Farmland in Southern England. *J. Appl. Ecol.* **1997**, *34*, 1462. [CrossRef]

99.   Murray, K.A. *Factors Affecting Foraging by Breeding Farmland Birds*; Open University: Buckinghamshire, UK, 2004.
100. Sozio, G.; Iannarilli, F.; Melcore, I.; Boschetti, M.; Fipaldini, D.; Luciani, M.; Roviani, D.; Schiavano, A.; Mortelliti, A. Forest management affects individual and population parameters of the hazel dormouse Muscardinus avellanarius. *Mamm. Biol.* **2016**, *81*, 96–103. [CrossRef]
101. Trout, R.C.; Brooks, S.E.; Rudlin, P.; Neil, J. The effects of restoring a conifer Plantation on an Ancient Woodland Site (PAWS) in the UK on the habitat and local population of the Hazel Dormouse (Muscardinus avellanarius). *Eur. J. Wildl. Res.* **2012**, *58*, 635–643. [CrossRef]
102. Goodwin, C.E.D.; Hodgson, D.J.; Al-Fulaij, N.; Bailey, S.; Langton, S.; McDonald, R.A. Voluntary recording scheme reveals ongoing decline in the United Kingdom hazel dormouse Muscardinus avellanarius population. *Mammal. Rev.* **2017**, *47*, 183–197. [CrossRef]
103. Goodwin, C.E.D.; Suggitt, A.J.; Bennie, J.; Silk, M.J.; Duffy, J.P.; Al-Fulaij, N.; Bailey, S.; Hodgson, D.J.; McDonald, R.A. Climate, landscape, habitat, and woodland management associations with hazel dormouse Muscardinus avellanarius population status. *Mammal. Rev.* **2018**, *48*, 209–223. [CrossRef]
104. Otto, C.R.V.; Roth, C.; Carlson, B.L.; Smart, M.D. Land-use change reduces habitat suitability for supporting managed honey bee colonies in the Northern Great Plains. *Proc. Natl. Acad. Sci. USA* **2016**, *113*, 10430–10435. [CrossRef]
105. Andrew, M.E.; Ustin, S. Habitat suitability modelling of an invasive plant with advanced remote sensing data. *Divers. Distrib.* **2009**, *15*, 627–640. [CrossRef]
106. Gomez, J.J.; Túnez, J.I.; Fracassi, N.; Cassini, M.H. Habitat suitability and anthropogenic correlates of Neotropical river otter (Lontra longicaudis) distribution. *J. Mammal.* **2014**, *95*, 824–833. [CrossRef]
107. Duro, D.; Franklin, S.; Dubé, M.G. A comparison of pixel-based and object-based image analysis with selected machine learning algorithms for the classification of agricultural landscapes using SPOT-5 HRG imagery. *Remote Sens. Environ.* **2012**, *118*, 259–272. [CrossRef]
108. Cord, A.F.; Brauman, K.; Chaplin-Kramer, R.; Huth, A.; Ziv, G.; Seppelt, R. Priorities to Advance Monitoring of Ecosystem Services Using Earth Observation. *Trends Ecol. Evol.* **2017**, *32*, 416–428. [CrossRef] [PubMed]
109. Cochran, F.; Daniel, J.; Jackson, L.; Neale, A. Earth observation-based ecosystem services indicators for national and subnational reporting of the sustainable development goals. *Remote Sens. Environ.* **2020**, *244*, 111796. [CrossRef] [PubMed]
110. Watmough, G.R.; Marcinko, C.L.J.; Sullivan, C.; Tschirhart, K.; Mutuo, P.K.; Palm, C.A.; Svenning, J.-C. Socioecologically informed use of remote sensing data to predict rural household poverty. *Proc. Natl. Acad. Sci. USA* **2019**, *116*, 1213–1218. [CrossRef] [PubMed]

 *sustainability*

Article

Are Global Environmental Uncertainties Inevitable? Measuring Desertification for the SDGs

Alan Grainger

School of Geography, University of Leeds, Leeds LS2 9JT, UK; a.grainger@leeds.ac.uk

Abstract: Continuing uncertainty about the present magnitudes of global environmental change phenomena limits scientific understanding of human impacts on Planet Earth, and the quality of scientific advice to policy makers on how to tackle these phenomena. Yet why global environmental uncertainties are so great, why they persist, how their magnitudes differ from one phenomenon to another, and whether they can be reduced is poorly understood. To address these questions, a new tool, the Uncertainty Assessment Framework (UAF), is proposed that builds on previous research by dividing sources of environmental uncertainty into categories linked to features inherent in phenomena, and insufficient capacity to conceptualize and measure phenomena. Applying the UAF shows that, based on its scale, complexity, areal variability and turnover time, desertification is one of the most inherently uncertain global environmental change phenomena. Present uncertainty about desertification is also very high and persistent: the Uncertainty Score of a time series of five estimates of the global extent of desertification shows limited change and has a mean of 6.8, on a scale from 0 to 8, based on the presence of four conceptualization uncertainties (terminological difficulties, underspecification, understructuralization and using proxies) and four measurement uncertainties (random errors, systemic errors, scalar deficiencies and using subjective judgment). This suggests that realization of the Land Degradation Neutrality (LDN) Target 15.3 of the UN Sustainable Development Goal (SDG) 15 ("Life on Land") will be difficult to monitor in dry areas. None of the estimates in the time series has an Uncertainty Score of 2 when, according to the UAF, evaluation by statistical methods alone would be appropriate. This supports claims that statistical methods have limitations for evaluating very uncertain phenomena. Global environmental uncertainties could be reduced by devising better rules for constructing global environmental information which integrate conceptualization and measurement. A set of seven rules derived from the UAF is applied here to show how to measure desertification, demonstrating that uncertainty about it is not inevitable. Recent review articles have advocated using 'big data' to fill national data gaps in monitoring LDN and other SDG 15 targets, but an evaluation of a sample of three exemplar studies using the UAF still gives a mean Uncertainty Score of 4.7, so this approach will not be straightforward.

Keywords: uncertainty evaluation; desertification; global change; Earth observation; planetary measurement; Land Degradation Neutrality; Sustainable Development Goals

Citation: Grainger, A. Are Global Environmental Uncertainties Inevitable? Measuring Desertification for the SDGs. *Sustainability* 2022, 14, 4063. https://doi.org/10.3390/su14074063

Academic Editors: Stephen Morse, Richard Murphy and Ana Andries

Received: 17 December 2021
Accepted: 24 February 2022
Published: 29 March 2022

1. Introduction

The present magnitudes of major global environmental change phenomena, such as forest area change, biodiversity loss and desertification, have been very uncertain for decades. Judged purely by the number of available estimates, one of the most uncertain of these phenomena is desertification, which is land degradation in dry areas. The annual rate of desertification has only been estimated once, for the 1970s [1], and estimates of the global extent of desertification show it contracting, not expanding: an estimate of the area of at least moderately desertified land in the 1970s [2] is over six times an estimate for the 1980s made by the World Atlas of Desertification [3,4]. That estimate has not been updated by the recently published Third Edition of the Atlas, since its authors claim that desertification cannot be mapped satisfactorily [5]. This is an important statement, for while the first two editions of the Atlas were produced by the United Nations Environment Programme, the third comes from the European Commission Joint Research Centre (JRC), a leading centre for global environmental monitoring using remote sensing data. In 2011, a report from a group of remote sensing scientists, coordinated by JRC, recommended that a Global Drylands Observing System be established to monitor desertification [6], but such a system is still awaited.

Continuing uncertainty about the extent and rate of change of desertification makes it difficult to assess the effectiveness of the United Nations Convention to Combat Desertification (UNCCD). Moreover, since drylands account for half of the Earth's land surface area [3], without accurate estimates of the extent and rate of change of their degradation, it will be impossible to reliably monitor whether the world offsets the rate of land degradation by the rate of restoration of degraded land by 2030, and so achieves Land Degradation Neutrality (LDN), which is Target 15.3 in the UN Sustainable Development Goal 15: "Life on Land" [7,8]. The other eight targets cover two other key global environmental change phenomena: forest area change (15.2) and biodiversity loss (15.1 and 15.4–15.9). According to Allen et al., the 17 Sustainable Development Goals (SDGs) "suffer from a lack of national data needed for effective monitoring and implementation. Almost half of the SDG indicators are not regularly produced and available datasets are often out of date" [9]. They, like Hassani et al. [10], identify satellite data and other sets of "big data" as a potential solution to this problem, but conclude that using these data will not be straightforward. Indeed, in the journal papers on using big data for monitoring SDGs which they review, SDG 15 accounts for the largest share of all papers but one of the smallest shares with *global* datasets cited in them [9]. This paper addresses these data deficiencies for land degradation in dry areas, but its analysis of global environmental uncertainties is also relevant to other targets in SDG 15.

Does the persistence of global environmental uncertainties mean that they are inevitable? At the other extreme of spatial scales, in 1927, Heisenberg deduced from the new theory of quantum mechanics an inequality which showed that for electrons and other sub-atomic particles, "the exact knowledge of one variable can exclude the exact knowledge of another" [11,12], since the disturbance involved in measuring the position of a particle, for example, affects the measurement of its momentum. Yet while Heisenberg's Uncertainty Principle was just a theoretical prediction in 1927, there is ample empirical evidence, for desertification and other phenomena, to show the persistence of global environmental uncertainties, despite all the planetary data collected in the 50 years since the first Landsat satellite was launched in 1972. Although sub-atomic physics may seem to have little in common with global change science, they both involve measuring phenomena with scientific instruments, and this paper is not the first to discuss potential parallels between Heisenberg Uncertainty and environmental uncertainties [13].

Are global environmental change phenomena equally uncertain? Global environmental uncertainties continue to inhibit governments from committing sufficient resources to tackling humanity's global impacts on the planet. So if science can differentiate between the uncertainties associated with different phenomena, this could lead to greater incentives to tackle them.

Surprisingly little research has been undertaken into global environmental uncertainties, despite their scientific and political importance. This may be because environmental uncertainties generally are too easily taken for granted: Brown even stated in 2010 that "there is no common understanding or consistent definition of uncertainty in environmental research" [14]. Neglect of uncertainty about the *natural* environment is apparent when Google Scholar searches for journal papers whose titles contain "environmental uncertainty" or "environmental uncertainties" generate results dominated by studies of organization theory [15] and control systems [16], which focus on the *business* environment, not the natural environment.

This paper aims to inspire fresh interest in environmental uncertainties by: (a) proposing an Uncertainty Assessment Framework (UAF) that can tackle the above questions about the inevitability and relative sizes of global environmental uncertainties, and indicate how they can be reduced by planetary measurement; and (b) applying the UAF to desertification and SDG Target 15.3. The UAF focuses on uncertainty about the magnitudes of environmental phenomena, rather than all knowledge about the latter. Instead of starting from a blank slate, it restructures sources of environmental uncertainty in two existing taxonomies [13,17] using an original conceptualization, dividing these sources into three categories linked to: (a) the features inherent in phenomena; (b) insufficient capacity to conceptualize phenomena; and (c) insufficient capacity to measure phenomena. It deals with *present* uncertainties, not *future* uncertainties and risk [18], uncertainties in modelling [19], or links between uncertainty and decision making [20].

This paper has four main sections. The first reviews previous research into environmental uncertainty. The second outlines the UAF, and the data and methods employed in the paper. The third applies the UAF to desertification, finding that it has a high inherent uncertainty and a persistently high present uncertainty. The fourth suggests how to reduce present uncertainty about desertification by planetary measurement, using an initial set of rules derived from the UAF for constructing reliable global environmental information, and shows that uncertainty about desertification is not inevitable. It also examines whether these rules are followed by a sample of papers, identified in recent reviews [9,10], which discuss using big data to monitor SDG Target 15.3.

2. Literature Review

2.1. Defining Uncertainty

Uncertainty is defined as "incomplete knowledge" by Böschen et al. [21], but is a contested term. For example, for Smithson, uncertainty is a type of *error* [22]; for Roth, it describes constraints on reproducing experimental procedures [23]; and for Brown, it is "a state of confidence" varying between certainty and irrelevance [14].

The relationship between uncertainty and *risk* is contentious too. Knight divided *ignorance* into risk, which can be assessed by probabilities, and uncertainty, which cannot [24]. Probabilities remain central to analysing future risk today [25], though Beck argued that prediction "is not reducible to . . . probability" [18].

Wynne distinguishes between uncertainty and risk when classifying "kinds of uncertainty" and proposes two more categories: *ignorance*, in which "we don't know what we don't know"; and *indeterminacy*, which is an inability to classify "things . . . as the same or different, [based on] specific properties or criteria" [26]. This views indeterminacy as a conceptualization limitation. Yet physicists treat it more explicitly as a measurement limitation, so parameters are known but cannot be properly measured [27]. Such different views illustrate the contributions made to uncertainty by conceptualization and measurement, and synergies between them.

2.2. The Sociology of Knowledge Accumulation

Uncertainty about any phenomenon is usually reduced as science systematically accumulates knowledge about it through observation, experiment and explanation. Isolated facts, or *data*, are collected and then processed within a conceptual framework into meaningful *information* [28]. After being verified and reported, information is synthesized into even more usable *knowledge*.

Science, however, is a social activity in which continuous development is punctuated by discontinuities as scientific communities switch from one dominant theoretical paradigm to another [29]. It also differentiates into an increasing number of subject-specific disciplines, each with its own language and rules [30] and authority and monopoly claims [31].

Planetary measurement uses instruments on satellites to collect *global data*, and then, with appropriate support from ground data, converts these data into *global information*. It is difficult to explain on purely technological grounds the limited amount of planetary measurement since the first Landsat satellite was launched in 1972, but much easier when allowing for the sociology of science, since different approaches are taken towards data collection and information production by remote sensing scientists, on the one hand, and scientists in other disciplines which study land cover change, on the other [32]. Ecologists, for example, have traditionally preferred to collect data by intensive measurements in small sample plots, and have been slow to make full use of remote sensing data [33]. Remote sensing scientists are skilled in processing the latter data but have taken time to convert them into global information. For example, the first global forest area map based on "wall-to-wall" Landsat data was not published until 2012 [34]; and of a sample of 96 papers published before this advance in the *International Journal of Remote Sensing* in 2009, only one focused on mapping at global scale (Supplementary Table S1).

Knowledge about global environmental change is gained not only by *scientific processes*, but also by intergovernmental *political processes* in which UN and other international organizations conceptualize phenomena and estimate their magnitudes. One example is the UN Commission for Sustainable Development process which led to the Sustainable Development Goals [8]. Intergovernmental processes often characterize global phenomena by *indicators*—measurable quantities that represent specific attributes of a given system [35]. If indicators are to generate meaningful information, they should ideally be chosen using coherent conceptual frameworks [36]; yet, in practice, these processes tend to rely on long lists of indicators with limited coherency [37]. Interactions between scientific processes and political processes vary in intensity [38].

2.3. Existing Approaches to Evaluating Very Uncertain Environmental Phenomena

The conventional quantitative approach taken by many peer-reviewed studies to evaluate uncertainty about environmental phenomena uses statistical methods to estimate errors. Yet it is claimed that this approach is less meaningful in cases of severe uncertainty [39,40], when "unquantifiable uncertainties ... dominate the quantifiable ones" [41]. Estimates of global environmental change phenomena are particularly prone to this, because many estimates are still not wholly based on measurements of the kind that scientists working at lower spatial scales take for granted, but often rely heavily on national statistics whose links to measurements are less robust [32].

One alternative to purely quantitative analysis of uncertainty is to combine it with *qualitative* evaluation. The Numerical Unit Spread Assessment Pedigree (NUSAP) system divides uncertainty into three "sorts": "technical", or random error; "methodological", or unreliable measurement; and "epistemological", or how well scientific theories fit the real world [42]. The first two sorts represent *measurement* and the third *conceptualization*. Van der Sluijs has added a "societal" category in which society influences scientific activity [41]. NUSAP identifies for any number its random error (Spread); reliability, linked to systematic errors (Assessment); and how the number is produced (Pedigree). Although NUSAP has been applied to various environmental phenomena, Spread seems less relevant to highly uncertain phenomena; and Pedigree indicators may change from one phenomenon to another, and give measurement uncertainties priority over conceptualization uncertainties.

Another approach is to only evaluate sources of environmental uncertainty qualitatively. Regan et al. distinguish between "linguistic sources", which limit *conceptualization*, and "epistemic sources", which include *natural variability* and *measurement* sources [13] (Table 1). Van Asselt and Rotmans separate "variability" in phenomena from "limited knowledge" (or measurement) sources, but exclude conceptualization sources (except "value diversity") [17] (Table 1). Both taxonomies neglect economic factors, which limit the size, frequency and resolution of large scale surveys [43]. They are also rather arbitrary and inconsistent in categorizing sources, and in sequencing them in each category (Tables S2 and S3). Yet their similarities suggest that, suitably modified, they could form the basis for a more coherent taxonomy which distinguishes more clearly between inherent, conceptualization and measurement sources, and this has inspired the approach taken here.

Table 1. Two taxonomies of sources of environmental uncertainty proposed in 2002 by Regan et al. [13] and Van Asselt and Rotmans [17] (detailed definitions are provided in Tables S2 and S3).

Regan et al.	Van Asselt and Rotmans
Linguistic	Variability
L1. Vagueness	V1. Inherent randomness
L2. Context dependence	V2. Value diversity
L3. Ambiguity	V3. (Irrational) human behaviour
L4. Underspecificity	V4. (Non-linear) societal dynamics
L5. Indeterminacy	V5. Technological surprises
Epistemic	Limited Knowledge
E1. Measurement error	K1. Inexactness
E2. Systematic error	K2. Lack of measurements
E3. Natural variation	K3. Practically immeasurable
E4. Inherent randomness	K4. Conflicting evidence
E5. Moral uncertainty	K5. Reducible ignorance
E6. Subjective judgement	K6. Indeterminacy
	K7. Irreducible ignorance

3. Methodology, Materials and Methods

3.1. Overview

Böschen et al.'s definition of uncertainty as "incomplete knowledge" [21] suggests that to conceptualize the origins of environmental uncertainty, it is necessary to first identify what determines *complete knowledge* of an environmental phenomenon (K_c), and then explain how the gap between this and *present knowledge* at any time t (K_t) is linked to restrictions on capacity to construct knowledge.

The Uncertainty Assessment Framework (UAF) proposed here therefore divides sources of uncertainty about any environmental phenomenon into three interacting categories (Figure 1) which are linked to:

(1)   The features inherent in the phenomenon.
(2)   Insufficient capacity to conceptualize the phenomenon.
(3)   Insufficient capacity to measure the phenomenon.

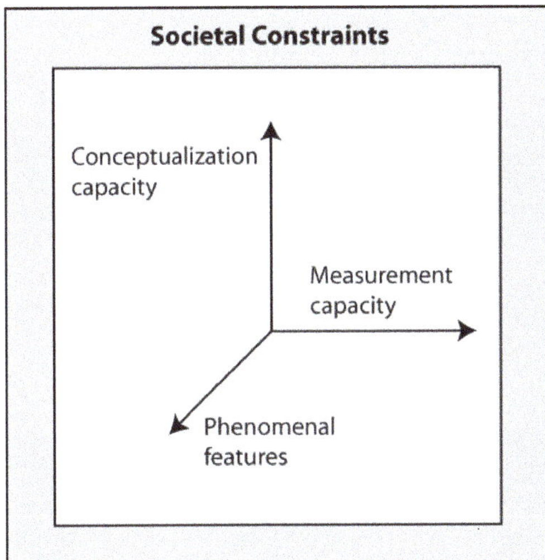

Figure 1. The Uncertainty Assessment Framework.

The *features* of a phenomenon determine what must be understood to have complete knowledge about it, and contribute to its inherent uncertainty. They include its: (a) spatial extent; (b) biophysical complexity, which depends on the minimum number of *attributes* needed to characterize its spatial distribution—attributes correspond to the different information layers which must be combined to map the phenomenon (see below); (c) spatio-temporal randomness, resulting from natural factors; and (d) human-environment complexity, which exacerbates biophysical complexity and natural randomness. The larger each feature is, the more knowledge is needed to understand the phenomenon, and the greater its *inherent uncertainty*.

The two *capacities* describe how improving technology, financial resources and people's skills (or 'Human Capital') can reduce uncertainty by constructing present knowledge about the phenomenon. The smaller the two capacities are, the larger the associated *difficulties* in conceptualization and measurement are likely to be.

If the difference between complete and present knowledge is represented by the sum of present *conceptualization uncertainties* (U_{ct}) and *measurement uncertainties* (U_{mt}) resulting from the associated capacity limitations at time t then:

$$K_c = K_t + U_{ct} + U_{mt} \tag{1}$$

Following Van der Sluijs [41], all three categories of sources are subject to *societal constraints*, which include political, economic and other social factors (Figure 1).

The UAF builds on previous research by restructuring the individual sources listed by Regan et al. [13] and Van Asselt and Rotmans [17], using the phenomenal features and measurement categories prominent in both taxonomies and the conceptualization category highlighted by Regan et al. [13] (Table 2).

Table 2. A taxonomy of sources of environmental uncertainty in the Uncertainty Assessment Framework (UAF) and corresponding terms in the taxonomies of Regan et al. [13] and Van Asselt and Rotmans [17].

UAF Taxonomy	Corresponding Terms in Other Taxonomies in Table 1 *
Phenomenal uncertainties	
P1. Spatial extent	—
P2. Biophysical complexity	RE3
P3. Spatio-temporal randomness	RE4; VV1
P4. Human-environment complexity	VV3, VV4, VV5
Conceptualization uncertainties	
C1. Terminological difficulties	RL1, RL3, RL5; VV2
C2. Underspecification	RL4
C3. Understructuralization	RL4, RE5
C4. Using proxies	—
Measurement uncertainties	
M1. Random errors	RE1; VK1
M2. Systematic errors	RE2; VK4, VK5
M3. Scalar deficiencies in measurement	RL2; VK2
M4. Using subjective judgment	RE6

* The second column lists the Linguistic (RL) and Epistemic (RE) categories of Regan et al. [13], and the Variability (VV) and Limited Knowledge (VK) categories of Van Asselt and Rotmans [17], with numbering as in Table 1.

3.2. Phenomenal Uncertainties

It is proposed that uncertainty inherent in an environmental phenomenon is associated with four of its features:

(1) Spatial extent (S). The greater the area of a phenomenon, the more difficult it is to measure, and the more spatially diverse its distribution is likely to be.

(2) Biophysical complexity (B), potentially involving many environmental *attributes*—each of which may be represented by at least one variable—and processes linking

these attributes. For example, forest area change involves change in just one forest attribute: area. In contrast, forest carbon change involves changes in at least two attributes: area and carbon density, each of which needs to be mapped. Biodiversity involves changes in at least three attributes: ecosystem diversity, species diversity and genetic diversity [44] (Table 3). In the two latter cases the number of attributes could be expanded to include intermediate ones, e.g., biomass density in the case of forest carbon change [32], but for simplicity, the minimum number of attributes is used here. Desertification is an even more complex phenomenon, with at least seven attributes, as discussed in Section 4.1.3.

(3) Randomness in spatial and temporal distributions (R), resulting from natural factors.

(4) Human-environment complexity (H), evident in multidirectional, multitemporal and multiscalar interactions between human systems and environmental systems. Often involving changeable, conflicting and inconsistent human behaviour in causing or responding to phenomena, these interactions can exacerbate biophysical complexity and natural randomness and shift the characteristics of phenomena outside previously recorded ranges.

Table 3. The multiple attributes of four global environmental change phenomena.

Phenomenon	Number of Attributes	Attributes
Forest area change	1	Area
Forest carbon change	2	Area
		Carbon density
Biodiversity loss	3	Ecosystem diversity
		Species diversity
		Genetic diversity
Desertification	7	Vegetation area
		Vegetation density
		Water erosion of soil
		Wind erosion of soil
		Soil compaction
		Waterlogging/salinization/ alkalinization of soil
		Rainfall variation

The last three features encompass but expand the scope of the "epistemic" sources 3 and 4 of Regan et al. [13] and the "variability" sources 1, 3, 4 and 5 of Van Asselt and Rotmans [17] (Table 2). Neither study recognizes the first feature, spatial extent, even though it is far more difficult to measure environmental change at global scale than at national and local scales [32].

The relationship between inherent uncertainty (U) and the four features of an environmental phenomenon listed above can be expressed algebraically by an *inherent uncertainty function*:

$$U = f\,(S, B, R, H) \tag{2}$$

If S is represented by the total area of the phenomenon (A_i), B is related to the minimum number of attributes required to characterize it (b_i), and R and H are jointly represented on the ground by the inverses of the smallest area (a_i) (areal variability) and shortest time period (t_i) (turnover time) over which the phenomenon varies, then U can also be expressed as:

$$U = g\,(A_i, b_i, 1/a_i, 1/t_i) \tag{3}$$

Ideally, there would be a close fit between these variables and the properties of the remote sensing system chosen to measure the phenomenon. Thus, A_i would be linked to the maximum area which a remote sensing system can measure in practice; a_i and t_i to the spatial and temporal resolutions of the system, respectively; and b_i to the minimum number of attributes which can be measured remotely and/or in situ.

3.3. Knowledge Construction Mechanisms

Identifying the social *mechanisms* which limit the conceptualization and measurement capacities of scientific groups and intergovernmental and other organizations, and lead to conceptualization and measurement uncertainties, can show how to restructure the sources listed in Table 1 to construct the more coherent taxonomy proposed in Table 2. The UAF assumes that conceptualization and measurement capacities can be linked to two characteristics of a group:

(1)   Its world view, or *discourse*, which frames conceptualization. Hajer [45] defines a discourse as "a specific ensemble of ideas, concepts, and categorizations that are produced, reproduced and transformed in a particular set of practices and through which meaning is given to physical and social realities." Ideas, concepts, and categorizations are ideally expressed in an internally consistent language which, starting with the smallest unit, or *term*, is used to construct increasingly complex *narratives*: sets of statements that give a meaningful totality of events [46].

(2)   Its set of repeated practices, or *institutions*, which comprise the methods used for measurement and constructing knowledge generally. Institutions are "enduring regularities of human action in situations structured by rules, norms and shared strategies, as well as by the physical world" [47]. They occur in 'organizations' but are not equivalent to them. Ostrom proposed that any social setting has multiple levels of institutions: "operational institutions", which may be varied easily, are embedded in the "collective choice institutions" of a particular group that change more slowly, and are framed by "constitutional choice institutions", consistent with national and international laws, that change even more slowly, and are nested in "metaconstitutional institutions", such as social norms, that rarely change [48].

Each scientific discipline has a set of common formal collective choice institutions for conceptualization and measurement that influence the operational institutions used by its members. All scientists can devise new conceptualizations and institutions. When new informal institutions are widely adopted by other members of a discipline, they may become formal institutions, and widespread adoption of a new conceptualization may change the dominant paradigm of a discipline [29].

Hajer's definition of "discourse", which is generic but was devised for environmental research, implies that reproducing discourse in conceptualization is inseparable from reproducing institutions in measurement [45]. *Synergistic interactions* between conceptualization and measurement are quite common in science: new theories are tested by comparing their predictions with empirical data, but new data may raise questions about existing theories and lead to better ones, and to more measurements to test these theories. Such interactions are not deterministic or predictable, and may have positive *and* negative effects on uncertainty.

3.4. Societal Constraints

The concepts of discourse and institutions can also explain societal constraints on groups that construct knowledge [41] (Figure 1), e.g., governments and intergovernmental organizations can impose their discourses and/or institutions on scientists working for them [49]. Science is also restricted by the operation of markets, but since governments frame the latter, by establishing and sustaining suitable constitutional choice institutions, they can also modify this restriction for social ends.

3.5. Conceptualization Uncertainties

Estimating the magnitude of an environmental phenomenon is constrained by insufficient capacity to conceptualize it, resulting in four main sources of conceptualization uncertainty that limit the clarity and coverage of statements about it (Table 2). If insufficient conceptualization capacity is linked to limitations in discourse and language, as proposed in Section 3.3, then these sources can be listed in order of the increasing *linguistic complexity* of the statements to which they refer:

(1)   Terminological difficulties, in which using unclear, poorly defined or group-specific *terms*, e.g., A and B, to name and represent a phenomenon or its attributes can create confusion or ambiguity. Every scientific discipline has a different dominant discourse, so the same term may mean different things to different disciplines [50], or to scientists and lay people.

(2)   Underspecification, which involves lack of *completeness* in statements that combine various terms, e.g., "A + B", to describe the multiple *attributes* of a phenomenon. Every discipline at any time only has sufficient common formal rules, and corresponding institutions, to combine some of the terms in its current discourse and theories into statements that describe a phenomenon at particular spatial scales. Statements made by different disciplines may be mutually inconsistent.

(3)   Understructuralization, in which the actual spatial distributions of the characteristics of a complex phenomenon are not fully represented by the *disaggregation* of combinations of terms and statements about relationships between multiple attributes, or states and flows related to these. Such combinations may include groups of symbolic statements (equations), e.g., "$aA + bB = C_1$, and $dA + eB = C_2$", and nested hierarchical taxonomies of attributes and states that structure multiscalar knowledge. Structural classifications of phenomena are called "ontologies" in geographical information science [51]. So two conceptualizations of a phenomenon may differ structurally (ontologically) as well as terminologically (semantically).

(4)   Using proxies, in which attributes are represented by indicators loosely linked to the ideal variables for measuring these attributes, or phenomena are represented by models constructed with easily quantified variables. This happens when it is difficult to: (a) identify more appropriate variables by conceptualization, or (b) collect empirical data for such variables even if they are known.

Conceptualization uncertainties impose very real constraints on the accuracy of estimates, as the analysis of desertification below will show. Our first three sources are included in Regan et al.'s "linguistic" sources of uncertainty [13] (Table 2) but are structured more coherently here. Terminological difficulties can influence other sources. Proxies are used in reaction to the first three sources, and can involve synergies between conceptualization and measurement. They are mentioned in NUSAP [42] but not by Regan et al. [13] or Van Asselt and Rotmans [17]. Limitations on conceptualization capacity are also analysed in other literatures, such as that on "vagueness" [52].

If conceptualization uncertainty (U_c in Equation (1)) is the sum of uncertainties resulting from terminological difficulties (U_{cte}), underspecificity (U_{cusp}), understructuralization (U_{cust}) and using proxies (U_{cpr}) then:

$$U_c = U_{cte} + U_{cusp} + U_{cust} + U_{cpr} \qquad (4)$$

Societal constraints on scientific conceptualization can exacerbate these uncertainties by: (a) *territorialization*, in which a scientific community is divided into 'insiders' and 'outsiders' when policy makers appoint 'expert' advisors who are unaccountable to other scientists, contrary to norms for good communication [53]; and (b) *scope shaping*, in which policy makers influence the scope of knowledge that these experts supply by imposing discourses and institutions on them [49].

3.6. Measurement Uncertainties

Estimating the magnitude of an environmental phenomenon is also restricted by insufficient capacity to measure it, leading to four main sources of measurement uncertainty which inhibit construction of quantitative statements. If insufficient measurement capacity is linked to institutional limitations, as proposed in Section 3.3, then these sources can be listed in order of increasing *institutional nesting*:

(1)   Random errors in measured data, resulting from deficient equipment and human error.

(2) Systematic errors in measured data, which are linked immediately to technical constraints, and through these to formal and informal institutions. For example, measurements of environmental phenomena may be biased by informal adoption of repeated practices which use: (a) equipment with insufficient resolution to observe a phenomenon reliably; and (b) inadequate sampling designs.

(3) Scalar deficiencies in measurement, which are linked more directly to institutional constraints. If the *formal* measurement institutions of a discipline do not specify all the scalar contexts that characterize an environmental phenomenon [54], then scientists may create ad hoc *informal* institutions for collecting and processing data. This can lead to errors in estimates that evade scrutiny in peer review.

(4) Using subjective judgment in making estimates, when data are lacking.

These measurement uncertainties combine in a more coherent way the "epistemic" sources 1, 2 and 6, and "linguistic" source 2 of Regan et al. [13]; and the "limited knowledge" sources 1, 2, 4 and 5 of Van Asselt and Rotmans [17] (Table 2). Subjective judgment is used in reaction to the other three uncertainties, and can involve synergies between conceptualization and measurement.

If measurement uncertainty (U_m in Equation (1)) is the sum of uncertainties resulting from random errors (U_{mr}), systematic errors (U_{msy}), scalar deficiencies (U_{msc}) and using subjective judgment (U_{msu}) then:

$$U_m = U_{mr} + U_{msy} + U_{msc} + U_{msu} \tag{5}$$

Societal constraints complicate measurement uncertainties when, for example: (a) scientists use global compilations of national statistics in the absence of planetary measurement, as when basing estimates of forest carbon change on national forest area statistics [55]; (b) governments ask scientific "experts" to use subjective judgment in making estimates for them, as in estimates of desertification evaluated below [49]; and (c) economic factors limit the size, frequency and resolution of surveys and hence the accuracy of estimates of phenomena characterized by the variables A_i, a_i, and t_i in Equation (3)—for example, market forces inhibited planetary measurement at appropriate spatial resolutions until the US government modified its institutions and made medium resolution Landsat images freely available in 2008.

3.7. Constructing the Uncertainty Fingerprint of an Estimate

The *Uncertainty Fingerprint* of an estimate combines its conceptual and measurement uncertainties in a row of a matrix, and is constructed by:

(1) Identifying which of the eight sources of conceptual and measurement uncertainties (Table 2) are associated with the estimate.

(2) Coding the uncertainties as follows:

    a. Conceptualization uncertainties: terminological difficulties (te); underspecification (usp); understructuralization (ust); and using proxies (pr).

    b. Measurement uncertainties: random errors (r); systematic errors (sy); scalar deficiencies (sc); and using subjective judgment (su).

(3) Calculating the total number of uncertainties in the fingerprint to give its *Uncertainty Score* (US), on a scale from 0 to 8.

3.8. Trends in Uncertainty over Time

Stacking the Uncertainty Fingerprints of successive estimates of an environmental phenomenon on top of each other in multiple rows in a matrix shows how the composition of its uncertainties changes over time. Among conceptualization uncertainties, ideally the use of proxies should end first (as estimates are increasingly based on appropriate measurements), followed by terminological difficulties, understructuralization and underspecification in a related manner. Among measurement uncertainties, reliance on subjective judgment should ideally end first, for the same reason as for proxies. Scalar deficiencies will decline as common rules for planetary measurement are devised, agreed and widely adopted, enabling reductions in random errors and systematic errors.

Assembling the trend in the Uncertainty Scores of successive estimates of a phenomenon in a stack gives its *Uncertainty Profile,* which can show if present uncertainty is persistent or not. If the Uncertainty Score falls to the *statistical threshold* value of US = 2, then ideally uncertainty should be dominated by two measurement uncertainties—random errors (U_{mr}) and systematic errors (U_{msy})—that can be evaluated by standard statistical methods alone, thereby showing continuity between the latter and the UAF (see also Supplementary Information). The Uncertainty Profiles of different phenomena can be used to compare trends in their present uncertainties.

The UAF only applies to information on the *magnitudes* of environmental phenomena. So gaining an accurate estimate of a phenomenon does not end the accumulation of knowledge about it. It is merely a precondition for allowing scientists to develop increasingly reliable explanations of the processes that cause and control it.

3.9. Rules for Constructing Reliable Global Environmental Information

The conceptualization uncertainties and measurement uncertainties listed in Table 2 and the inherent uncertainty function (Equation (3)) lead to seven rules for constructing reliable global environmental information by planetary measurement:

(1)   Define a phenomenon clearly and appropriately.
(2)   Specify the minimum number of attributes to measure, to completely characterize a phenomenon.
(3)   Disaggregate measurement of a phenomenon, to represent the full diversity of its spatial distribution.
(4)   Minimize spatial systematic errors, by using sensors whose spatial resolution matches the areal variability of a phenomenon and whose spectral resolution matches its most distinctive property.
(5)   Minimize temporal systematic errors, by choosing a monitoring frequency consistent with the turnover time of a phenomenon.
(6)   Minimize the systematic and random errors associated with the method used to classify satellite images, e.g., supervised classification, unsupervised classification, crowd classification etc., supported by ground data.
(7)   Minimize the systematic and random errors associated with the algorithm used to combine estimates of the various attributes of a phenomenon.

The first three rules will avoid terminological difficulties (1), underspecification (2), understructuralization (3), and using proxies. Rules 4–7 will avoid using subjective judgment, and reduce random and systematic errors and scalar deficiencies.

3.10. Methods

The inherent uncertainty of desertification was assessed using the components of the inherent uncertainty function (see Equations (2) and (3)).

Individual estimates of the extent of desertification were evaluated to identify the presence of conceptualization and measurement uncertainties, produce their Uncertainty Fingerprints, and calculate their Uncertainty Scores (US). The US values of five global estimates were combined to give the Uncertainty Profile of desertification. Underlying mechanisms which limit conceptualization and measurement capacities and generate uncertainties were also identified.

The rules proposed here for constructing global environmental information were applied to suggest how to reduce uncertainty about desertification by planetary measurement, and to inform the Uncertainty Fingerprinting of methods proposed to use 'big data' to monitor SDG Target 15.3.

3.11. Data

A time series of five estimates of the global extent of desertification, estimated by scientists working within the framework of intergovernmental (UN) institutions [1–3,56,57], was analysed using the UAF, together with methods proposed by scientific groups to use big data to monitor SDG Target 15.3 in seven papers identified in two recent reviews [9,10]. A sample of 96 papers in the *International Journal of Remote Sensing* in 2009 was examined to identify topics given priority in remote sensing science (see Supplementary Table S1). Another 50 papers on assessing dryland degradation, published in *Land Degradation and Development* from 2006 to 2010, were analysed to identify the scalar preferences, and diversity of discourses and institutions, of dryland scientists (see Tables S4, S5, S8 and S9). To avoid bias, both samples precede the start of global forest measurement using Landsat satellite data [34], and exclude special issues.

4. Results

To illustrate how the Uncertainty Assessment Framework (UAF) can be used in practice this section applies it to desertification. After examining the inherent uncertainty of desertification it identifies present conceptualization and measurement uncertainties in a time series of five estimates of the global extent of desertification, and then assembles the Uncertainty Fingerprints of these estimates and the overall Uncertainty Profile of desertification.

4.1. The Inherent Uncertainty of Desertification

4.1.1. Definition

Desertification is defined in the United Nations Convention to Combat Desertification (UNCCD) as "land degradation in arid, semi-arid and dry sub-humid areas resulting from various factors, including climatic variations and human activities" [58]. Countering it by the *restoration* of degraded land is necessary to achieve the Land Degradation Neutrality Target 15.3 of UN Sustainable Development Goal 15 [7,8] in dry areas.

4.1.2. Spatial Extent

Desertification affects the drylands, which, according to the UN Environment Programme World Atlas of Desertification [3], cover 6147 million hectares (Mha) in the hyper-arid, arid, semi-arid and dry sub-humid zones. All of this area except for 978 Mha of hyper-arid land (natural desert) is vulnerable to desertification [59] and so 5169 Mha should be measured to determine its extent.

4.1.3. Biophysical Complexity

Desertification is a complex phenomenon in which the degradation (or reduction in quality) of vegetation and soil, and the corresponding decline in their collective ecological functions, is influenced by variation in climate [2]. Long-term human degradation of land can accelerate when drought reduces land productivity and human impacts intensify. It involves continuous transitions between different degrees of degradation, and is usually reversible by restoration up to a threshold degree of degradation [59].

Desertification has multiple *attributes*. Thus, each type of dryland ecosystem has a particular *area*, within which its multiple layers of grasses, shrubs and trees grow at varying *densities*. Degradation through overuse causes each of these types of plants and their species (including crops) to decline in density, which makes soil more vulnerable to degradation by: (a) *water erosion*; (b) *wind erosion*; (c) *compaction* by animals and machinery; and (d) *salinization, alkalinization and waterlogging*—three related forms of degradation to which irrigated cropland is especially susceptible. Desertification therefore has *at least* six terrestrial attributes plus rainfall variation, for which vegetation maps must be corrected to avoid misleading inferences about vegetation change [60] (Table 3).

4.1.4. Spatio-Temporal Randomness

Desertification is highly dispersed and spatially variable, owing to variation in soil erosivity [61], and how the irregular timing and location of rainfall influence vegetation growth in dry areas and human responses to it.

4.1.5. Human-Environment Complexity

Biophysical complexity and natural randomness are exacerbated by how complex underlying social, economic and political driving and controlling forces [62,63] can lead to cross-scalar relationships [64] and coupled relationships with multiple feedback loops [65].

Consequently, areal variability (Equation (3)) may be as little as 0.1 ha, since tree density is low in dry open woodlands, and gullies caused by soil erosion may only be a few metres wide, even in advanced stages of erosion [66]. A turnover time of 2 years fits the great fluctuation in rainfall and short-term vegetation and human responses to this [67] within long-term cycles.

4.1.6. The Relative Inherent Uncertainty of Desertification

Desertification is one of the most inherently uncertain of all global environmental change phenomena. For example, in terms of the components of the inherent uncertainty function (Equation (3)), it has seven times as many attributes as forest area change (Table 3), and the area potentially affected is three times the area of forest in the tropics (Table 4), where forest area is currently changing most rapidly. An areal variability of as little as 0.1 ha is just a fifth of that of tropical forest area change (0.5 ha): the smallest agricultural clearances in tropical moist forest are usually of the order of 1 ha, but this overall tropical mean allows for the greater spatial complexity of tropical dry forest change. The turnover time of desertification (2 years) is slightly less than that of the 3 years for tropical forest area change (Table 4).

Table 4. Values of components of the inherent uncertainty function for two global environmental change phenomena.

Phenomenon	Potentially Affected Area (Mha)	No. of Attributes	Areal Variability (ha)	Turnover Time (yrs)
Desertification	5169	7	0.1	2
Tropical forest area change	1770	1	0.5	3

4.2. Conceptualization Uncertainties of Estimates of Desertification

This assessment of present uncertainty about desertification begins by checking to see if the four sources of present conceptualization uncertainty in Table 2—terminological difficulties, underspecification, understructuralization and using proxies—occur in the time series of five estimates of the global extent of desertification in Table 5.

4.2.1. Terminological Difficulties

Terminological difficulties lead to uncertainty about what a number refers to, and to inconsistency between estimates of what may appear to be the same variable.

The first four estimates of the extent of at least moderately desertified land were prepared for the United Nations Environment Programme (UNEP), which convened the UN Conference on Desertification (UNCOD) in 1977 and coordinated implementation of the Plan of Action to Combat Desertification agreed there [68]. The estimates vary greatly, from 4002 Mha [2] to 3272 Mha [1] and 3475 Mha [56] in the 1970s to 608 Mha [3,4] in the 1980s (Table 5). Counter-intuitively, they appear to show desertified land *contracting*, not expanding, over time. The rate of desertification has only been estimated once, for the 1970s (20 Mha.a^{-1}) [1].

These estimates have no terminological difficulties as they all assume that desertification includes a range of soil and vegetation degradation, much of it dispersed and reversible, with only the most severe degradation leading to new desert. This is consistent with how UNCOD defined the term as: "an aspect of the widespread deterioration of ecosystems under the combined pressure of adverse and fluctuating climate and excessive exploitation . . . [involving] the diminution or destruction of the biological potential of land, and can lead ultimately to desert-like conditions" [59]. The fourth estimate was reported in the UNEP World Atlas of Desertification [3] and included in its Second Edition too [4], though this used instead the more compact definition in the UN Convention to Combat Desertification [58] (see Section 4.1.1).

Table 5. Estimates of the global extent of desertification (Mha).

Estimate	Primary Variable	Period	Magnitude (Mha)	Notes
Dregne (1977) [2]	Area of at least moderately desertified land	1970s	4002	–
Dregne (1983) [1]	Area of at least moderately desertified land	1970s	3272	–
Mabbutt (1984) [56] *	Area of at least moderately desertified land	1970s	3475	–
Middleton and Thomas (1992, 1997) [3,4]	Area of at least moderately desertified land	1980s	608	UNEP World Atlas of Desertification
LADA (2008) [57]	Degrading area	1981–2003	771	From Bai et al. [69]

* This paper also included an estimate of 1942 Mha that omitted unused rangelands.

The fifth estimate in Table 5, 771 Mha, *does* have terminological difficulties. It comes from a "preliminary [global] map of land degradation" published by the Land Degradation Assessment in Drylands (LADA) project of another UN agency, the Food and Agriculture Organization [57]. It refers not, as FAO states, to the area of "degraded land", but to a proxy variable of *"degrading area"* [69]. The estimate is based on a drop in biomass growth from 1981 to 2003 estimated from satellite data. So here conceptualization is affected by the practicalities of measurement.

Table 5 contains no estimate for the Third Edition of the World Atlas of Desertification, published in 2018 not by UNEP, but by the Joint Research Centre of the European Commission (JRC). JRC is a leading centre for planetary measurement, and a new map of desertification based on remote sensing data could have provided a more robust estimate than those in earlier editions, which relied on subjective judgment. Yet the Atlas states that: "'desertification' or 'land degradation' cannot be captured in global maps in a way that satisfies all stakeholders. Instead, [the Atlas] illustrates the geographic distribution of coincident patterns of issues that may indicate potential land degradation" [5].

Difficulties in "satisf[ying] stakeholders" in the new Atlas reflect the different perceptions of the governments of developing countries, who are concerned about the impacts of drought (a natural hazard) on economic development, and those of developed countries, who are more concerned about land degradation (a human-made hazard) [49]. The term "desertification" is also contested by scientists, as its original meaning of frontier-like desert expansion [70] is not how UNCOD understood desertification [71]. UNEP 'territorialized' the drylands science community (see Section 3.5) into 'insiders', who advised it for UNCOD [59] and later initiatives and accepted its discourse, and 'outsiders' (other scientists), many of whom did not. Thus, in our sample of 50 papers that assess dryland degradation, only 36% mention the term "desertification" in the text and just 4% include it in their titles (Table S4). The focus of the new Atlas on *"potential* land degradation" is consistent with a scientific discourse within which maps of potential desertification *hazard* are generated by biophysical models [72,73]. A "World Map of Desertification" was the most widely publicized of four maps presented to UNCOD, though it only showed potential land degradation hazard, not the actual current *status* of desertification [74].

4.2.2. Underspecification

Underspecification limits the *completeness* of estimates in covering all attributes of a phenomenon.

All the estimates of the extent of desertification in Table 5 are underspecified. In 1977, Dregne was the first to specify desertification as a combination of vegetation degradation and soil degradation [2], and used this approach to produce for UNCOD the first world map of current desertification *status* (Figure 2) [75]. This subjective estimate divides soil degradation into wind erosion, water erosion and salinization, but omits soil compaction (Figure 3a). Two later estimates by Dregne in 1983 [1] and Mabbutt in 1984 [56] are even less complete, as they only refer to total soil erosion (Figure 3b).

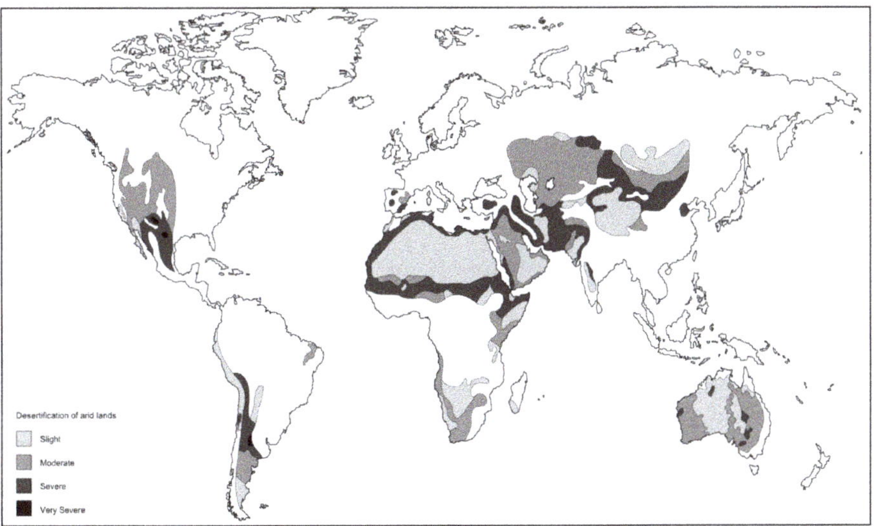

Figure 2. The first world map of desertification status (Based on [75]).

The UNEP World Atlas of Desertification estimate is underspecified too, since it treats soil degradation as a proxy for all desertification (Figure 3c). The estimate is well specified in soil degradation, covering all soil attributes, but it omits *vegetation* degradation. The Atlas acknowledges this limitation, and includes a map combining soil and vegetation degradation, but no estimate based on this map [3]. This conceptualization was influenced by measurement practicalities, since UNEP used the dryland component of an existing soil degradation map based on subjective estimates by a large team of scientists [76], instead of commissioning a special survey of desertification.

In contrast, the LADA estimate is underspecified because it omits *soil* degradation and uses a decline in vegetation productivity as a proxy for all land degradation [57] (Figure 3d). Yet vegetation productivity corresponds to just one of 11 indicators ("vegetation activity") in LADA's own comprehensive taxonomy of land degradation indicators, the other ten covering climate, soil and water [77] (Table S6). This proxy also involves a synergy between conceptualization and measurement since LADA used an existing map of vegetation change originally produced for another purpose [69]. Underspecification in the LADA and UNEP World Atlas of Desertification estimates contributes to their values being lower than the earlier estimates, since both omit a major group of attributes.

a. Dregne (1977)

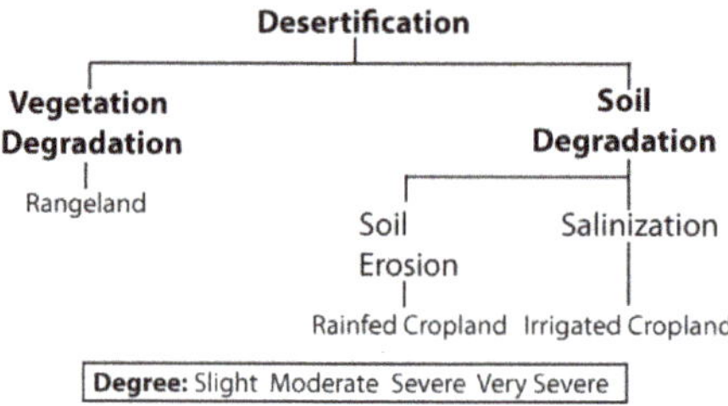

b. Dregne (1983), Mabbutt (1984)

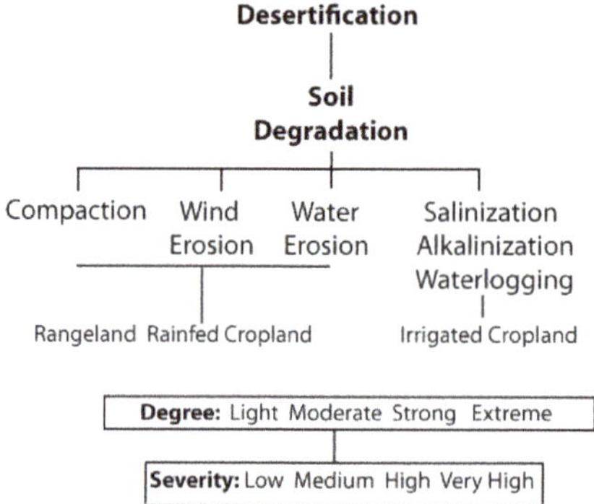

c. UNEP World Atlas of Desertification (Middleton and Thomas, 1992)

d. Land Degradation Assessment in Drylands (LADA) (Bai et al., 2008)

Figure 3. Alternative conceptual structures for specifying the attributes of desertification in five global estimates of the extent of desertification by Dregne [1,2], Mabbutt [56], UNEP [3] and LADA [57], also showing their disaggregation by land use type and the scales used for ranking the degree and severity of desertification.

4.2.3. Understructuralization

Understructuralization limits the extent to which an estimate is *disaggregated* to represent the actual distribution of a phenomenon.

Estimates of the extent of desertification would ideally be disaggregated by types of land use, aridity and degradation of irrigated cropland. Only the Dregne (1977) and LADA estimates in Table 5 are understructuralized by land use type [2,57]. The Dregne (1983) and Mabbutt (1984) estimates divide areas of land by degree of desertification, e.g., slight, moderate, severe and very severe, for the three main uses of drylands: rainfed cropping, livestock raising and irrigated cropping (Figure 3b) [2,56]. The UNEP World Atlas of Desertification estimate takes a different approach, by focusing on the causes of desertification, but it identifies areas in which soil is degraded by "overgrazing" and "agricultural activities". The latter include both rainfed cropping and irrigated cropping, whose degraded area is listed separately (Figure 3c) [3].

Only the UNEP World Atlas of Desertification [3] and LADA [57] estimates are disaggregated between the aridity zones within which desertification can occur according to the UN [58,59], though the LADA estimate combines the arid and hyper-arid zones (Table S7). The other estimates are understructuralized and this limits their spatial resolution.

The UNEP World Atlas of Desertification estimate is also fully disaggregated between the different types of degradation of irrigated cropland: salinization, alkalinization and waterlogging [3] (Figure 3c). The other estimates are understructuralized since they merely list the area of all degraded irrigated cropland under the heading of "salinization or waterlogging", as with the estimates by Dregne [1] and Mabbutt [56], or aggregate degraded irrigated cropland with other degraded land, as with the estimates by Dregne [2] and LADA [57].

4.2.4. Using Proxies

All estimates of the extent of desertification in Table 5 use proxies, indicating their tenuous foundation on measured variables and/or data. Dregne only uses one proxy in his two estimates–an economic indicator (crop yield) to represent salinization of irrigated cropland [1,2] though his second estimate does include electrical conductivity equivalents [1]); but Mabbutt [56] relies on economic proxy indicators (crop and livestock yields) for all three of his attributes (Table S6).

The UNEP World Atlas of Desertification uses soil degradation as a proxy for desertification. It assesses different types of soil degradation using quantifiable indicators, and then converts these into the extent of desertification by using four *additional* proxy indicators: "changes in agricultural suitability", "decline in agricultural productivity", the quality of the terrain, and intactness of "biotic functions" and the ease of restoring these [3] (Table S6). The LADA map uses "degrading area" as a proxy for "degraded land" [57], though the map's original authors [78], and later LADA itself [79], recognized that this did not properly represent land degradation observable on the ground.

4.3. Measurement Uncertainties of Estimates of Desertification

This section reports the presence in the time series of estimates of the four sources of measurement uncertainty listed in Table 2: random errors, systematic errors, scalar deficiencies and using subjective judgment.

4.3.1. Random and systematic errors

Systematic errors can be evaluated in relation to areal variability and turnover time in the inherent uncertainty function (Equation (3)). They are analysed here with random errors since both are high in all the estimates of desertification in Table 5. Systematic errors are difficult to assess for the first four estimates, owing to the limited empirical data on which these are based, but are more easily traced in LADA's map of lands, where, according to the Normalized Difference Vegetation Index calculated from satellite data, biomass growth fell from 1981 to 2003 [57]. Drylands only account for 22% of the global total of this "degrading area" (Table S7), and since in Africa the latter is concentrated below the Equator, the estimate is biased as it excludes degradation of drylands immediately to the south of the Sahara. Desertification, by definition, can lead to "the diminution or destruction of the biological potential of land ... ." [59], but it is not *equivalent* to a reduction in net primary productivity, as this can also occur because of lack of rainfall [80]. Systematic errors also result from the gap between the 8 km resolution of satellite data used for this map and the much higher resolutions needed to monitor the areal variabilities of the different attributes of desertification reliably (see Section 4.1.4) [81].

4.3.2. Scalar Deficiencies

All estimates of desertification in Table 5 have scalar deficiencies owing to limitations of the informal institutions devised to produce them. 'Insider' scientists who worked within UN institutions to make subjective global estimates of the extent of desertification for UNEP devised informal institutions to do this, since few local ground data were available [49]. Studies by autonomous scientists have scalar deficiencies too, e.g., LADA's global map of "degrading area" relies on another ad hoc set of institutions [57]. None of our sample of 50 papers on assessing dryland degradation shows evidence for the use of conceptual frameworks and formal measurement institutions suited to global and regional scales. Only 4% of papers even produce national information using national conceptual frameworks (Table S8).

4.3.3. Using Subjective Judgment

It is difficult to evaluate properly the reliability of subjective estimates by referring to the methods and/or data on which they are based. Only the LADA estimate does not rely on subjective judgment [57].

4.4. Fingerprinting the Sources of Uncertainty about Desertification

The Uncertainty Fingerprints of the estimates of the extent of desertification by Dregne (1983) [1] and Mabbutt (1984) [56] show that the estimates are limited by underspecification, understructuralization by irrigated cropland and climate, random errors, systematic errors, scalar deficiencies and using proxies and subjective judgment (Figure 4). The Dregne (1977) estimate is also understructuralized by land use type [2]. The least uncertain estimate, by the UNEP World Atlas of Desertification [3], lacks understructuralization, but resembles the preceding three estimates in being underspecified, relying on proxies and subjective judgment, and having no terminological difficulties. The LADA estimate is not based on subjective judgment but does have terminological difficulties [57].

Estimate	Conceptualization				Measurement				Score
	Terminological difficulties	Under-specification	Understructuralization	Using proxies	Random errors	Systematic errors	Scalar deficiencies	Using subjective judgment	
Dregne (1977)	-	usp	ust_c ust_{lu} ust_{ir}	pr	r	sy	sc	su	**7**
Dregne (1983)	-	usp	ust_c ust_{ir}	pr	r	sy	sc	su	**7**
Mabbutt (1984)	-	usp	ust_c ust_{ir}	pr	r	sy	sc	su	**7**
Middleton & Thomas (1992)	-	usp	-	pr	r	sy	sc	su	**6**
LADA (2008)	te	usp	ust_{lu} ust_{ir}	pr	r	sy	sc	-	**7**

Figure 4. A stack of Uncertainty Fingerprints to show changes over time in the conceptualization and measurement uncertainties associated with five estimates of the global extent of desertification and in their Uncertainty Scores (ust_c = understructuralization by climate; ust_{lu} = understructuralization by land use, and ust_{ir} = understructuralization by irrigated cropland degradation).

Measurement uncertainties exceed conceptualization uncertainties in the first four estimates in Table 5, yet conceptualization uncertainties still account for over 40% of all sources of uncertainty (Figure 5). This supports claims by Van der Sluijs [41] and others that statistical methods alone have limitations for evaluating very uncertain phenomena.

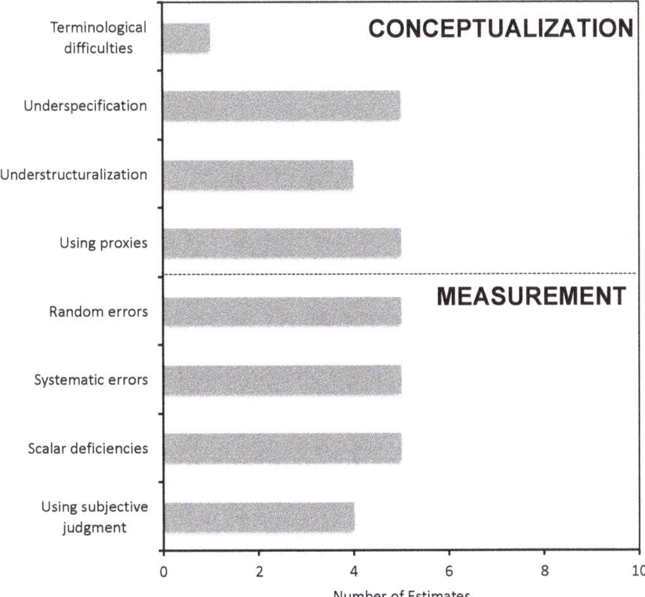

Figure 5. The total numbers of the eight main sources of conceptualization and measurement uncertainties found in a time series of five estimates of the global extent of desertification.

Stacking the fingerprints on top of each other to give the Uncertainty Profile of desertification shows that uncertainty about it is high and persistent. The first three estimates, by Dregne [1,2] and Mabbutt [56], all have Uncertainty Scores of 7 on a scale from 0 to 8. This drops to 6 for the UNEP World Atlas of Desertification estimate [3], but returns to 7 for the LADA estimate [57] (Figure 6). The mean score of 6.8 is far above the *statistical threshold* of 2, when only random and systematic errors are expected and statistical evaluation alone is appropriate, according to the UAF, so this also supports the claim of Van der Sluijs [41].

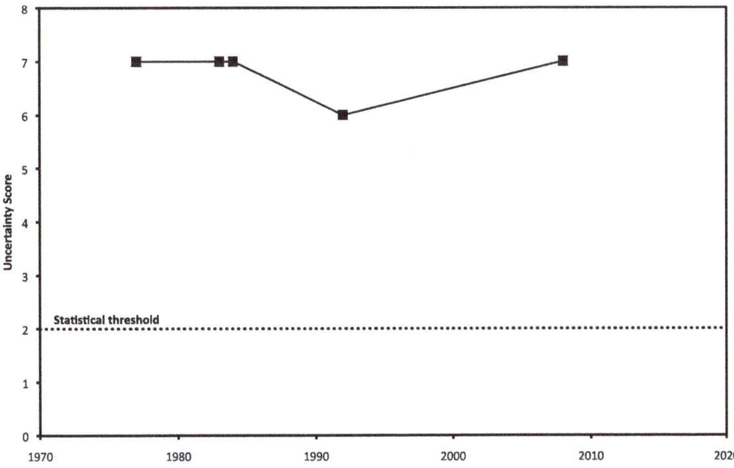

Figure 6. The Uncertainty Profile of desertification, based on the Uncertainty Scores of five estimates of the global extent of desertification made between 1977 and 2008 [1–3,56,57].

4.5. The Underlying Mechanisms of Global Environmental Uncertainties

The UAF can explain *why* uncertainties about estimates persist, by linking trends in uncertainties, as in the Uncertainty Profile in Figure 6, to underlying discursive and institutional constraints on conceptualization and measurement capacities in the monitoring systems that produce the estimates (see Section 3.3).

Intergovernmental discourses responding to societal influences have framed conceptualization in all estimates of desertification evaluated here, allowing the use of proxies (Table 6).

Table 6. Numbers of conceptualization and measurement uncertainties associated with five estimates of the global extent of desertification and their underlying mechanisms (I = intergovernmental, S = scientific, Y = present, and − = absent).

	Conceptualization Uncertainties	Measurement Uncertainties	Uncertainty Score	Discourse	Formal Institutions	Informal Institutions	Conceptualization–Measurement Synergies
Dregne (1977) [2]	3	4	7	I	I	S	−
Dregne (1983) [1]	3	4	7	I	I	S	−
Mabbutt (1984) [56]	3	4	7	I	I	S	−
Middleton and Thomas (1992) [3]	2	4	6	I	I	S	Y
LADA (2008) [57]	4	3	7	IS	I	S	Y
Mean			6.8				

Uncertainty is also influenced by the institutions of intergovernmental and governmental organizations, and by scientific institutions. Formal intergovernmental institutions are linked here to large uncertainties in monitoring desertification, but they have allowed scientists to devise informal institutions to make estimates (Table 6).

Negative synergies between conceptualization and measurement can promote uncertainty too (Table 6), as when ease of access to existing maps of soil degradation and vegetation change led to underspecification in the UNEP World Atlas of Desertification estimate [3] and LADA estimate [57], respectively. So while in Heisenberg Uncertainty, one measurement disturbs another [11], in environmental uncertainty it seems that how a phenomenon is 'measured' can disturb how it is conceptualized.

5. Measuring Desertification

The results presented in the previous section, which show that uncertainty about desertification has been persistently high for decades, imply that global environmental uncertainties are indeed inevitable, and so support the statement in the Third Edition of the World Atlas of Desertification that the global extent of desertification cannot be mapped satisfactorily [5]. However, this evidence is not conclusive. This section applies the seven rules for constructing reliable global environmental information through planetary measurement, derived from the UAF in Section 3.9 (Table 7), to examine if it is technically feasible to measure desertification reliably at global and national scales, for example, to quantify the indicator for Target 15.3 of the Sustainable Development Goals: "proportion of land that is degraded over total land area" [8]. It then examines if these requirements are met by a sample of seven papers, identified in recent reviews [9,10], which propose using "big data" to monitor SDG Target 15.3.

Table 7. Seven rules for constructing reliable global environmental information.

---

1. Define a phenomenon clearly and appropriately.
2. Specify the minimum number of attributes to measure, to completely characterize a phenomenon.
3. Disaggregate measurement of a phenomenon, to represent the full diversity of its spatial distribution.
4. Minimize spatial systematic errors, by using sensors whose spatial resolution matches the areal variability of a phenomenon and whose spectral resolution matches its most distinctive property.
5. Minimize temporal systematic errors, by choosing a monitoring frequency consistent with the turnover time of a phenomenon.
6. Minimize the systematic and random errors associated with the method used to classify satellite images.
7. Minimize the systematic and random errors associated with the algorithm used to combine estimates of the various attributes of a phenomenon.

5.1. Conceptualizing Desertification

Conceptualization frames the design of data collection, the analysis of data, and presentation of the resulting information, and is the subject of the first three rules in Table 7.

5.1.1. Define a Phenomenon Clearly and Appropriately

If desertification is defined as in either the UNCOD or UNCCD definitions (see Sections 4.1.1 and 4.2.1) then this should avoid terminological difficulties.

5.1.2. Specify the Minimum Number of Attributes to Measure

An estimate of the extent of desertification will be fully specified if all six attributes of vegetation degradation and soil degradation in Table 3 are measured, and their estimates are adjusted to remove misleading signals caused by rainfall variation.

5.1.3. Disaggregate Measurement of a Phenomenon

To avoid understructuralization, any measurement of desertification should be disaggregated to represent the actual diversity of its spatial distribution by estimating the degree of degradation for all types of land use, aridity and degradation of irrigated cropland. Past experience, reviewed in Section 4.2.3, shows how to do this. Disaggregating by aridity requires that a digital map of climatic zones is overlaid on a map of desertification. As changes in global climate will shift climatic zones [82], existing maps of the latter should be revised using ground-based climate measurements. To disaggregate by land use, it is necessary to map land use *before* measuring degradation, so that measurements can incorporate criteria appropriate to each land use [1]. Mapping land use is also a prerequisite for mapping degradation of irrigated cropland, as specific measurement methods, discussed below, are required for this too.

5.1.4. Avoiding Other Types of Conceptualization Uncertainties

Using remote sensing data, supported by ground data, does not prevent the use of proxies (see Section 4.2.4), but proxy uncertainty should be absent if planetary measurement is properly conceptualized and carried out at appropriate spatial and temporal resolutions.

5.2. Measuring Desertification

Measurement involves collecting data and converting them into meaningful information. It is the subject of the last four rules in Table 7.

5.2.1. Minimize Spatial Systematic Errors

Matching the spatial resolution of a sensor to the *areal variability* (smallest area of variation) of each attribute of a phenomenon, and the sensor's spectral resolution to the most distinctive property of each attribute, will minimize spatial systematic errors. Desertification has at least six terrestrial attributes plus rainfall variation. Each is now discussed in turn.

(1)　*Vegetation area.* Mapping vegetation cover in dry areas is challenging since dryland ecosystems commonly involve trees scattered at low density over grasslands. This is difficult to measure with the medium (20–100 m) resolution optical satellite sensors used to map changes in the area of the much denser forests in humid areas with reasonable accuracy [83]. The first global map of tree cover in drylands based on very high ($\leq$ 1 m) resolution satellite images was not published until 2017, and led to a much higher estimate of dry forest area than earlier estimates using lower resolution images [84]. The correlation which that study found between dry forest area and the spatial resolution of sensors (Figure 7) supports the relationship between spatial resolution and areal variability (a_i) in the inherent uncertainty function (Equation (3)).

(2)   *Vegetation density*. Measuring vegetation degradation in dry areas, e.g., by a decline in tree and grass density, is even more challenging than measuring vegetation cover [83]. Very high resolution satellite images are suitable for this too, but measurement is complicated by: (a) the maintenance of vegetation cover when invasive species proliferate on degraded land; (b) the lack of an absolute benchmark for 'non-degraded' ecosystems in the drylands [85]; and (c) the temporal dimension, e.g., tree and grass density vary with rainfall, and so apparent trends should be corrected for this (see below).

(3)   *Water erosion*. Medium resolution (Landsat) images have been used to measure trends in areas suffering from water erosion based on their spectral properties [86]. They can also identify large- and medium-sized gullies but cannot track their development over time [66]. Very high resolution satellite images are therefore needed for comprehensive measurements of the features of water erosion. Research has found that as spatial resolution rises, so too does the number of gullies identified. For example, 9, 15 and 30 gullies were mapped in an area in Tunisia by automated classification of images from SPOT multispectral (10 m resolution), SPOT panchromatic (5 m resolution) and Quickbird (0.6 m resolution) sensors, respectively [87]. This also supports the relationship between the spatial resolution of sensors and areal variability (a_i) in the inherent uncertainty function (Equation (3)).

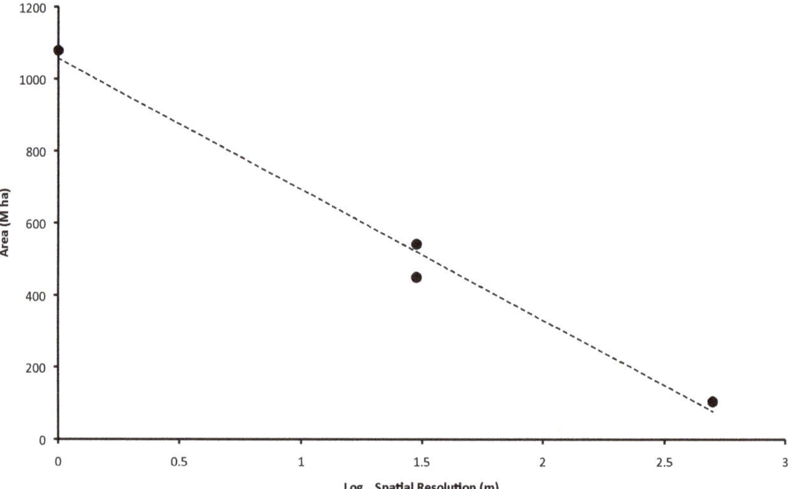

Figure 7. The expansion of estimates of dry forest area [84,88–90] as the spatial resolution of satellite sensors used for measurement gets closer to the areal variability of dry forest.

Radar sensors and light detection and ranging (LiDAR) sensors can be used to measure water erosion too, e.g., gullies below forest canopies have been mapped by an airborne LIDAR sensor [91].

(4)   *Wind erosion*. The spatial distribution of wind erosion has not yet been directly measured using satellite images, possibly because of the absence of the same large

physical artefacts seen in water erosion. One way to overcome this problem, discussed in Section 5.2.2, currently suffers from temporal resolution issues. Most estimates of the rate of wind erosion are currently made using mathematical models that incorporate meteorological factors, such as wind speed, and the susceptibility of soil to erosion, with the use of satellite images confined to mapping land use and land cover and how these change over time [92].

Landsat images, on the other hand, can measure trends in sandy areas, showing that while in some parts of northern China, for example, sandy areas are contracting, elsewhere they are expanding [93–95].

(5)   *Soil compaction*. A literature search using Google Scholar found no studies which measured soil compaction using optical satellite sensors. LIDAR and radar sensors might be suitable for this purpose, however.

(6)   *Salinization, alkalinization and waterlogging of irrigated cropland*. The spectral signatures of salinized and waterlogged areas differ sufficiently from those of non-affected areas for them to be separated by medium resolution optical satellite images [96], but best results are obtained by using ground and laboratory data too [97]. Areas affected by salinization and alkalinization can also be distinguished using medium resolution images [98]. Measuring the *degree of salinization* using satellite sensors was previously thought to be too difficult, owing to sensor limitations and variable spectral responses [99–101]. Yet recent research in Morocco and Turkey shows that the degree of salinization can be measured by soil salinity indices constructed using reflectance characteristics in the visible and near infrared bands of Landsat images [102] and high (10 m) resolution Sentinel 2 images [103]. So desertification maps based on satellite images can be disaggregated by the type *and* degree of degradation of irrigated cropland.

(7)   *Rainfall variation*. The role of rainfall variation is discussed in (2) above and in the next section.

5.2.2. Minimize Temporal Systematic Errors

Ensuring that monitoring frequency is consistent with the shortest time period over which a phenomenon varies (*turnover time*) will minimize temporal systematic errors. The temporal resolution at which desertification generally is measured should ideally match its turnover time, set above at 2 years, while allowing correction of misleading signals due to the seventh attribute, rainfall variation, over longer periods.

Without appropriate correction, cyclical rainfall patterns make it difficult to determine if a reduction in vegetation cover is caused by land degradation or declining rainfall, or if a greater profusion of vegetation is the result of land restoration, the spread of invasive species, or simply a rise in rainfall [104,105]. Confusion over this issue has previously led to incorrect estimates of the rate of desertification and, in turn, to scepticism about whether desertification actually exists [71]. For example, in 1977 UNEP reported that comparing aerial survey observations with an 18-year-old map of the Sahara Desert's southern border implied that the desert was moving south at over 5 km per annum [106]. Scientific scepticism about the existence of desertification grew in the late 1980s [71], after analysis of low spatial resolution satellite images showed that while the boundary between the Sahara Desert and the Sahelian region shifted south in 1981, it moved north in 1985 when rainfall returned [60,107,108]. So rainfall measurements at long-term monitoring stations are indispensable for correcting for the variation of vegetation growth with rainfall, and for future changes in climatic zone boundaries resulting from global climate change [82].

Annual rainfall variation is used here as the climate attribute because it is important for analysing satellite data on land cover. Other climatic variables contribute to understanding desertification but in different ways, and so are not listed here for measuring actual desertification status. For example, prolonged droughts have a *causative* role in accelerating actual desertification [59], and so would be independent variables in future models in which the measured extent of desertification is the dependent variable. Mean dry season length and the mean annual number of extreme precipitation events could be used in a similar way.

Research suggests that measuring wind erosion by combining satellite data and ground data will be challenging for temporal reasons. The origins and paths of dust storms can in principle be measured using optical satellite images, but dust storms are often not detected from space due to high cloud cover, and even on cloudless days the temporal resolution of satellite sensors may not match the relevant turnover time (t_i in Equation (3)). For example, using ground-based cameras to collect images in the Mojave Desert every 15 min over six years recorded major dust events on 68 days each year, on average. Yet none of these events was identified in images from the low (250 m) spatial resolution MODIS sensor, despite its high temporal resolution (daily image collection), as the timing of dust storms did not coincide with when cloud-free images were collected [109]. National ground-based networks are vital for measuring airborne dust transport but are still few in number, and even the US network has only 13 measurement sites [110]. Furthermore, according to Webb et al., such networks generally "do not address which areas are eroding, and why, with enough accuracy to inform management" [111].

5.2.3. Minimize Errors Associated with the Method Used to Classify Satellite Images

It is also important to minimize the systematic and random errors associated with the method used to classify satellite images, supported by ground data, since planetary measurement methods are still embryonic. Thus, the first global "wall-to-wall" map of forest area based on Landsat images, published only in 2012, relied on a major innovation in *semi-automated* supervised classification software [34]. The first global wall-to-wall map of forest area change based on Landsat images followed a year afterwards and appeared to use *automated* classification [112].

Since these innovations for classifying medium resolution satellite images are so recent, corresponding innovations for the reliable automated or semi-automated supervised large-area classification of very high resolution satellite images will take time to emerge. This is why the first very high resolution map of tree cover in the drylands used crowd-based visual classification [84], and why the same method is likely to be used to measure desertification at very high resolution for the first time.

5.2.4. Minimize Errors Associated with the Algorithm Used to Combine Estimates of the Various Attributes of a Phenomenon

When the multiple attributes of desertification have been measured, it is necessary to use an algorithm to combine the resulting estimates to map spatial variation in the overall degree of desertification. The choice of algorithm may lead to systematic and random errors and limit comparability between different estimates.

In the early estimates evaluated in Section 4, algorithms are only employed to allow for the contextuality of desertification [61], so it may occur in some parts of an area but not in others [113,114]. Thus, the UNEP World Atlas of Desertification first assesses the *degree* of desertification from Light to Extreme, and then uses an algorithm to designate the *severity* of desertification in areas on another four-point scale from Low to Very High, according to the percentage incidence of Light, Moderate, Strong and Extreme desertification in that area [3].

5.2.5. Avoiding Other Types of Measurement Uncertainties

Planetary measurement of desertification should prevent uncertainty due to the use of subjective judgment. Scalar deficiencies will be minimized if a robust set of planetary measurement rules, such as those proposed here, are employed. Gaining a consensus in the global change science community for a common set of rules will take time. However, the seven rules in Table 7 could provide a foundation on which initial theoretical discussions can build, so that the variety of informal planetary measurement institutions now in use can become increasingly consistent.

5.3. The Prospects for Reducing Uncertainty about Desertification

This section has presented an optimistic view of the technical feasibility of using planetary measurement to reduce uncertainty about desertification, but has also indicated that *current* state-of-the-art remote sensing methods still impose limits on the extent of this reduction. For instance, the Uncertainty Score for estimates is unlikely to fall below 3 soon, because of continuing underspecification owing to the lack of measurement of wind erosion and soil compaction.

Measuring the extent of desertification at global scale must be organizationally feasible as well as technically feasible. Thus, measuring global forest area using a wall-to-wall survey of Landsat images was, arguably, technically feasible in the 1970s but it did not become organizationally feasible until 2012 [34]. A similar organizational advance is needed to reduce uncertainty about desertification. For a Global Drylands Observing System, which was advocated in various studies in the late 2000s, Verstraete et al. proposed a nested hierarchy of monitoring centres covering all scales from global to local [6]. Bastin et al. later found that tree cover in drylands could be measured at global scale by crowd-based classification of very high resolution satellite images in regional centres [84]. This could provide the basis for planetary measurement of desertification, though this section has shown that ground-based measurements, especially of wind erosion, soil compaction and rainfall, may also be needed for the foreseeable future.

5.4. Recent Proposals to Use "Big Data" to Monitor SDG Target 15.3

The measurement approach proposed here can be used to quantify the indicator for SDG Target 15.3 listed in the Sustainable Development Goals: "proportion of land that is degraded over total land area" [8]. In the absence of sufficient *national* data to allow countries to monitor progress in meeting the SDGs, two recent reviews have advocated using *global* sets of "big data" (including satellite data) instead [9,10]. Yet since analysis earlier in this paper has shown that existing global information on desertification is inadequate, this section uses the UAF to evaluate the reliability of the methods proposed to monitor Target 15.3 in a sample of seven of the papers that are cited as exemplars of the big data approach in these two review studies.

Only one of the seven papers, by Christian et al. [115], specifically aims to measure the actual *status* of desertification, in a 144,368 ha area of Rajasthan State in India. While it has no terminological difficulties, it is understructuralized since it is only disaggregated by land use types and climatic zones (even though salinization is a major problem in Rajasthan [116]), and is also underspecified since it merely maps a 25 year (1991–2016) trend in vegetation degradation and water erosion, with vegetation degradation only being assessed on land with natural ecosystems. Random and systematic errors are relatively high, because 30 m resolution satellite data are employed as standard, with 5.8 m resolution data only used for 2016, and temporal resolution (≥ 9 years) is also rather low. A second paper, by Wang et al. [117], measures the status of "land degradation" in the whole of Mongolia, but since this is in a dry area it is equivalent to desertification. The method has no terminological difficulties and corrects informally for rainfall variation, but it is underspecified as it effectively uses vegetation degradation (between non-degraded land, desert steppe, sand, desert and barren land) as a proxy for land degradation as a whole, and does not measure soil degradation as such. It is also understructuralized by aridity zones, land use types and degradation of irrigated cropland. Spatial systematic errors are relatively high, because 30 m resolution satellite data are used as standard, though temporal systematic errors are relatively low since the highest temporal resolution is 5 years.

Two more papers merely use models to predict the *potential* hazard of desertification [118,119], following the approach of the UNCOD "World Map of Desertification" [74] described in Section 4.2.1, so they are not evaluated here. Nor is another modelling study which predicts soil organic carbon content and other soil properties at global scale, using a network of sample plots and low (250 m) resolution optical satellite data on land cover and other land properties [120].

In the two remaining papers, a global study by Giuliani et al. [121] discusses how to assess land degradation in a range of climatic zones while Mitri et al. focus on a 140,800 ha area in Lebanon [122]. Both studies are framed by three UNCCD desertification indicators—land cover, land productivity, and soil organic carbon stocks—that have been proposed to substitute for the single SDG indicator [123], since the UNCCD is coordinating implementation of the LDN target. As discussed in Section 4.3.1, land cover change is an inadequate proxy for vegetation degradation. Satellite-based measurement of change in the net primary productivity of areas stratified by land cover type may be used to estimate vegetation degradation, but it is an inadequate proxy for land degradation as a whole. The same is true for estimates of changes in soil organic carbon content, which should be derived from direct measurements of soil carbon density and the different types of soil degradation (Table 3), and not used as a substitute for them. A full critique of the UNCCD indicators requires a separate study [124], but they and other indicators have been critically evaluated by a group of experts appointed by the UNCCD [125]. As Giuliani et al. only aim to provide a "proof of concept" of accessing different data sources, their global study lacks sufficient methodological detail to be evaluated here, though it does recognize the need to use high spatial and temporal resolution data, and appreciates the limitations of the soil organic carbon indicator [121]. The Lebanon study is disaggregated by climatic zones and land use/land cover types, but not by degradation of irrigated cropland. It is underspecified, as it uses vegetation degradation (estimated using the change in net primary productivity for forest, grassland and cropland) as a proxy for all land degradation, and land use and land cover change to predict changes in soil organic carbon content, rather than measuring soil degradation directly. Temporal systematic errors are high, as the measurement period is 13 years. Spatial systematic errors are substantial, owing to the use of data from satellite sensors with resolutions ranging from 5 m to 1000 m. Despite being framed by the UNCCD indicators, it uses an original algorithm to estimate the degree of overall land degradation by a weighted sum of the magnitudes of land cover change, land productivity trend, change in net primary productivity, soil organic carbon content, erosion risk, soil fertility and rainfall [122]. Yet since these parameters and their weights are not justified in the study, this incurs further systematic errors (Table 7).

This evaluation of three of the seven exemplar big data studies complements the evaluation of the five global UN studies in Section 4 by showing how the UAF can be used to assess uncertainties in studies by scientific groups, and how ranking random and systematic errors in Uncertainty Fingerprints can be informed by the last four rules for constructing reliable global environmental information in Table 7. While the Uncertainty Scores of the five UN estimates vary between 6 and 7 (Figure 4) and have a mean of 6.8, these three studies have a lower mean of 4.7: the studies of Lebanon [122] and Mongolia [117] have scores of 5 while that of India [115] has a score of 4 (Figure 8). None of the three studies has terminological difficulties or uses subjective judgment. Only the Lebanon study by Mitri et al. uses an algorithm to provide an overall estimate of the degree of land degradation [122], and this has systematic errors associated with it. However, it is important to note that all three studies lack scalar deficiencies since they are limited in spatial scope.

So while there is clearly potential to use big data to substitute for inadequate national data when monitoring SDG Target 15.3, such measurements require a more careful selection of methods than those used in the three recent studies assessed in Figure 8 if Uncertainty Scores are to decline substantially. Two of the other four studies [118,119] illustrate the continuing popularity among scientific groups of estimating potential desertification hazard, rather than actual desertification status. Allen et al. are therefore justified in arguing that substituting big global datasets for national data will face challenges.

Estimate	Conceptualization				Measurement				Score
	Terminological difficulties	Under-specification	Understructuralization	Using proxies	Random errors	Systematic errors	Scalar deficiencies	Using subjective judgment	
Christian et al. (2018)	-	usp	ust_{ir}	-	r	sy	-	-	**4**
Wang et al. (2020)	-	usp	ust_c ust_{lu} ust_{ir}	pr	r	sy	-	-	**5**
Mitri et al. (2019)	-	usp	ust_{ir}	pr	r	sy	-	-	**5**

Figure 8. A stack of Uncertainty Fingerprints to show the conceptualization and measurement uncertainties associated with three recent estimates of the extent of land degradation based on 'big data' sources and their Uncertainty Scores (ust_c = understructuralization by climate; ust_{lu} = understructuralization by land use; and ust_{ir} = understructuralization by irrigated cropland degradation).

6. Conclusions

Fifty years after the first remote sensing satellite was launched to collect global data, estimates of the magnitudes of global environmental change phenomena remain very uncertain, since global data collected by these satellites have not been fully converted into global information. This paper has built on two previous taxonomies of the sources of environmental uncertainty [13,17] to propose an Uncertainty Assessment Framework (UAF) for evaluating very uncertain environmental phenomena, and has applied it to study the magnitude and persistence of global uncertainty about desertification and suggest how this may be reduced.

This paper has demonstrated, using the UAF, that desertification is one of the most uncertain of all global environmental change phenomena. Based purely on their relative complexities, estimated using the number of attributes needed to measure them, the *inherent* uncertainty of desertification, which has at least seven attributes, is much greater than that of forest area change, which has just one attribute. *Present* uncertainty about desertification is high too: the five available global estimates have a mean Uncertainty Score of 6.8 out of a maximum score of 8, corresponding to four conceptualization uncertainties and four measurement uncertainties.

Another finding is that uncertainty about desertification is persistent. The Uncertainty Score (US) is a more objective measure of the persistence of uncertainty than the mere frequency of estimates mentioned in Section 1, and using the UAF to evaluate the five available global estimates of desertification shows that the US has remained at 7 since the 1970s, except for a dip to 6 in the 1980s.

In none of the estimates of desertification evaluated here has the Uncertainty Score therefore fallen to the threshold of 2 when, according to the UAF, statistical evaluation of uncertainties alone is appropriate. This, and the finding that conceptualization uncertainties account for over 40% of all sources of uncertainty about desertification, support claims that standard statistical methods are inadequate for evaluating very uncertain phenomena [39–41].

While global environmental uncertainties are persistent, they are not inevitable like Heisenberg Uncertainty [11]. This paper has also shown how the UAF can be used to devise an initial set of seven rules for constructing reliable global environmental information. Contrary to a statement in the Third Edition of the World Atlas of Desertification [5], applying these UAF rules shows that even the large uncertainty about the extent of desertification could be substantially reduced if surveys are properly conceptualized, and involve measurements using sensors with appropriate spatial, temporal and spectral resolutions. Yet while it is technically feasible to measure most attributes of desertification at global scale using currently available remote sensing methods, this does not mean that uncertainty about it will diminish quickly. Translating the *technical* potential of Earth observation into practice is often hindered by *organizational* constraints [126], and until remote sensing methods become available to monitor two particularly challenging attributes of desertification—wind erosion and soil compaction—estimates are likely to remain underspecified, ensuring that the US value does not fall below 3.

These findings have two implications for measuring compliance at national scale in dry areas with the Land Degradation Neutrality Target 15.3 of the UN Sustainable Development Goal 15 "Land and Life". First, within the limits of underspecification mentioned in the last paragraph, it is technically feasible to monitor national progress in complying with the official indicator of "proportion of land that is degraded over total land area" listed in the Sustainable Development Goals [8], provided that measurements are properly conceptualized and use both medium and very high resolution satellite images, supported by ground data. While very high resolution satellite images are still not yet widely used in national environmental monitoring, FAO has made the Collect Earth software it used to map dry forests [84] freely available, and government use of this software is increasing. Second, however, Allen et al. are right to caution that using "big data" to fill gaps in national data to monitor SDG Target 15.3 will not be straightforward [9]: (a) the five existing UN global estimates of desertification are out of date and our analysis has shown that they were very uncertain when they were made; and (b) although the uncertainty associated with the methods used in three recent studies of the potential to use 'big data' for this purpose is, according to our analysis, lower (with a mean Uncertainty Score (US) of 4.7) than that of the five UN estimates (US = 6.8), it is still substantial, owing to limitations in conceptualization and measurement.

The UAF can differentiate between different degrees of high inherent and present uncertainty about different phenomena. It complements the use of statistical methods for uncertainty evaluation and is consistent with them at the limits of their reliability. This is because it identifies sources of uncertainty that are missed by statistical methods and which are particularly important for complex multiple attribute global environmental change phenomena, such as desertification. The UAF can also show how to reduce uncertainty to a level where it can be estimated by statistical methods alone. The UAF is consistent with, but more coherent than, previous taxonomies of sources of environmental uncertainty because it synthesizes the sources using a novel theoretical approach to linking conceptualization and measurement.

The simplicity of the UAF is another of its advantages, but it also leads to disadvantages. For example, it is convenient to compare the uncertainty of different environmental phenomena, and different estimates of the same phenomenon, using the Uncertainty Score (US) on a common scale from 0 to 8, but the presence of different degrees of individual conceptualization uncertainties in different estimates may not be reflected in the corresponding US values. Thus, an estimate of desertification is ranked: (a) as understructuralized if it has one form of understructuralization or all three; and (b) as using proxies whether this occurs for just one attribute or all of them. One way to tackle this is to extend the scale when comparing the uncertainties of multiple estimates of the same phenomenon. Wider application of the UAF will lead to further critical evaluation of its advantages and disadvantages, and to refinements to counter the latter.

While the Earth is a "small planet" [127], it is worrying that current estimates of the magnitudes of global environmental change phenomena continue to be so uncertain. This is of particular concern now that human impacts on the planet have reached global proportions [82] and the world's governments have agreed on ambitious Sustainable Development Goals which include a considerable environmental component [8]. To address this shortcoming, it is vital to give greater priority to fundamental research into the origins of global environmental uncertainties and how to evaluate them. Using the UAF more extensively to evaluate present uncertainty about other global environmental change phenomena, e.g., forest area change, forest carbon change, and biodiversity loss, will enable their US values to be compared with the mean of 6.8 reported here for desertification and inform the monitoring of other targets in SDG 15. Another priority is to devise new rules for constructing reliable global environmental information, so disparities between different planetary measurements using different methods can be reduced. The initial set of seven rules derived from the UAF that are proposed in this paper could provide a starting point for this work.

More research of this kind will benefit global environmental governance, and humanity's capacity to tackle its global impacts. Politicians often wrongly assume that scientists provide them with 'certain' knowledge. Countering this assumption remains a challenge, but scientists could also do more to evaluate the uncertainty of information about global environmental changes which they communicate to politicians, and to reduce this uncertainty by realizing the full potential of planetary measurement.

Supplementary Materials: The following supporting information can be downloaded at: https://www.mdpi.com/article/10.3390su14074063/s1, Supplementary Information, Modelling the transition to quantifiable uncertainty; Figure S1, The gap between complete knowledge and present knowledge at limiting uncertainty; Table S1, Dominant topic areas of 96 papers in *International Journal of Remote Sensing* Volume 30, Issues 17–18 and 21–24 in 2009; Table S2, A taxonomy of sources of environmental uncertainty proposed by Regan et al. (2002) [13]; Table S3, A taxonomy of sources of environmental uncertainty proposed by Van Asselt and Rotmans (2002) [17]; Table S4, Key features of 50 papers published in *Land Degradation and Development* between 2006 and 2010; Table S5, Mapping, modelling and linguistic preferences in a sample of 50 papers in *Land Degradation and Development* between 2006 and 2010 on assessing land degradation; Table S6, Six sets of indicators used to specify desertification; Table S7, Sizes of "degrading area" in drylands by climatic zone in the original study [69] on which the Land Degradation Assessment in Drylands (LADA) "preliminary [global] map of land degradation" [57] is based; Table S8, Scalar foci of a sample of 50 papers published in *Land Degradation and Development* between 2006 and 2010; Table S9, Scalar preferences in a sample of 50 papers in *Land Degradation and Development* between 2006 and 2010 on assessing land degradation.

Funding: This research was supported by the Science and Technology Facilities Council under grant ST/K006738/1, and by the European Space Agency/National Remote Sensing Centre of China Dragon 5 Programme under grant 59313.

Conflicts of Interest: The author declares no conflict of interest.

References

1. Dregne, H.E. *Desertification of Arid Lands*; Harwood Academic Publishers: London, UK, 1983.
2. Dregne, H.E. Desertification of arid lands. *Econ. Geogr.* **1977**, *53*, 322–331. [CrossRef]
3. Middleton, N.J.; Thomas, D.S.G. (Eds.) *World Atlas of Desertification*; Arnold: London, UK, 1992.
4. Middleton, N.J.; Thomas, D.S.G. (Eds.) *World Atlas of Desertification*, 2nd ed.; Arnold: London, UK, 1997.
5. Cherlet, M.; Hutchinson, C.; Reynolds, J.; Hill, J.; Sommer, S.; von Maltitz, G. (Eds.) *World Atlas of Desertification*, 3rd ed.; Publication Office of the European Union: Luxembourg, 2018.
6. Verstraete, M.M.; Hutchinson, C.F.; Grainger, A.; Stafford Smith, M.; Scholes, R.J.; Barbosa, P.B.; Léon, A.; Mbow, C. Towards a Global Drylands Observing System, observational requirements and institutional solutions. *Land Degrad. Dev.* **2011**, *22*, 198–213. [CrossRef]
7. Grainger, A. Is land degradation neutrality feasible in dry areas? *J. Arid. Environ.* **2015**, *112*, 14–24. [CrossRef]
8. UN. *Sustainable Development Goals*; United Nations: New York, NY, USA, 2015.
9. Allen, C.; Smith, M.; Rabiee, M.; Dahmm, H. A review of scientific advancements in data sets derived from big data for monitoring the Sustainable Development Goals. *Sustain. Sci.* **2021**, *16*, 1701–1716. [CrossRef]
10. Hassani, H.; Huang, X.; MacFeely, S.; Entezarian, M.R. Big data and the United Nations Sustainable Development Goals (UN SDGs) at a glance. *Big Data Cogn. Comput.* **2021**, *5*, 28. [CrossRef]
11. Heisenberg, W. Über den anschulichen Inhalt der quantentheoretischen Kinematik und Mechanik. *Z. Phys.* **1927**, *43*, 172–198. [CrossRef]
12. Heisenberg, W. *The Physics of the Atomic Nucleus*; Taylor and Francis: London, UK, 1952.
13. Regan, H.M.; Colyman, M.; Burgman, M.A. A taxonomy and treatment of uncertainty for ecology and conservation biology. *Ecol. Appl.* **2002**, *12*, 618–628. [CrossRef]
14. Brown, J.D. Prospects for the open treatment of uncertainty in environmental research. *Prog. Phys. Geogr.* **2010**, *34*, 75–100. [CrossRef]
15. Wong, C.Y.; Boon, S.; Wong, W.Y. The contingency effects of environmental uncertainty on the relationship between supply chain integration and operational performance. *J. Oper. Manag.* **2011**, *29*, 604–615. [CrossRef]
16. Niksefat, N.; Sepehri, N. Designing robust force control of hydraulic actuators despite system and environmental uncertainties. *IEEE Control Syst. Mag.* **2001**, *21*, 66–77.
17. Van Asselt, M.B.A.; Rotmans, J. Uncertainty in integrated assessment modelling. *Clim. Change* **2002**, *54*, 75–105. [CrossRef]
18. Beck, U. Living in the world risk society. *Econ. Soc.* **2006**, *35*, 329–345. [CrossRef]
19. Beven, K. A manifesto for the equifinality thesis. *J. Hydrol.* **2006**, *320*, 18–36. [CrossRef]

20. Polasky, S.; Carpenter, S.R.; Folke, C.; Keeler, B. Decision-making under great uncertainty: Environmental management in an era of global change. *Trends Ecol. Evol.* **2011**, *26*, 398–404. [CrossRef]
21. Böschen, S.; Kastenhofer, K.; Rust, I.; Soentgen, J.; Wehling, P. Scientific nonknowledge and its political dynamics. *Sci. Technol. Hum. Values* **2010**, *35*, 783–811. [CrossRef]
22. Smithson, M. *Ignorance and Uncertainty: Emerging Paradigms*; Springer: Berlin/Heidelberg Germany, 1989.
23. Roth, W.-M. Radical uncertainty in scientific discovery work. *Sci. Technol. Hum. Values* **2009**, *34*, 313–336. [CrossRef]
24. Knight, F.H. *Risk Uncertainty and Profit*; Kelley: New York, NY, USA, 1921.
25. Hermansson, H. Defending the conception of "objective risk". *Risk Anal.* **2012**, *32*, 16–24. [CrossRef]
26. Wynne, B. Uncertainty and environmental learning: Reconceiving science and policy in the preventive paradigm. *Glob. Environ. Chang.* **1992**, *2*, 111–127. [CrossRef]
27. Eddington, A.S. *The Nature of the Physical World*; Cambridge University Press: Cambridge, UK, 1928.
28. Aamodt, A.; Nygard, M. Different roles and mutual dependencies of data information and knowledge-an AI perspective on their integration. *Data Knowl. Eng.* **1995**, *16*, 191–222. [CrossRef]
29. Kuhn, T. *The Structure of Scientific Revolutions*; University of Chicago Press: Chicago, IL, USA, 1962.
30. Robertson, D.W.; Martin, D.K.; Singer, P.A. Interdisciplinary research. *BMC Med. Res. Methodol.* **2003**, *3*, 20–25. [CrossRef]
31. Sillitoe, P. Interdisciplinary experiences: Working with indigenous knowledge in development. *Interdiscip. Sci. Rev.* **2004**, *29*, 6–23. [CrossRef]
32. Grainger, A. Uncertainty in constructing global knowledge about tropical forests. *Prog. Phys. Geogr.* **2010**, *34*, 811–844. [CrossRef]
33. Leimgruber, P.; Christen, C.A.; Laborderie, A. The impact of Landsat satellite monitoring on conservation biology. *Environ. Monit. Assess.* **2005**, *106*, 81–101. [CrossRef] [PubMed]
34. Townshend, J.R.; Masek, J.G.; Huang, C.; Vermote, E.F.; Gao, F.; Channan, S.; Sexton, J.O.; Feng, M.; Narasimhan, R.; Kim, D.; et al. Global characterization and monitoring of forest cover using Landsat data: Opportunities and challenges. *Int. J. Digit. Earth* **2012**, *5*, 373–397. [CrossRef]
35. Gallopin, G.C. Indicators and their use: Information for decision-making. In *Sustainability Indicators Report on the Project on Indicators of Sustainable Development*; Moldan, B., Bilharzia, S., Eds.; John Wiley: Chichester, UK, 1997; pp. 13–27.
36. Gudmundsson, H. The policy use of environmental indicators-learning from evaluation research. *J. Transdiscipl. Environ. Stud.* **2003**, *2*, 1–12.
37. Grainger, A. The role of forest sustainability indicator systems in global governance. *Glob. Environ. Chang.* **2012**, *22*, 147–160. [CrossRef]
38. Grainger, A. Reducing uncertainty about hybrid lay-scientific concepts. *Curr. Opin. Environ. Sustain.* **2010**, *2*, 444–451. [CrossRef]
39. Halpern, B.S.; Regan, H.M.; Possingham, H.P.; McCarthy, M.A. Accounting for uncertainty in marine reserve design. *Ecol. Lett.* **2006**, *9*, 2–11. [CrossRef]
40. Regan, H.M.; Ben-Haim, Y.; Wilson, W.G.; Lundberg, P.; Andelman, S.J.; Burgman, M.A. Robust decision-making under severe uncertainty for conservation management. *Ecol. Appl.* **2005**, *15*, 1471–1477. [CrossRef]
41. Van Der Sluijs, J.P. Anchoring Amid Uncertainty: On the Management of Uncertainties in Risk Assessment of Anthropogenic Climate Change. Ph.D. Thesis, Utrecht University, Utrecht, The Netherlands, 1997.
42. Funtowicz, S.O.; Ravetz, J.R. *Uncertainty and Quality in Science for Policy*; Kluwer: Dordrecht, The Netherlands, 1990.
43. Fisher, A.C. *Resource and Environmental Economics*; Cambridge University Press: Cambridge, UK, 1981.
44. UN. *Convention on Biological Diversity*; United Nations: New York, NY, USA, 1992.
45. Hajer, M.A. *The Politics of Environmental Discourse: Ecological Modernization and the Policy Process*; Clarendon Press: Oxford, UK, 1995.
46. Barton, E. Sanctioned and non-sanctioned narratives in institutional discourse. *Narrat. Inq.* **2000**, *10*, 341–375. [CrossRef]
47. Crawford, S.E.; Ostrom, E. A grammar of institutions. *Am. Political Sci. Rev.* **1995**, *89*, 582–600. [CrossRef]
48. Ostrom, E. *Governing the Commons: The Evolution of Institutions for Collective Action*; Cambridge University Press: Cambridge, UK, 1990.
49. Grainger, A. The role of science in implementing international environmental agreements: The case of desertification. *Land Degrad. Dev.* **2009**, *20*, 410–430. [CrossRef]
50. Bracken, L.J.; Oughton, E.A. "What do you mean?" The importance of language in developing interdisciplinary research. *Trans. Inst. Br. Geogr.* **2006**, *31*, 371–382. [CrossRef]
51. Guarino, N.; Giaretta, P. Ontologies and knowledge bases, towards a terminological clarification. In *Towards Very Large Knowledge Bases: Knowledge Building and Knowledge Sharing*; Mars, N., Ed.; IOS Press: Amsterdam, The Netherlands, 1995; pp. 25–32.
52. Bennett, B. What is a forest? On the vagueness of certain geographic concepts. *Topoi* **2001**, *20*, 189–201. [CrossRef]
53. Cash, D.W.; Clark, W.; Alcock, F.; Dickson, N.; Eckley, N.; Guston, D.; Jager, J.; Mitchell, R. Knowledge systems for sustainable development. *Proc. Natl. Acad. Sci. USA* **2003**, *100*, 8086–8091. [CrossRef]
54. Star, S.L. Scientific work and uncertainty. *Soc. Stud. Sci.* **1985**, *15*, 391–427. [CrossRef]
55. Grainger, A. Difficulties in tracking the long-term global trend in tropical forest area. *Proc. Natl. Acad. Sci. USA* **2008**, *105*, 818–823. [CrossRef]
56. Mabbutt, J.A. A new global assessment of the status and trends of desertification. *Environ. Conserv.* **1984**, *11*, 100–113. [CrossRef]

57. Bai, Z.G.; Dent, D.L.; Olsson, L.; Schaepman, M.E. *Global Assessment of Land Degradation and Improvement: 1. Identification by Remote Sensing*; GLADA Report 5; Land Degradation Assessment in Drylands Project and International Soil Reference and Information Centre University of Wageningen; UN Food and Agriculture Organization: Rome, Italy, 2008.

58. UN. *Elaboration of an International Convention to Combat Desertification in Countries Experiencing Serious Drought and/or Desertification Particularly in Africa*; United Nations: New York, NY, USA, 1994.

59. UN. *Overview. UN Conference on Desertification*; UN Environment Programme: Nairobi, Kenya, 1977.

60. Tucker, C.J.; Choudhury, B.J. Satellite remote sensing of drought conditions. *Remote Sens. Environ.* **1987**, *23*, 243–251. [CrossRef]

61. Warren, A. Land degradation is contextual. *Land Degrad. Dev.* **2002**, *13*, 449–459. [CrossRef]

62. Geist, H.J.; Lambin, E.F. Dynamic causal patterns of desertification. *Bioscience* **2004**, *54*, 817–829. [CrossRef]

63. Reynolds, J.F.; Stafford Smith, D.M.; Lambin, E.F.; Turner, B.L., II; Mortimore, M.; Batterbury, S.P.J.; Downing, T.E.; Dowlatabadi, H.; Fernández, R.J.; Herrick, J.E.; et al. Building a science for dryland development. *Science* **2007**, *316*, 847–851. [CrossRef] [PubMed]

64. Grainger, A. Characterization and assessment of desertification processes. In *Desertified Grasslands: Their Biology and Management, Papers Presented at an International Symposium, Linnean Society, London, UK, 27 February–1 March 1991*; Chapman, G.P., Ed.; John Wiley: Chichester, UK, 1992; pp. 17–33.

65. Liu, J.; Dietz, T.; Carpenter, S.R.; Alberti, M.; Folke, C.; Moran, M.; Pell, A.N.; Deadman, P.; Kratz, T.; Lubchenco, J.; et al. Complexity of coupled human and natural systems. *Science* **2007**, *317*, 1513–1516. [CrossRef] [PubMed]

66. Vrieling, A. Satellite remote sensing for water erosion assessment. *Catena* **2006**, *65*, 2–18. [CrossRef]

67. Bell, M.A.; Lamb, P.J. Integration of weather system variability to multidecadal regional climate change, the West African Sudan-Sahel zone 1951–1998. *J. Clim.* **2006**, *19*, 5343–5365. [CrossRef]

68. UN. *Draft Plan of Action to Combat Desertification*; UN Environment Programme: Nairobi, Kenya, 1977.

69. Bai, Z.G.; Dent, D.L.; Olsson, L.; Schaepman, M.E. *Global Assessment of Land Degradation and Improvement 1. Identification by Remote Sensing Report*; World Soil Information Project, International Soil Reference and Information Centre, University of Wageningen: Wageningen, The Netherlands, 2008.

70. Aubréville, A. *Climats Forêts et Désertification de l'Afrique Tropicale*; Société d'Edition Géographiques Maritimes et Coloniales: Paris, France, 1949.

71. Thomas, D.S.G.; Middleton, N.J. *Desertification: Exploding the Myth*; John Wiley: Chichester, UK, 1994.

72. Greco, M.; Miranda, D.; Squicciarino, G.; Telesca, V. Desertification risk assessment in southern Mediterranean areas. *Adv. Geosci.* **2005**, *2*, 243–247. [CrossRef]

73. Salvati, L.; Zitti, M. Regional convergence of environmental variables: Empirical evidence from land degradation. *Ecol. Econ.* **2008**, *68*, 162–168. [CrossRef]

74. FAO; UNESCO; WMO. *World Map of Desertification*; UN Food and Agriculture Organization, UN Educational Scientific and Cultural Organization and World Meteorological Organization: Nairobi, Kenya, 1977.

75. Dregne, H.E. *Map of The Status of Desertification in the Hot Arid Regions*; UN Environment Programme: Nairobi, Kenya, 1977.

76. Oldeman, L.R.; Hakkeling, R.T.A.; Sombroek, W.G. *World Map of the Status of Human-Induced Soil Degradation, an Explanatory Note*; University of Wageningen: Wageningen, The Netherlands, 1990.

77. LADA. *Biophysical Indicator Toolbox Technical Report No 2, Land Degradation Assessment in Drylands Project*; UN Food and Agriculture Organization: Rome, Italy, 2005.

78. Bai, Z.G.; Dent, D.L.; Olsson, L.; Schaepman, M.E. Proxy global assessment of land degradation. *Soil Use Manag.* **2008**, *24*, 223–234. [CrossRef]

79. LADA. *Land Degradation Assessment in Drylands. LADA. Project Findings and Recommendations*; UN Environment Programme, UN Food and Agriculture Organization: Rome, Italy, 2011.

80. Milich, L.; Weiss, E. GAC NDVI images: Relationship to rainfall and potential evaporation. *Int. J. Remote Sens.* **2000**, *21*, 261–280. [CrossRef]

81. Turner, M.D. Methodological reflections on the use of remote sensing and geographic information science in human ecological research. *Hum. Ecol.* **2003**, *31*, 255–279. [CrossRef]

82. Grainger, A. The prospect of global environmental relativities after an Anthropocene tipping point. *For. Policy Econ.* **2017**, *79*, 36–49. [CrossRef]

83. Lambin, E.F. Monitoring forest degradation in tropical regions by remote sensing: Some methodological issues. *Glob. Ecol. Biogeogr.* **1999**, *8*, 191–198. [CrossRef]

84. Bastin, J.F.; Berrahmouni, N.; Grainger, A.; Maniatis, D.; Mollicone, D.; Moore, R.; Patriarca, C.; Picard, N.; Sparrow, B.; Abraham, E.A.; et al. The extent of forest in dryland biomes. *Science* **2017**, *356*, 635–638. [CrossRef] [PubMed]

85. Eyre, S.R. *Vegetation and Soils*; Arnold: London, UK, 1968.

86. Dube, T.; Mutanga, O.; Sibanda, M.; Seutloali, K.; Shoko, C. Use of Landsat series data to analyse the spatial and temporal variations of land degradation in a dispersive soil environment: A case of King Sabata Dalindyebo local municipality in the Eastern Cape Province.; South Africa. *Phys. Chem. Earth A/B/C* **2017**, *100*, 112–120. [CrossRef]

87. Desprats, J.F.; Raclot, D.; Rousseau, M.; Cerdan, O.; Garcin, M.; Le Bissonnais, Y.; Slimane, A.B.; Fouche, J.; Monfort-Climent, D. Mapping linear erosion features using high and very high resolution satellite imagery. *Land Degrad. Dev.* **2013**, *24*, 22–32. [CrossRef]

88. Miles, L.; Newton, A.C.; DeFries, R.S.; Ravilious, C.; May, I.; Blyth, S.; Kapos, V.; Gordon, J.E. A global overview of the conservation status of tropical dry forests. *J. Biogeogr.* **2006**, *33*, 491–505. [CrossRef]

89. Lindquist, E.J.; D'annunzio, R.; Gerrand, A.; Macdicken, K.; Achard, F.; Beuchle, R.; Brink, A.; Eva, H.D.; Mayaux, P.; San-Miguel-Ayanz, J.; et al. *Global Forest Land-Use Change 1990–2005*; UN Food and Agriculture Organization: Rome, Italy, 2010.

90. Achard, F.; Beuchle, R.; Mayaux, P.; Stibig, H.J.; Bodart, C.; Brink, A.; Carboni, S.; Desclée, B.; Donnay, F.; Eva, H.D.; et al. Determination of tropical deforestation rates and related carbon losses from 1990 to 2010. *Glob. Chang. Biol.* **2014**, *20*, 2540–2554. [CrossRef]

91. Parker, C.; Thorne, C.; Bingner, R.; Wells, R.; Wilcox, D. *Automated Mapping of Potential for Ephemeral Gully Formation in Agricultural Watersheds*; NSL, Technical Research Report No. 56; National Sedimentation Laboratory: Oxford, MS, USA, 2007.

92. Chi, W.; Zhao, Y.; Kuang, W.; He, H. Impacts of anthropogenic land use/cover changes on soil wind erosion in China. *Sci. Total Environ.* **2019**, *668*, 204–215. [CrossRef]

93. Jabbar, M.R.; Chen, X. Land degradation assessment with the aid of geo-information techniques. *Earth Surf. Process. Landf.* **2006**, *31*, 777–784. [CrossRef]

94. We, B.; Ci, L.J. Landscape change and desertification development in the Mu Us Sand land.; northern China. *J. Arid. Environ.* **2002**, *50*, 429–444.

95. Zhang, L.; Yule, L.; Via, B. The study of land desertification in transitional zones between the Mu Us desert and the loss plateau using RS and GIS-a case study of the Yulin region. *Environ. Geol.* **2003**, *44*, 530–534. [CrossRef]

96. Karavanova, E.I.; Shrestha, D.P.; Orlov, D.S. Application of remote sensing techniques for the study of soil salinity in semi-arid Uzbekistan. In *Response to Land Degradation*; Bridge, E.M., Ed.; Science Publishers: Enfield, UK, 2001; pp. 261–273.

97. Metternicht, G.; Zinck, J.A. Remote sensing of soil salinity: Potentials and constraints. *Remote Sens. Environ.* **2003**, *85*, 1–20. [CrossRef]

98. Metternicht, G.; Zinck, J.A. Spatial discrimination of salt-and sodium-affected soil surfaces. *Int. J. Remote Sens.* **1997**, *18*, 2571–2586. [CrossRef]

99. Rao, B.R.M.; Sankar, T.R.; Dwivedi, R.S.; Thammappa, S.S.; Venkataratnam, L.; Sharma, R.C.; Das, S.N. Spectral behaviour of salt-affected soils. *Int. J. Remote Sens.* **1995**, *16*, 2125–2136. [CrossRef]

100. Dwivedi, R.S.; Sreenivas, K.; Ramana, K.V. Inventory of salt-affected soils and waterlogged areas: A remote sensing approach. *Int. J. Remote Sens.* **1999**, *20*, 1589–1599. [CrossRef]

101. Dwivedi, R.S.; Sreenivas, K. Delineation of salt-affected soils and waterlogged areas in the Indo-Gangetic plains using IRS-1C LISS-III data. *Int. J. Remote Sens.* **1998**, *19*, 2739–2751. [CrossRef]

102. El Harti, A.; Lhissou, R.; Chokmani, K.; Ouzemou, J.; Hassouna, M.; Bachaoui, E.M.; El Ghmari, A. Spatiotemporal monitoring of soil salinization in irrigated Tadla Plain (Morocco) using satellite spectral indices. *Int. J. Appl. Earth Obs. Geoinf.* **2016**, *50*, 64–73. [CrossRef]

103. Gorji, T.; Yildirim, A.; Hamzehpour, N.; Tanik, A.; Sertel, E. Soil salinity analysis of Urmia Lake Basin using Landsat-8 OLI and Sentinel-2A based spectral indices and electrical conductivity measurements. *Ecol. Indic.* **2020**, *112*, 106173. [CrossRef]

104. Olsson, L.; Eklundh, L.; Ardö, J. A recent greening of the Sahel-trends.; patterns and potential causes. *J. Arid. Environ.* **2005**, *63*, 556–566. [CrossRef]

105. Herrmann, S.M.; Anyamba, A.; Tucker, C.J. Recent trends in vegetation dynamics in the African Sahel and their relationship to climate. *Glob. Environ. Chang.* **2005**, *15*, 394–404. [CrossRef]

106. Lamprey, H.F. Report on the Desert Encroachment Reconnaissance in Northern Sudan, 21 October–10 November 1975. *Desertif. Control Bull.* **1988**, *17*, 1–7.

107. Tucker, C.J.; Dregne, H.E.; Newcomb, W.W. Expansion and contraction of the Sahara desert from 1980 to 1990. *Science* **1991**, *253*, 299–301. [CrossRef] [PubMed]

108. Tucker, C.J.; Newcomb, W.W.; Dregne, H.E. AVHRR data sets for determination of desert spatial extent. *Int. J. Remote Sens.* **1994**, *15*, 3547–3565. [CrossRef]

109. Urban, F.E.; Goldstein, H.L.; Fulton, R.; Reynolds, R.L. Unseen dust emission and global dust abundance: Documenting dust emission from the Mojave Desert USA by daily remote camera imagery and wind-erosionmeasurements. *J. Geophys. Res. Atmos.* **2018**, *123*, 8735–8753. [CrossRef]

110. Webb, N.P.; Herrick, J.E.; Van Zee, J.W.; Courtright, E.M.; Hugenholtz, C.H.; Zobeck, T.M.; Okin, G.S.; Barchyn, T.E.; Billings, B.J.; Boyd, R.; et al. The National Wind Erosion Research Network: Building a standardized long-term data resource for aeolian research.; modeling and land management. *Aeolian Res.* **2016**, *22*, 23–36. [CrossRef]

111. Webb, N.P.; Kachergis, E.; Miller, S.W.; McCord, S.E.; Edwards, B.L.; Herrick, J.E.; Karl, J.W.; Leys, J.F.; Metz, L.J.; Smarik, S.; et al. Indicators and benchmarks for wind erosion monitoring.; assessment and management. *Ecol. Indic.* **2020**, *110*, 105881. [CrossRef]

112. Hansen, M.C.; Potapov, P.V.; Moore, R.; Hancher, M.; Turubanov, S.A.; Tyukavina, A.; Thau, D.; Stehman, S.V.; Goet, S.J.; Loveland, T.R.; et al. High-resolution global maps of 21st-Century forest cover change. *Science* **2013**, *342*, 850–853. [CrossRef]

113. Rasmussen, K.; Fog, B.; Madsen, J.E. Desertification in reverse? Observations from northern Burkina Faso. *Glob. Environ. Chang.* **2001**, *11*, 271–282. [CrossRef]

114. Tiffen, M.; Mortimore, M. Questioning desertification in dryland sub-Saharan Africa. *Nat. Resour. Forum* **2002**, *26*, 218–233. [CrossRef]

115. Christian, B.A.; Dhinwa, P.S. Long term monitoring and assessment of desertification processes using medium high resolution satellite data. *Appl. Geogr.* **2018**, *97*, 10–24. [CrossRef]
116. Rathore, V.S.; Kumar, M.; Yadava, N.D.; Yadav, O.P. Dryland agriculture and secondary salinization in canal commands of arid Rajasthan. *J. Soil Salin. Water Qual.* **2017**, *9*, 30–46.
117. Wang, J.; Wei, H.; Cheng, K.; Ochir, A.; Davaasuren, D.; Li, P.; Chan, F.K.S.; Nasanbat, E. Spatio-temporal pattern of land degradation from 1990 to 2015 in Mongolia. *Environ. Dev.* **2020**, *34*, 100497. [CrossRef]
118. Yaojie, Y.; Min, L.; Lin, W.; Xing, Z.A. A data-mining-based approach for aeolian desertification susceptibility assessment: A case-study from Northern China. *Land Degrad. Dev.* **2019**, *30*, 1968–1983. [CrossRef]
119. Salvati, L.; Kosmas, C.; Kairis, O.; Karavitis, C.; Acikalin, S.; Belgacem, A.; Gungor, H. Unveiling soil degradation and desertification risk in the Mediterranean basin: A data mining analysis of the relationships between biophysical and socioeconomic factors in agro-forest landscapes. *J. Environ. Plan. Manag.* **2015**, *58*, 1789–1803. [CrossRef]
120. Hengl, T.; Mendes De Jesus, J.; Heuvelink, G.B.; Ruiperez Gonzalez, M.; Kilibarda, M.; Blagotić, A.; Shangguan, W.; Wright, M.N.; Geng, X.; Bauer-Marschallinger, B. Soil Grids 250m: Global gridded soil information based on machine learning. *PLoS ONE* **2017**, *12*, e0169748. [CrossRef]
121. Giuliani, G.; Mazzetti, P.; Santoro, M.; Nativi, S.; Van Bemmelen, J.; Colangeli, G.; Lehmann, A. Knowledge generation using satellite earth observations to support sustainable development goals (SDG): A use case on land degradation. *Int. J. Appl. Earth Obs. Geoinformatics* **2020**, *88*, 102068. [CrossRef]
122. Mitri, G.; Nasrallah, G.; Gebrael, K.; Nassar, M.B.; Abou Dagher, M.; Nader, M.; Masri, N.; Choueiter, D. Assessing land degradation and identifying potential sustainable land management practices at the subnational level in Lebanon. *Environ. Monit. Assess.* **2019**, *191*, 567. [CrossRef]
123. UNCCD. *Expert Meeting on a Land Degradation Indicator (SDG target 15.3), Washington, DC, USA, 25–26 February 2016*; Summary of Main Outcomes; UN Convention to Combat Desertification: Bonn, Germany, 2016.
124. Grainger, A. Monitoring desertification after the Ad Hoc Advisory Group of Technical Experts. In Proceedings of the 5th International Conference on Drylands, Deserts and Desertification, Beersheba, Israel, 17–20 November 2014; Ben-Gurion University.
125. UNCCD. Advice on How Best to Measure Progress on Strategic Objectives 1, 2 and 3 of The Strategy. Refinement of the Set of Impact Indicators on Strategic Objectives 1, 2 and 3. Recommendations of the ad hoc Advisory Group of Technical Experts. Note by the Secretariat. Conference of the Parties/Committee on Science and Technology, Eleventh Session, Windhoek, Namibia, 17–20 September 2013. Document ICCD/COP(11)/CST/2. UN Convention to Combat Desertification, Bonn, 2013. Official Documents CST 11, Windhoek, 2013. Available online: http://www.unccd.int/sites/default/files/sessions/documents/ICCD_COP11_CST_2/cst2eng.pdf (accessed on 10 October 2021).
126. Grainger, A. Citizen observatories and the new Earth observation science. *Remote Sens.* **2017**, *9*, 153. [CrossRef]
127. Ward, B.; Dubos, R. *Only One Earth: The Care and Maintenance of a Small Planet*; Andre Deutsch: London, UK, 1972.

Article

Can Current Earth Observation Technologies Provide Useful Information on Soil Organic Carbon Stocks for Environmental Land Management Policy?

Ana Andries [1,*], Stephen Morse [1], Richard J. Murphy [1], Jim Lynch [1], Bernardo Mota [2] and Emma R. Woolliams [2]

1 Centre for Environment and Sustainability, Faculty of Engineering and Physical Sciences, University of Surrey, Guildford GU2 7XH, UK; s.morse@surrey.ac.uk (S.M.); rj.murphy@surrey.ac.uk (R.J.M.); j.lynch@surrey.ac.uk (J.L.)

2 Climate and Earth Observation Group, National Physical Laboratory, Teddington, Middlesex TW11 0LW, UK; bernardo.mota@npl.co.uk (B.M.); emma.woolliams@npl.co.uk (E.R.W.)

* Correspondence: a.andries@surrey.ac.uk

Abstract: Earth Observation (EO) techniques could offer a more cost-effective and rapid approach for reliable monitoring, reporting, and verification (MRV) of soil organic carbon (SOC). Here, we analyse the available published literature to assess whether it may be possible to estimate SOC using data from sensors mounted on satellites and airborne systems. This is complemented with research using a series of semi-structured interviews with experts in soil health and policy areas to understand the level of accuracy that is acceptable for MRV approaches for SOC. We also perform a cost-accuracy analysis of the approaches, including the use of EO techniques, for SOC assessment in the context of the new UK Environmental Land Management scheme. We summarise the state-of-the-art EO techniques for SOC assessment and identify 3 themes and 25 key suggestions and concerns for the MRV of SOC from the expert interviews. Notably, over three-quarters of the respondents considered that a 'validation accuracy' of 90% or better would be required from EO-based techniques to be acceptable as an effective system for the monitoring and reporting of SOC stocks. The cost-accuracy analysis revealed that a combination of EO technology and in situ sampling has the potential to offer a reliable, cost-effective approach to estimating SOC at a local scale (4 ha), although several challenges remain. We conclude by proposing an MRV framework for SOC that collates and integrates seven criteria for multiple data sources at the appropriate scales.

Keywords: monitoring; verification; reporting; soil organic carbon; soil organic matter; Earth Observation

Citation: Andries, A.; Morse, S.; Murphy, R.J.; Lynch, J.; Mota, B.; Woolliams, E.R. Can Current Earth Observation Technologies Provide Useful Information on Soil Organic Carbon Stocks for Environmental Land Management Policy? *Sustainability* **2021**, *13*, 12074. https://doi.org/10.3390/su132112074

Academic Editor: Jamal Jokar Arsanjani

Received: 30 September 2021
Accepted: 28 October 2021
Published: 1 November 2021

Publisher's Note: MDPI stays neutral with regard to jurisdictional claims in published maps and institutional affiliations.

1. Introduction

Soil organic carbon (SOC) is one of the components of soil organic matter (SOM), which also includes other elements such as hydrogen, oxygen, and nitrogen, as well as fresh (living) and decomposed plant/animal (dead) materials [1]. SOM consists of several components in varying proportions, including microorganisms (10–40%) and stable organic matter (40–60%) (also referred to as humus) [1]. Most productive agricultural soils have between 3% and 6% organic matter, with the organic matter contributing to soil productivity in many ways. Although the carbon content of SOM varies considerably, the soil remains one of the key components of the carbon cycle [2]. A mean value of 1500 gigatonnes for the first 1 m depth of soils has been estimated from many studies of global carbon stocks, with about half of this amount present in the top 30 cm of the soil [3,4]. SOC is critical for soil health and biological, chemical, and physical functions such as nutrient cycling, water retention, and maintaining soil structure and crop yield [5]. As a result, carbon storage (C-storage) in soils has gained increasing international attention as a means of mitigating climate change. Initiatives such as '4p1000' [6], the FAO's Global assessment of carbon sequestration by soils (e.g. the GSOCseq programme) [7], and the UN's Sustainable

Development Goals (SDGs) all recognise the importance of C-storage in soils [8]. As part of its action on soils, the UK has signed up for the '4p1000' soil carbon initiative to increase the soil carbon levels by 0.4% per year [6]. Similarly, the UK 25 Year Environment Plan aims to enhance the natural capital of air, water, soil, and ecosystems that support all forms of life. This is viewed as being essential for economic growth and productivity over the long term. This includes developing a soil health index and restoring and protecting peatlands [9].

The UK Department for Environment, Food and Rural Affairs (DEFRA)'s Environmental Land Management (ELM) initiative is a national programme of intensive policy development, for implementation post-2025, to help farmers and other land managers to improve the environment with associated public money used to help deliver 'public goods' (e.g. clean air and water, healthy soils, climate change mitigation, wildlife protection, and the promotion of landscape beauty and heritage). As part of achieving this, cost-effective and innovative solutions will be needed to deliver a robust and flexible system for the monitoring, reporting, and verification (MRV) of soil status and change [9].

Certain agricultural management practices that result in land use change, such as agroforestry, tillage/residue, manure/biosolid management, grazing land management/pasture improvement, land restoration, and livestock management, can maintain or increase SOC stocks [10,11]. Using such practices, it has been suggested that carbon sequestration in soils could be a cost-effective, environmentally friendly strategy for greenhouse gas removal [12]. Increasing soil organic matter content and thereby improving soil quality can also enhance other soil-based ecosystem services, such as agricultural production, clean water supply, and biodiversity [13].

In terms of the current large-scale soil assessments (including SOC levels), regional and country scale systems have been developed by the FAO and the International Union of Soil Societies (IUSS) (Table 1). In the case of the UK, the National Soil Inventory (NSI) and Countryside Survey (CS) are the two major soil surveys (Table 1). Bellamy et al. [14] reported changes in SOC (up to 30 cm) during the 1980s and 1990s in England and Wales based on NSI and observed gains in OC in some areas but large losses in others and, overall, noted a net loss of about 4 million tonnes of C per year. More recent results reported by Carey et al. [15], Emmett et al. [16], and Norton et al. [17] based on the CS show only a negligible change in OC over the 1978–2007 period. The drivers of SOC variation are discussed in further detail in [18] and are considered to include environmental change and habitat-specific trends as influencing soil processes. However, Bellamy et al. [14] showed a linear relationship between the rate of change in soil carbon and the soil carbon content, indicating that the loss was greater in soils with large carbon contents while soils with small carbon contents gained carbon. Because the losses occurred across all land use types, they considered that climate change might play a role in this, in addition to the effects of changes in land use and agricultural management activities.

The reasons for the differences in SOC between the NSI and CS results are not clearly understood. However, Kirk et al. [19] considered that they were not comparable because of differences in the sampling designs, with the CS using a hierarchical stratified sampling scheme and single samples from point locations, whereas the NSI used a fixed grid with bulked samples centred on the target point. Hence, the answer obtained about the SOC of a field can depend to some extent on the sampling methodology.

Table 1. Existing and emerging large-scale soil assessment projects (global, Europe and UK scale).

Network	Scale	Data/Method	Reference
FAO soil map	Global	Derived from soil profile data.	[20]
IUSS Global soil map	Global	Different mapping methods based on existing data.	[21]
Global Soil Organic Carbon map (GSOCMap)	Global	Employs Earth Observation (EO) imagery to derive prediction factors for global soil organic carbon mapping.	[7,20]
European Soil Database (ESDB) map	Europe	This database includes the Soil Geographical Database of Eurasia at scale 1:1,000,000 (SGDBE), Pedotransfer Rules Database (PTRDB), Soil Profile Analytical Database of Europa (SPADBE), and the Database of Hydraulic Properties of European Soils (HYPRES).	[22]
Land Use/Cover Statistical Area Frame Survey (LUCAS)	Europe	LUCAS project monitors landscape diversity, land cover, land-use changes, and soil chemical data, mainly on agricultural land.	[23]
National Soil Inventory (NSI) database	England and Wales	Results are based on in situ soil sampling. Between 1978 and 1981, the topsoil (0–15 cm) was sampled at every 1 km^2, 5 samples in 6127 points in England and Wales, and analysed for a number of soil parameters, including SOC. About a third of the points were resampled in 1995.	[24]
Countryside Survey (CS)	UK	The CS is an integrated national monitoring program that makes measurements of vegetation; topsoil physical, chemical, and biological characteristics (0–15 cm); water quality; and land cover across the UK in 1978, 1998, and 2007.	[25]
UK Soil Observatory	UK	Users of the UK Soil Observatory and the mySoil mobile app play a role in citizen science to build up a soil dataset. They crowdsource information about the soil properties in their area and send photographs and measurements to gather vital new information on soil science and help to improve soil maps of the UK. By June 2021, over 2000 contributions have been recorded.	[26]

The increasing recognition of soil carbon as a key component in policy and action on climate change, as well as for ecosystem health, highlights the need for MRV systems that can track and update SOC status and change estimates at scales ranging from local to global. At present, the available large-scale soil surveys (c.f. Table 1) represent major resourcing efforts due to the complex and intensive sampling requirements to address the spatial variability of SOC, even within a single field. The FAO [1] has outlined recommendations for determining SOC stock and monitoring its dynamics and spatial variability. For instance, to determine SOC stocks between 0 and 30 cm depth at a particular site, the FAO specifies as a minimum requirement the need to quantify the following: (i) SOC concentration in the fine earth (<2 mm size) soil fraction, (ii) the fine earth (<2 mm) and coarse mineral fraction content (>2 mm size) of the soil, and (iii) the soil bulk density of fine earth mass. To capture the spatial variability of SOC within an agricultural field (land parcel), with an acceptable level of accuracy, it is required that the site be assessed using the following factors: soil and land use type, climate, grazing practices, field size, fertiliser use, historical management practices, and ploughing frequency [27,28]. In addition, a large number of soil samples is often required.

For instance, Garten and Wullschleger [29] recommended the collection of more than 100 soil samples (to detect 2–3% change in SOC stocks), whereas Vanguelova et al. [30] suggested between 4 and 25 soil samples for 0.25 ha with more or equal 5 m between sampling points to eliminate spatial autocorrelation. SOC measurement relies on appropriate study designs and sampling protocols to deal with high spatial variability [31]. SOC content is typically expressed as the (mass) percentage dry weight of soil, once stones, etc., have been removed.

Laboratory techniques for assessing SOC are commonly based on approaches such as dry combustion using an elemental analyser (prior acidification of sample will remove inorganic carbon), loss on ignition, the Walkley–Black method, soil respiration, and active carbon tests. The most accurate standard laboratory test for soil carbon is dry combustion using an elemental analyser. These instruments heat a small sample (usually a fraction of a gram) of dry pulverized soil to around 900 degrees C and measure the carbon dioxide gas that is a combustion product. In terms of loss of ignition, a change in weight after heating a soil sample (ignition test) is observed, although complications can arise if the sample has carbon in an inorganic form (e.g. calcium carbonate), which will also generate carbon dioxide at high temperatures. The latter can be accommodated by comparing the change in weight of a soil sample divided into two parts, with one part first treated with acid to remove the calcium carbonate while the other is subjected to combustion. Comparing changes in the two provide a measure of the SOC for the sample, although there is clearly room for error. Nonetheless, knowledge of the potential sources of error and uncertainty in SOC stock estimation, coupled with sampling strategies that integrate lateral and vertical variability of SOC stock, enable SOC and its change to be determined with an acceptable level of uncertainty to help guide management, albeit often requiring intensive sampling regimes [1].

Smith et al. [28] classified approaches to assess SOC measurements into "direct"—those that could be performed through intensive soil sampling—and "indirect methods"—measurements which infer SOC stock changes from flux measurements (based on the gross primary production) and/or from the comparison of spectral reflectance images of soil samples with consolidated spectral reflectance libraries derived from measurements on reference soil samples of known properties (determined by traditional laboratory methods). The reflectance-based measurements can be obtained at a high level of spatial detail from hyperspectral sensors mounted on unmanned aerial systems (UASs) or multispectral sensors onboard satellites and airborne platforms.

EO satellite imagery also can be used to estimate changes in SOC through proxies such as Net Primary Production (NPP) or Gross Primary Productions (GPP) [32], land use management [33], land cover change [34], and land degradation indices [35]. Conant et al. [36] described approaches such as statistical upscaling via in situ samplings (providing greater accuracy and lower errors but expensive at large scales) and geostatistical upscaling through combining initial sampling with covariates (obtaining spatial maps of SOC status at the whole field and large scale at low cost, but with low accuracy).

Many published studies [37–45] show that remote sensing imagery (both EO satellite and UAS) can be used to estimate SOC stock and its change, by building statistical relationships between the measured remotely sensed spectra and in situ soil samples, to make estimations of the SOC status. As the traditional methods based on soil sampling and laboratory analysis can be very expensive and time consuming, especially when estimating SOC over large areas, the potential of EO technologies to provide timely, repeatable, and potentially cost-effective estimates of SOC stocks and their spatial and temporal variability is of considerable interest [46,47]. However, of course, much depends on the relationship between estimates of SOC made using EO and those made by using more 'traditional' (i.e. well-established) approaches of taking soil samples that are assessed in the laboratory. While, as noted above, the 'traditional' approaches to SOC assessment have their own uncertainty in terms of the sampling regime and laboratory methods, they arguably provide the standard to which EO assessments have to be compared. Thus, the term 'accuracy' is typically used to describe how the EO-based assessment of SOC matches (within uncertainties) that derived from traditional methods. Inevitably there are trade-offs here for those concerned with using the estimates of SOC to manage the land. Given that EO-based approaches have the advantages noted above, including perhaps a lower cost, then they may provide an attractive option for those concerned with land management, but what level of agreement with the more traditional assessments of SOC would be acceptable to them? Would it be acceptable, for example, if the EO-based estimates of SOC gave values that matched those using traditional techniques for half the number of plots or does it need to be higher?

The research described in this paper was designed to assess the current state of the art on the use of EO for SOC stock measurement and its potential for the current and future MRV of SOC. The following objectives are addressed (i) to assess (from literature) the capabilities of EO techniques for assessing SOC stocks at local to national scales (throughout this paper the term EO is used to encompass remote sensing technologies such as satellite, aerial photography, Unmanned Aircraft Systems (UAS), etc.), (ii) to evaluate expert perspectives on the relevance/importance of SOC estimation by EO and the acceptable levels of accuracy required for application to UK (and other) policy implementation derived from such approaches, and (iii) to perform a cost-accuracy analysis using different data sources for estimating SOC.

2. Materials and Methods

2.1. EO for SOC—Literature Acquisition and Evaluation

A literature search was restricted to peer-reviewed papers published between 2000 and 2020 which assessed the use of EO to estimate SOC stocks. This literature was also used to refine questions and topics for exploration in the expert interviews. Three sequential steps were undertaken to select literature for analysis: (i) search for peer-reviewed articles containing one or more of the following keywords: "Earth Observation", "Remote Sensing", "Airborne", "Lidar", "Satellite", "UAS", along with either "Soil Organic Carbon" or "Soil Organic Matter"; (ii) selecting relevant work that included the keywords; and (iii) review and analysis. These steps identified 54 relevant peer-reviewed articles.

2.2. Expert Interviews

Primary data collection was conducted via a set of semi-structured interviews between March and August 2020 with 13 specialists in soil health (highly cited researchers) and environment/agricultural policymaking (top relevant institutions), most with more than 15 years of experience. An email invitation was sent with a brief explanation of the request and a detailed description of this study. In addition, the framework established from the analysis of the literature was provided to all participants and discussed during the interviews. Following the initial seven interviews, several suggestions for further potential respondents were made and followed up in a 'snowballing' approach. Each interview lasted for between 20 and 45 min. Twelve were carried out through online conversations with permitted recording, and one was undertaken in writing as the expert preferred that format.

The structure of the interviews was led by open-ended questions acting to promote and produce rich information which requires a high level of attention and interpretation. We consider that the sample size was large enough to allow understanding of the issue under study to unfold but small enough that rich qualitative data could be obtained.

The interviewees represented the following range of sectors and roles: academics and researchers (5), consultants (4), and policymakers (4) (Table 2). All of the respondents were based in the UK.

Table 2. Participants in the semi-structured interviews by sector and institution.

Sector	Institution Name	Number of Respondents
Academia	University of Aberdeen	1
	University of Reading	1
	University of Surrey	1
Research Centre	Rothamsted Research	1
	Soil and Agrifood Institute at Cranfield University	1
Consultancy	Climate Solutions Exchange	2
	ADAS	1
	Farm Africa	1
Governmental Organisation	Department for the Environment, Food and Rural Affairs (DEFRA)	3
	Environment Agency	1

In the opening section of the interview, a detailed introduction was given regarding the study's objectives and the framework derived from the literature. The set of 12 questions in Sections 2–4 of Table 3 was then explored. Section 2 comprised six questions about the importance of SOC in the policy context and the current use and frequency of SOC/SOM assessment within agricultural policy. The five questions in Section 3 sought to explore the levels of accuracy that would be required from EO techniques in order to provide reliable data for monitoring and reporting SOC levels to benefit ELM. Finally, in Section 4, the respondents were given the opportunity to raise any further points or observations. This structure enabled consistent coverage of the topics whilst allowing for the respondents to provide their suggestions and comments based on their area of expertise.

Table 3. Format and questions of the semi-structured interview.

Section	Description	Questions Addressed
1. Introduction	Introduction to the objectives of the study and framework derived from the literature analysis.	No questions.
2. SOC/SOM in agricultural policy	Six questions attempt to understand SOC/SOM in the agricultural policy context.	1. Please provide a brief background of your experience in the field of agricultural research/policy. 2. Do you think SOC is currently used within agricultural policy/do you think it will be? 3. Do you think SOC is important in agricultural policy? If it is important then why? 4. How often is SOC assessed in the current policy—do you think this is adequate/ If there are plans to assess SOC in the policy then how do you think this should best be done? 5. (optional) Are you familiar with the SDGs? If yes, how do the policies on SOC relate to the SDGs? Has there been an influence of the SDG framework on the policy? 6. (optional-depending on the above answers) If SOC is considered to be important but is not currently/or planned to be a part of the policy, why do you think that is?

Table 3. *Cont.*

Section	Description	Questions Addressed
3. Potential for using EO to monitor/ predict SOC	Five questions focusing on the expected/required accuracy of SOC predictions from remote sensing methods	7. Are you aware that EO satellite-derived data can be used to assess SOC? 8. EO has the potential to estimate soil organic carbon under the right circumstances, although this is at an early stage of development. As part of our research, we are also investigating this concerning potentially using EO to help assess trends in SOM of various types of managed and unmanaged land and it is a useful approach for monitoring and reporting, e.g. for ELMS. Please provide your estimate as a percentage of the accuracy when EO technology is used to measure SOC and provide a short statement indicating why you consider this to be acceptable. 9. Studies have shown that EO derived data is not a perfect predictor for SOC contents so what would be the level of 'accuracy' for making agricultural policy decisions when SOC contents are predicted via EO data relative to the traditional methods? 10. What do you think could be done to improve the assessment via EO? 11. What other information would be needed to supplement the assessment via EO and how might it be collected?
4. Round up	Any other comments by the respondent.	12. Do you have any other comments/observations?

The transcribed interviews were subjected to manual content analysis. A series of open and axial coding techniques were used to identify themes and build the theory. The first step was 'open coding', in which tentative labels were assigned to the interview transcription data. Secondly, in the axial coding, codes were selected to focus on the analysis of the core categories to identify relationships among the open codes. Thus, categories/themes were identified in the data using the underlying objectives and structure of the questions posed, together with insights emerging from the respondents.

2.3. Cost-Accuracy Analysis

In Andries et al. [48], a structured framework was described for evaluating the usefulness of various EO-based approaches to inform the UN Sustainable Development Goal (SDG) indicators. One premise that was considered as part of the framework (based on recommendations from expert interviews) was the "cost-effectiveness" of the EO solution when compared with more traditional techniques. It was noted that the existing literature rarely discussed the cost-effectiveness of a particular approach, and the inclusion of this information in studies was recommended. Therefore, in the present research, we have taken into consideration information on the cost of the data type, methods, prediction, and accuracy. This cost information is expected to be relevant for decision making in land management and the design of cost-effective MRV systems for SOC within schemes such as the UK ELM scheme.

For the selected 4 ha reference site (Figure 1), we assumed that the land manager or ELM verification body would wish to obtain a map of SOC within the delimited field site. This would be obtained by propagating data from passive and active sensors on satellites or UAS platforms through a mathematical model that has been derived from in situ measurements, either from the field site or from existing libraries of in situ observations.

Figure 1. Trial site for cost-accuracy analysis (source—Google Earth Pro).

We obtained the prices for satellite data supplied by Earth-i, a global space company that exploits commercial satellite imagery and cutting-edge artificial intelligence and machine learning techniques. In terms of Unmanned Aircraft Systems (UAS), we have used the cost analysis conducted by Aldana-Jague et al. [49] and the prices from ADAS, a commercial operator that provides services including digital data collection by UAS and aerial photography, use of software for data processing, and expertise for estimating SOC. For conventional laboratory tests for SOC analysis, we used pricing obtained from a leading UK soil analysis company (preferred to remain anonymous in this study).

3. Results

3.1. Literature Analysis of EO-Based Approaches for SOC Measurement

There are various studies in the literature which seek to assess how EO-based assessments of variables thought to be related to SOC (such as land cover and topography) relate to measures of SOC based on ground-level soil sampling. These variables (covariates) are used as inputs to a statistical model that is used to estimate SOC. It should be noted that the more traditional assessments using soil samples and laboratory analysis are deemed to be the best estimates of SOC. Figure 2 presents our synthesis of the main steps that appear in most literature reports and are needed for estimating SOC content from the field to the national scale.

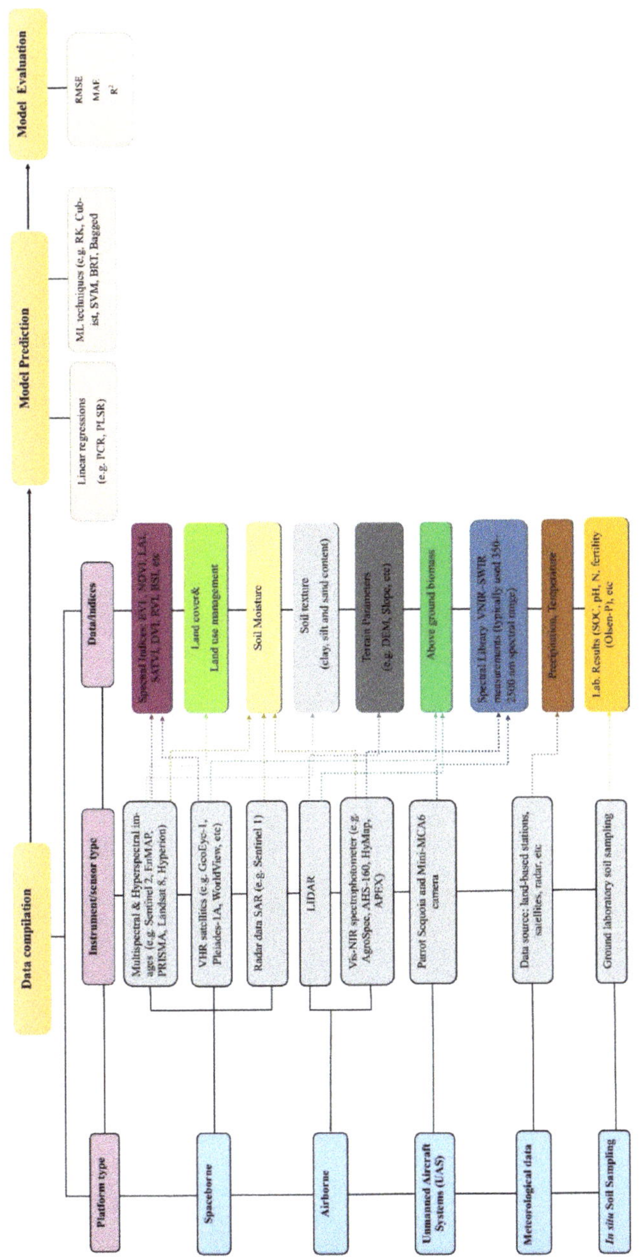

Figure 2. A conceptual framework for digital mapping of SOC/SOM—based on literature review.

3.1.1. Selection of Environmental Variables (Covariates) and Data Compilation

Different methods use different covariates in developing SOC estimation models. Covariates used in methods described in the literature include land cover and land use maps (e.g. CEH Land cover 2015, CORINE), meteorological data (e.g. temperature, precipitation, evapotranspiration), vegetation index (e.g., spectral indices), terrain parameters (e.g., digital elevation model (DEM) and its derivates, slope degree (SD), aspect, depth of sinks (m) (DoS), compound topogrphic index (CTI), and profile curvature (PC)). As these different quantities are provided by data providers with different spatial resolutions and on different grids, data compilation requires the covariate dataset to be resampled to a common resolution.

The covariate data sets are collated from a wide variety of data sources. These data sources range from imagery obtained by passive or active sensors onboard remote sensing platforms (UASs, airborne and spaceborne), and meteorological data measured by land-based stations. Our literature review showed that the following are the most common EO data sources used in the development of SOC estimation models (note that some methods use a combination of data sources):

i. Satellite multispectral (e.g., Sentinel 2, Landsat 8 OLI, Very High resolution (VHR) images) and hyperspectral (e.g., EnMAP, PRISMA, and HyspIRI) remote sensing reflectance images, and laboratory spectral measurements of soil samples [50,51];

ii. Spectral indices derived from multispectral images such as Normalized Difference Vegetation Index (NDVI) [52], Enhanced Vegetation Index (EVI) [50], Soil Adjusted Total Vegetation Index (SATVI) [53], Brightness Index (BI) and Greenness Index (GI) [52], Vegetation Temperature Condition Index (VTCI) [52,54,55], Ratio Vegetation Index (RVI) (also for aboveground biomass estimation) [56], soil moisture [57], and soil texture [58];

iii. Topographic data derived from active radar sensors such as SRTM, stereophotogrammetric images from ASTER [47], and LiDAR capable of providing a measure of elevation to derive topographic-based parameters such as slope, aspect, depth of sinks, etc. [59];

iv. VHR satellite data and LiDAR, which is used to estimate above-ground biomass by assessing tree composition and height, both of which are useful proxies in predicting SOC stock [60];

v. Data derived from multispectral sensors mounted on UAS, for example, by combining hyperspectral and high-resolution images of bare soil to obtain libraries of soil spectral reflectance for training machine learning techniques [49,61,62]. The model development also relies on traditional laboratory SOC measurements taken from samples collected in the field.

3.1.2. SOC Model

The soil organic mapping approaches mentioned above adopt a wide variety of statistical techniques to relate SOC retrieved from soil samples (dependent variable) to the predictor variables (independent variables, e.g., spectral libraries/measurements, spectral indices, terrain parameters). A range of statistical and machine learning techniques have been applied to develop models to estimate SOC, and some had to accommodate the issue of co-linearity whereby a number of the independent variables could be related to each other. These techniques include principal component regression (PCR) and partial least square regression (PLSR) [37,38,43,44,51,63], regression-kriging (RK) [55], penalized-spline signal regression (PSR) [38], random forest (RF) [47,55,64,65], cubist (CB) [44,65,66], support vector machines (SVM) [38,42,47], boosted regression trees (BRT) and bagged CART [47], and convolutional neural networks (CNNs) [67].

This wide variety includes methods based on linear regression, where a linear relationship is assumed between soil properties and predictor variables, and more sophisticated machine learning (ML) techniques that can provide highly non-linear mapping and classification approaches. Since the relationship between soil properties and predictor variables is complex and nonlinear [68], these more sophisticated approaches can achieve higher accuracy in terms of matching the observed values of SOC and outperform more conventional methods [69].

3.1.3. Model Evaluation

The performances of the statistical models presented above are commonly tested under different calibration/training (model development) and validation (model evaluation) steps. In general, soil samples are divided into two sets that are randomly selected, one for calibration/training (70% to 80% of the available data) and the other for validation (20% to 30%) [44,52,55,60,65,68,70,71]. The calibration dataset is used to develop the model, including selecting the relevant covariates and deriving the model parameters. The evaluation dataset is used independently to assess the model performance [40]. To provide an additional degree of robustness to the validation, an n-fold cross-validation approach can be adopted. This approach builds on repeated train/test splits into the observed data, ensuring that each data point occurs at least once in the test set [72,73]. In addition, to repeat, training and testing subsets can also be subject to resampling techniques such as bootstrapping. This has been shown to be efficient in assessing accuracy (i.e., the ability to correctly predict observed values of SOC based on field measurements) when a small data set is available [60].

The error metrics often used to compare these models to the validation datasets are the root mean square error (RMSE), the mean absolute error (MAE), and the mean error (ME). These metrics all represent different information associated with the differences between the model predicted values and the observed values from the reference (usually in situ) data set. These are all given in the same units as the measurement, here as g/kg^{-1} org/kg. The ME is the simple mean of all such differences and can be positive or negative. For unbiased results, it will take the value 0; some of the more sophisticated results can give a ME that is different from 0. The MAE is the mean of the unsigned differences, and the RMSE gives the standard deviation. All of these metrics must be interpreted with respect to the amount of SOC measured. The coefficient of determination, R^2, is, in contrast, a relative, normalised quantity and the proportion or percentage of variation explained by each model. The R^2 is commonly reported in regression studies and provides a measure of the ability of the model to predict the dependent variable but fails to provide information on the model error and bias. An R^2 close to 1 alongside lower values of RMSE, MAE, and, where relevant, ME, together provide more information on model quality. Much depends, of course, on the quality of the data used in the statistical analysis. All of these methods rely on being able to calculate an error based on a reliable in situ data set that is representative of the region observed by remote sensing methods and do not account for uncertainties associated with those measurements or site variability and other representation uncertainties. None, particularly quoted without context, can be considered a pure "accuracy".

3.1.4. Summary of Methods

Several of the above studies and their claimed 'accuracy metrics' are summarized in Table 4. It should be noted here that 'accuracy' in this context is defined in terms of how well the statistical model is able to predict the observed measures of SOC based on field samples (dependent variable). The table includes values of the 'accuracy metrics' for the model development and model evaluation phases of the analysis. The accuracy metrics quoted may not represent the full model accuracy for the reasons given above and do not necessarily apply when the method is extrapolated to new areas. The study that obtained the best set of accuracy metrics in predicting SOC was performed at a single-field scale and used very high spatial resolution (12 cm) and multispectral (6 bands) reflectance data from a UAS that was processed with a non-linear machine learning regression trained with samples from the same site. This method achieved an R^2 value of 0.98 in calibration (spread of calibration points) and $R^2 = 0.95$ in validation [49]. The studies with a poorer set of 'accuracy metrics' used free satellite data (e.g., Sentinel 1 & 2, Landsat, etc.) and covered much broader areas, with less detailed in situ data and more natural variability. High-resolution studies are generally limited to small areas due to UAS flight duration for covering large areas and legal barriers to initiate flights, and it is significantly more expensive to do such detailed campaigns. Free satellite data can be used to cover a much broader area and provide information for national and transnational studies.

3.2. Expert Interviews

This section presents the three main themes that emerged from the analysis of the interviews. The key concerns and suggestions (KCS) associated with these themes are presented in Tables 5–7, together with the most relevant quotes from the respondents (with the respondent number indicated as R*x*).

3.2.1. Theme 1: SOC/SOM within Agricultural Policy

SOC was discussed by most respondents as being a critical component in soil science and soil quality, alongside other soil properties (Table 5: KCS 1.1 recognised by 77% of respondents). In addition, many participants considered that soil health has been moved up the political agenda after initiatives such as '4 per 1000' and the UN SDGs increased awareness at the global scale (Table 5: KCS 1.2, 55% of respondents). In 2018, the UK Environment Secretary and Farming Minister both promised to put soil health at the heart of the 25 Year Environment Plan, emphasising the UK Government's ambition to improve the approach to soil management and restating the commitment to all of the UK's soils being managed sustainably by 2030 (Table 5: KCS 1.3, 46% of respondents).

Likewise, some respondents (Table 5: KCS 1.4, 23% of respondents) also raised the crucial role of tackling pressures (e.g., climate change, different agricultural practices, drainage schemes, intensive sheep grazing, etc.) on the land in the long term as important additional factors affecting SOC.

Interestingly, some respondents noted (Table 5: KCS 1.5–1.7, 38%, 31%, and 25%, respectively) that SOC has been neglected so far in the UK policy framework, with this coming from two different perspectives. The one coming from soil scientists was that, due to a lack of interest and knowledge, policymakers have postponed addressing soil health for far too long because priority was given to maximising the production of cheap food. For policymakers, the explanation for a lack of focus on SOC, as well as soil health, was perceived to be due to a lack of resources, rather than from any intentional avoidance of integrating SOC into policy.

3.2.2. Theme 2: EO for SOC and Its Level of Accuracy

In this theme, the main concerns and suggestions that refer to the potential that EO technology could have to monitor SOC and the level of accuracy that would be needed are presented in Table 6. It needs to be noted here that 'accuracy' in the context of the interviews was interpreted in terms of how often the results from an EO-based assessment match those obtained via the 'traditional' methodology of taking field samples and using laboratory analysis. The highest level of consensus from respondents (77% of respondents) was KCS 2.1 and 2.2—providing strong support for EO being relevant and valuable for recognising different land use, agricultural practices, etc., and the linkage of these characteristics with SOC. An advantage for EO approaches was acknowledged to be their ability to estimate SOC stock over large spatial scales using proxies such as vegetation indices, slope degree, precipitation, temperature, etc. About half of the respondents (54%) noted that a direct determination of SOC from EO requires bare soil and that the spectral signatures derived from EO imagery can be constrained by cloud coverage (KCS 2.3). In addition, there was a strong consensus that EO could be useful for estimating the overall stock of SOC but did not have the consistency and accuracy required to detect its detailed change over time (KCS 2.4).

Table 4. Example studies for digital soil mapping (SOC estimation) that used remote sensing techniques.

Data Used [Reference]	Spatial Resolution	Study Area	Predictor (Independent) Variables	Number of Soil Samples (Dependent Variable)	Model Prediction and Results of the Evaluation Methods	SOC Contents Estimates (Ranging across the Entire Area)
Sentinel 1 & 2 [47]	10 m	Southern part of Central Europe (~700 km^2)	NDVI; EVI; SATVI	179	RF (RMSE—0.63–0.70%) BRT (0.57–0.68%) Bagged CART (RMSE—0.59–0.70%) SVM (RMSE-0.57–0.70%)	0.47% to 43.91% SOC
SRTM & ASTER [47]	25 m		Terrain parameters (e.g., DEM, slope, and Wetness Index)			
Meteorological data [47]	1 km		Average annual precipitation and annual temperature			
Airborne imaging spectrometer (APEX) [74]	2 m	Northern part of Wallonia, Belgium (240 km^2)	NDVI CAI (cellulose absorption index) NBR2 (Normalized Burn Ratio2)	104	PLSR (RMSE—2.13 g kg^{-1})	0.75% to 2.02% SOC
Sentinel 2 [74]	10/20/60 m					
LISS-III images, Landsat TM, Soil map [52]	24 m–30 m	Central India (20,558 km^2)	Land use, NDVI, soil types, VTCI	210	RK (RMSE—0.23%)	0.028% to 3.195% SOC
ASTER GDEM [52]	25 m		Terrain parameters (e.g., DEM, slope)			
Sentinel-2 [52]	10/20/60 m	Field scale (1.45 km^2)	Soil Composite Mapping Processor (SCMaP)	50	RF (RMSE—0.20–0.27%) CB (RMSE—0.26–0.34%) SVM (RMSE—0.20–0.28%) PLS (RMSE—0.27%)	0.84% to 2.62% SOC
Landsat-8 OLI [52]	30 m (15 m PAN)		Spectral Library of VNIR–SWIR reflectance values			
PlanetScope [52]	3.5–4 m		Spectral Library of VNIR reflectance values			
Multispectral Parrot Sequoia [52]	Variable (cm)		DTM and surface reflectance values			
Spectrometer-XDS™ Rapid Content Analyzer [49]	N/A	Field scale (590 m^2)	Samples scanned with a diffuse reflectance spectrometer	109	SVM (RMSE—0.21%)	0.5% to 4.7% SOC

Table 5. Theme 1: SOC/SOM within agricultural policy.

Key Concerns and Suggestions (KCS)	Respondents %	Associated Quotes
1.1 SOC is a crucial aspect of soil science and soil quality alongside other soil properties	77%	"SOC is the key parameter for soil health; however, it comes with other types of components, that are important for its functioning. It is obvious that all are very interlinked. We cannot just focus on one component because then we won't be doing robust and trustworthy policies." (R6) "Soil nutrients are highly interlinked, and if you ignore one of them, then it affects the others. So, you can't just focus only on one aspect. You have to focus on the whole system, in order to come up with something sensible that covers the practicalities of managing soils within the constraints that the agricultural production or land use puts on it." (R8) "When soil loses SOC, then it produces an imbalance of the soil structure and function, and it won't be able to support food productivity and biodiversity." (R7) "We believe that soil health and soil management are hugely important for the future of British agriculture and that currently, we are seeing significant soil degradation taking place in many key areas across the country. The impact of soils loss on rivers, aquatic organisms, and water abstraction has a substantial cost to the nation's economy." (R19)
1.2 '4 per 1000' and the UN SDGs recently increased awareness	55%	"The need of increasing SOC levels to be included into the agricultural policy started to be mentioned at the international scale at the international initiative "4 per 1000" and later in the SDGs. UK signed up for both, hence more recently it has been included in 25-year Environment Plan." (R12)
1.3 SOC recently considered in the current national policy—25 Year Environment Plan	46%	"DEFRA ultimately set the agenda on including SOC, because of Michael Gove and George Eustice policies. They worked on the integrated agricultural policy and, better management of the land. And when you talk in those terms, you are more likely to need information on SOC. Since then, I think we have seen a big increase in interest in the way in which land is managed." (R11)
1.4 Potential impact of climate change on SOC	23%	"Before we work together to improve SOC levels, we also should focus on tackling climate change and stopping the wrong agricultural practices." (R9)
1.5 Agricultural policies have been focused on maximizing the production of cheap food so far	38%	"It has been very difficult, to demonstrate food production in the last 50 years. So, the agricultural policies so far have largely been about maximising the production of cheap food. It is much easier to maximise the production of cheap food by growing crops, over large areas of land using a whole raft of easily applicable fertilisers, which are usually inorganic fertilisers." (R4)

Table 5. *Cont.*

	Key Concerns and Suggestions (KCS)	Respondents %	Associated Quotes
1.6	Lack of a soil health monitoring methodology due to the difficulties in measuring SOC change	31%	"There has been a lack of interest and probably knowledge as well, mostly because the policy message has not been put over clearly enough." (R1) "Soil health is a multifaceted and difficult thing to define and measure; generally speaking, there is concern about the methods of measurement of SOC level change." (R5)
1.7	Lack of resources (staff, funding, data)	23%	"We have not been able to dedicate the resources required to actually address SOC. So, it is not from a lack of willingness. But we are definitely not neglecting to consider SOC or soil health in our current policy. It is something that we have been continuously rising, with other teams, who do have an impact on soils, for example, our fertiliser, the agri-environment, and the climate teams. It is exceedingly difficult to make progress on programs with limited resources, like funding people and data. However, we are currently working on a national program that develops the soil monitoring system." (R6)

Table 6. Theme 2: EO for SOC and its level of accuracy.

	Key Concerns and Suggestions (KCS)	Respondents %	Associated Quotes
2.1	Using EO can identify different land uses and verify different agricultural practices	77%	"If a farmer reports a certain practice that improves/maintains soil health, for example, leaving residues on the field, you can verify that with remote sensing to check if that is actually happening. So, EO can play a role not for measuring the soil carbon specifically, but for verifying that an agricultural management practice has been put in place to increase soil carbon." (R2)
	Increasing SOC level through different agricultural practices		"You can use the EO system to look at the type of land uses. People are being encouraged to introduce more mixed farming for all kinds of reasons. For instance, bringing grass into the rotation means that you have obtained naturally, more organic matter into the system, but you cannot measure the quantity of SOM levels." (R12)
2.2	Estimating SOC level via proxies and a spectral library	77%	"The importance of land use management can be an indicator for the age of soil organic matter. I think that could be a real key indicator. That is where EO can become quite important because it can show the land use. You could speculate how much organic matter is going into the soil." (R10) "We have truly little information about the variation change and the exact content of organic carbon in the soil of 10 hectares field, for instance. But if you want to measure carbon sequestration of a forest via remote sensing technology, that you can do, measuring the increase in the size of trees and record land use and vegetable and biomass production. So, I think that you could potentially infer the likely benefit to soil organic carbon on a particular piece of land by looking at how the land has been managed during the years." (R18) "So, the direct measurement of soil carbon from EO systems, it is not possible, and I can't foresee in the near future that it will be possible, but you can get some proxies that can be used to estimate soil organic carbon or develop a spectral library on bare soil." (R16) "You cannot measure soil organic matter content directly by looking at the satellite imagery, but you can infer it from a variety of observations that you are able to get using satellites." (R13)

Table 6. *Cont.*

	Key Concerns and Suggestions (KCS)	Respondents %	Associated Quotes
2.3	Bare soil necessity	54%	"We have got the issues around the ground cover because there is a need for bare soil that can be available only late autumn and winter but then you have high coverage of clouds." (R5)
2.4	EO can predict the SOC state but not the change	69%	"With satellite data, you can identify differences in land use and NDVI. Absolutely you can identify differences in slope and all this sort of thing. But you cannot measure the change of SOC levels. The measurements have to be consistent in the way that they are done. And they have to be done over a relatively long period of time." (R9) "EO can cover large areas, quite quickly, possibly multiple times, but of course, you need to have a clear atmosphere, no clouds." (R7)
2.5	EO is a cost-efficient source of data	54%	"Then as long as you are not making a policy decision or payment decision, if you like an implementation decision based on insufficient data, could do harm. But satellite imagery does have the advantages that you can essentially use this application, in a way that doesn't massively increase your costs." (R13)
2.6	Level of accuracy acceptable for an efficient EO-based system to provide data on SOC levels for monitoring and reporting		
	1. 90–100%	77%	"The accuracy of SOC estimates should have 95–100%. And the reason for saying this is because changes in soil organic carbon are very slow and you get a very small change against a very large background carbon stock. So, to detect change, you need to have quite accurate measurements in situ. You can maybe detect the stock with a little bit less accuracy if you are only interested in knowing the SOC state whereas the remote sensing measurements could feed into that process [. . . . ...] Because one of the advantages of the satellite measurements is, of course, the cost and that you can take measurements frequently." (R7)
	2. 70–90%	23%	"I think looking at it from a policy perspective, you would like to have high confidence in the data. I would have thought over 80% accuracy because if you are going to make decisions that are going to affect people's payments and avoid legal challenges, you need to have high accuracy. The obvious pay off from using EO is that you can reduce the number of actual inspections, people going out onto farms doing the sampling, and that's the kind of payoff benefit which we know that EO has." (R9)
2.7	Using a variety of data types through machine learning	31%	"I believe that a model combining earth observation and ground-truthing through machine learning techniques would considerably increase accuracy." (R13) "Of course, there might be advantages if you can connect satellite imagery with things like UAS, drones, Lidar, and ground-truthing. But I think the technology needs developing methods that integrate all these types of data." (R7)

Table 7. Theme 3: Designing a MRV SOC framework.

Key Concerns and Suggestions (KCS) Criteria Released		Respondents %	Associated Quotes
3.1.	Directness	85%	"It is not possible to measure SOM content directly by looking at the satellite or aerial photography imagery, the only way to get direct measurements is to perform a soil sampling, but you can infer a variety of these in situ observations with EO technology, sort of reflectance, soil colour, bare soil attributes as a set of covariates that can help to measure indirectly SOM/ SOC stock." (R2)
3.2	SOC metric standards	38%	"Attention should be first on the grounds and then using other technologies, such as satellites or UAS. And we currently do not have a good active program of in-situ soil data collection. There is a bigger problem that affects soil health monitoring. As we know, SOC can be unevenly distributed over varying soil depths, thus there is a need of creating a universal and consistent fieldwork protocol for the assessment of SOC dynamics in order to properly capture SOC on a landscape scale and to identify small SOC changes in a highly variable pool. SOC monitoring programs need to liaise with long-term soil experiments. This might offer a baseline for the SOC pool and can comprise a set of sites where targeted research on soil processes and their impacts on soil C can be performed. They can serve as a backbone for SOC monitoring. And then we can discuss the alternative data that can help us to estimate SOC stock at a larger scale. So, you need that as well as any new technology." (R8)
3.3	Scalability	77%	"It is important to know what's your universe. So, if you are looking, at the global level and you are establishing a kind of relationship that would have a certain level of errors. If you perform a model for a field scale then your prediction accuracy would be different, likely with a higher level of errors, but this would depend on the data used." (R9)
3.4	Accuracy	100%	"Accuracy of the method used strictly depends on the scale, but first we should define the SOC metrics for national and local scale." (R5) "What is suitable for making a national scale estimate of a change in soil carbon would be different from an individual farm scale. The amount of money required to do that accurately with various methods is pretty minor compared with the importance of the soil health." (R3) "Before we design a soil monitoring system, we should find an answer to these questions: How accurate is your field sampling? How accurate are the models used? What would be the approach that provides the highest accuracy and reliability?" (R8) "I think the accuracy of EO imagery will be partly determined by the accuracy of the ground truth data itself." (R7) "I think the challenge we would have is that if you are developing methods that have flaws, it is a PR disaster in terms of how the system works. You know, it could mean people getting rewarded for not doing enough. And the people are getting under rewarded for doing a lot of work and not, they are not getting the benefit of that. I guess there is a question on the type of technology used and the sensors. You know, if we have a high degree of confidence saying that we are using a variety of hyperspectral sensors could be used. I don't know what you would actually use this, but we need to get a high degree of confidence." (R2)
3.5	Variability	69%	"In terms of SOC variation, do you find that carbon content changes over the year, or does it change over a number of years, and are there things that can happen that can enrich and improve it? How can we monitor the change? We should find answers to these questions before we design a soil monitoring system that includes the usage of EO technologies." (R2) "In terms of the variability, you are going to see in soils, it is far too dispersed. Is there any literature on how soil sampling itself can provide an accurate figure of how much soil organic carbon has been increased/decreased on a piece of land?" (R6)

Table 7. *Cont.*

Key Concerns and Suggestions (KCS) Criteria Released		Respondents %	Associated Quotes
3.6	Consistency	62%	"Consistency is very important. In terms of measurements of soil, either in situ or remote, they must be consistent, otherwise we just can't have faith in methods used." (R13) "An ideal soil monitoring system should follow an approach that could be consistent and applicable across the country for doing a national coverage." (R12)
3.7	Cost-effectiveness analysis	54%	"If you want to make a monitoring system of carbon sequestration then you must consider the satellite data and other technologies because this would significantly drop the cost. For instance, we can use free satellite data, such Sentinels, to model spectral indices or soil moisture." (R10) "A system that could monitor soil at a national scale and identify innovative methods and cost-effective ways could deliver support for agri-environment policy and possibly enhance existing monitoring programmes." (R13)

Interestingly, more than half of the respondents noted that EO could provide a cost-effective alternative to the use of field measurements (KCS 2.5). It should be remembered, however, that most respondents indicated that they did not consider that EO-based imagery could provide a sufficiently accurate estimation of SOC directly, and, by inference, this cost-effectiveness refers to the value of EO in providing complementary data supporting SOC determinations. Concerning the accuracy requirement for SOC estimations by EO, it was clear that an accuracy of >90% was seen as a requirement by nearly 70% of respondents. This means, for example, that out of 100 fields, the EO-based measurement would need to match (within uncertainties) the assessments of SOC obtained via traditional methods for at least 90 fields. Similarly, only 15% of respondents (2 out of 13) thought that an accuracy between 70% and 90% was 'acceptable', confirming a strong concern over the impacts of using potentially inaccurate assessments (KCS 2.6).

Most of the respondents recognised that methods solely utilising the current remote sensing technologies would not be capable of substituting direct field sampling methods. About one-third of respondents considered that multiple combinations of data types, e.g., remotely sensed, ground-truthing measurements, multispectral imaging (derived from UAS), LiDAR, citizen science, etc., all assimilated by machine learning, could significantly increase the accuracy (KCS 2.7). Furthermore, it was also noted that for mapping the state of SOC using remote sensing data, it would be possible to devise efficient sampling schemes to achieve a specified accuracy based on the analysis of a number of ground-based samples with associated analyses (e.g., machine learning, regressions), thus reducing current physical survey costs.

3.2.3. Theme 3: Criteria for the Design of an MRV Framework for SOC

The third theme of measurement/monitoring, reporting, and verification (MRV) emerged from the analysis of the interviews. This extended beyond the previous two themes, which the semi-structured interviews were designed to explore with the respondents. The basis for the emergence of this MRV theme potentially relates to the goal ofLand Management (ELM) schemes to incentivise good soil management practices through the application of current technologies, enhancing the ability to deliver environmental benefits via improving or maintaining soil quality. Currently, most agri-environment schemes have to rely on limited national monitoring capabilities and a wider understanding of soil health and improved capability for its monitoring would be advantageous. Therefore, Table 7 provides seven criteria and associated quotes that resulted from the interviews.

A directness criterion (KCS 3.1) was mentioned by 85% of participants and refers to the ways that the SOC MRV system is comprised of direct measurements of SOC (field sampling) and indirect/proxy measures based on a spectral library and indices, topographic parameters, etc. (see Figure 2). A few respondents (38%) recognised that direct measurements should be first constructed on a universal protocol that should be consistent over all fields. Thus, respondents raised the importance of defining SOC metric standards as a criterion for a robust structured survey of soil sampling (KCS 3.2). About 77% of respondents considered scalability an important criterion to consider in designing a soil monitoring system. Scale plays a role as a driver for determining the accuracy of the methods. All participants admitted that the level of accuracy is a vital criterion to understand, and it would differ for different scales and data used. In terms of the variability in the SOC stock, many respondents (69%) recognised this as a challenge, mostly because SOC levels vary considerably across the same field and with soil depth (KCS 3.5). A consistency criterion was raised by 62% of the respondents through two different perspectives, one in terms of the standardization of the in situ soil sampling and the other in the methods used that are closely linked to scalability (KCS 3.6). Finally, more than half of the respondents (54%) consider that satellite and other technologies could enhance the cost-effectiveness as a criterion for a national monitoring system of SOC (KCS 3.7).

3.3. Cost-Accuracy Analysis

To evaluate the cost-accuracy of using different approaches to estimate SOC across our 4 ha (0.04 km^2) site, we combined information about the cost of obtaining and analysing different data sources (Table 8) for different methods (Table 9).

Table 8. Methods of measuring SOC—reference costs for different data sources.

Data Type	Cost		
	Raw Data	**Processing Data (e.g., Calibration)**	**Analytics (e.g., Applying ML)**
VHR satellite data	VHR satellite data [1] Multi spectral <= 50 cm spatial resolution [1] = 23 €/km^2 Multi spectral >50 cm = 12 €/km^2 Archived VHR satellite data [1] Multispectral <= 50 cm spatial resolution = 7 €/km^2 Multi spectral > 50 cm spatial resolution = 3.5 €/km^2	EUR 1.20/km^2	EUR 7/km^2
Medium and high resolution (e.g., EnMap, MODIS, Sentinel, Landsat)	Free	EUR 0.20/km^2	EUR 7/km^2
Multi-spectral imagery (480–1000 nm) UAS (prices can vary depending on the camera type and capabilities)	The cost analysis performed by Aldana-Jague et al. [49] estimated the total cost is between EUR 160 to EUR 400 per 0.01 km^2. ADAS UK provides services for a cost between EUR 600 to EUR 950 for a set of images (at least 0.04 km^2).		
In situ collection and laboratory analysis	The price of the SOC laboratory analysis may vary depending on the company accredited for soil testing. The cost for one soil sample would include: • sample disposal (per sample)—EUR 2.50; • soil sample for measuring the total organic carbon—EUR 12 (a minimum order charge of EUR 120 + VAT is applied); • postal charges for collection and delivery per total order—EUR 60. There will be additional costs relating to the collection of the soil samples from the field.		

[1] minimum order applied.

Table 9. Calculation based on the prices provided in Table 8 for 4 ha.

Method [Reference]	Prediction Accuracy Metrics (Based on Validation Phase)	Methods Cost for 4 ha	Total Cost for 4 ha
Remote sensing satellite data (Sentinel-2, Landsat-8 OLI, and VHR PlanetScope) [65]		Sentinel-2 and Landsat-8 OLI = free PlanetScope [1] = EUR 0.5 Analytics = EUR 0.3	
UAS multispectral sensor Parrot Sequoia [65]	$R^2 = 0.56–0.81$		EUR 750–1000
Hyperspectral Airborne Imaging CASI 1500 sensor and SWIR (960–2440 nm) SASI 600 sensor [65]		EUR 600–950 (including data capture and processing)	

Table 9. *Cont.*

Method [Reference]	Prediction Accuracy Metrics (Based on Validation Phase)	Methods Cost for 4 ha	Total Cost for 4 ha
Collection of Soil Sampling Data Study (50 soil samples across 145 ha—80% for training set and 20% for validation set) [65]		11 soil samples = EUR 150 + EUR 60 (delivery and collection charge)	
Remote sensing satellite data (Sentinel-1 and -2 images) [47]			
DEM derivatives from SRTM and ASTER GDEM [47]			
In situ soil for satellite calibration and validation (179 topsoil samples were collected from the Land Use and Coverage Area Frame Survey (LUCAS) topsoil dataset provided by the European Soil Data Centre [2] [47]	$R^2 = 0.32–0.44$	Processing and interpreting satellite data EUR 0.35	~EUR 0.35
Multi-spectral imagery (480–1000 nm) UASs, and 30 reference samples collected from the field site to build the model [49]	$R^2 = 0.98$	Purchased camera EUR 600–1600	EUR 600–1600
		Hiring camera (including data capture and processing) EUR 600–950	EUR 600–950
Laboratory analysis (between 6 and 42 samples, depending on the accuracy and carbon variability required) [27]	N/A		EUR 90–600 + EUR 60 (Postal charges)

[1] According to our correspondent from Earth-i, the most affordable way is to pay for a full year of access to the PlanetScope data captured over an area. [2] LUCAS2015_SOIL data are the property of the European Union, represented by the European Commission, represented by the Directorate General-Joint Research Centre (hereafter referred to as the JRC), which authorise third parties to use these data free of charge.

In Table 8, different types of data are presented that could be used to estimate SOC, including multispectral VHR satellite data, high- and medium-resolution satellite imagery, UAS-based imagery, and the laboratory analysis of soil samples collected from each site.

Table 9 provides the approaches selected from the literature that can be applied to the defined 4 ha study field to estimate SOC from some combination of the data sources given in Table 8. For each approach, the cost and accuracy are considered.

Žížala et al. [65] used free (medium resolution) and commercial (VHR) satellite data, along with UAS and in situ sampling to develop a bespoke model for the trial site. For this approach, the total cost for measuring SOC for our trial site would be between EUR 750 and EUR 1000 (number of samples scaled from their larger site to our smaller site) with an mboxemph$R^2 = 0.81$. Zhou et al. [47] used only free satellite data trained to a general European model (from the free LUCAS dataset), and here, the cost is extremely low at EUR 0.35 for 4 ha (price only for the processing and interpreting the dataset), obtaining an $R^2 = 0.44$ for their model fit. Aldana-Jague et al. [49] used UAS reflectance spectroscopy and a model trained on 30 reference samples from the field site. Including sample collection and processing, image acquisition, processing, and analysis costs, they estimated the total cost to be between EUR 160 and EUR 400 per ha, hence, for a 4-ha field, the cost would be between EUR 600 and EUR 1600, with an $R^2 = 0.98$. Their price considers the initial investment costs for purchasing a UAS, a multispectral camera, and software (estimated at circa EUR 25,000). Commercial operators charge approximately EUR 600–EUR 950 for services based on UAS imagery, where includes UAS and aerial photography for digital data collection, processing, and manipulation.

For comparison, the Gehl and Rice [27] approach is given for sample collection and laboratory analysis only. The range of prices depends on the types of laboratory tests performed to quantify SOC, the different methods include the Walkley–Black (WB) method, mass loss on ignition (LOI), automated dry combustion (ADC), and humic matter (HM). The most accurate standard laboratory test for soil carbon is ADC and is often cheaper than other tests [75]. Consequently, the laboratory price list for ADC is presented in Table 8, and the price has been applied on the 4-ha field (Table 9). According to Vanguelova et al. [30], the ideal number of samples is between 4 and a maximum of 25 samples per 0.25-ha plot, thus, for our trial site, the total cost could be between EUR 90 and EUR 600, for a sampling range between 6 and 42.

These hypothetical estimations are made recognising several limitations. The only accuracy metric given is the R^2, which is only one of the metrics of model fit, which may underestimate the uncertainty, and the costs are based on limited information. Likewise, for using UASs, there are two options available, to hire or to purchase, both of which have benefits and drawbacks.

4. Discussion

This paper has evaluated the potential contribution that EO technology could provide for SOC estimation. We have collated input from experts in soil science and agri-environment policymaking to identify key concerns and opportunities in this domain. In particular, the inclusion of expert opinion delivers a unique degree of insight on the acceptable level of accuracy that an EO-based soil MRV system would need to achieve to be appropriate for policy implementation. It is noteworthy that all respondents mentioned the benefit of such a study as an important input to the current DEFRA ELM scheme for developing a UK national soil MRV system.

According to the literature, EO imagery combined with ML techniques trained on soil samples representative of the location of interest can provide valuable information on the state of soils and SOC estimations in a given location. Such analysis involves the selection of variables (covariates), data compilation, model development, model evaluation, and model application. Similarly, interviewees discussed the potential of EO technology to estimate SOC, which they recognised relied on indirect measurements (proxies/covariates) and a derived model based on simple or more complex machine learning. The respondents were concerned about the use of EO technology and emphasised that it would have to be highly accurate (>90%) to allow fair payments if this was the basis of an MRV system. The meaning of 'accuracy' in this context was expressed mostly in terms of how well the EO-based assessments match (as far as practical) those assessments of SOC based on more traditional methods, although the meaning of this 'accuracy' is not universally defined.

The main advantages of using EO technologies have been highlighted in many studies. Our respondents described the advantages in the gathering of information on SOC over large and/or inaccessible areas, provision of information at different spatial and temporal resolutions, use of low-cost or free data, and cost-effective assessments. However, they also considered it of equal importance that the higher accuracy EO-based methods generally came at a higher cost.

Several substantial unanswered questions remain over how feasible current remote sensing methods are for estimating SOC and the levels of accuracy (i.e., how well the models predict observed values of SOC obtained via field measurement) that can be achieved, the amount of cross-validation or sampling that might be needed to assure such accuracy and, ultimately, what level of accuracy is required for SOC determinations for policy and decision making in support of sustainable SOC management. Our analysis has also discussed that there is a need to find a consistent definition of 'accuracy' that goes beyond simple metrics such as R^2, which can only ever describe how well models fit data, and not model biases or reference (in situ) data uncertainties. Other drawbacks noted in the use of EO-based data are that such methods are limited by cloud cover (for optical satellite sensors) and can only estimate SOC for the thin top layer of bare soil, while some agricultural practices that can increase SOC (cover of perennial rotation cropping, no-till planting in rotation) increase soil cover and thus compromise the window of bare soil. In addition, UAS have limited flight duration and hence spatial coverage.

It appears to be quite widely assumed that EO technology can be a cost-efficient tool in monitoring SOC. However, few studies have performed a cost-effectiveness analysis to test this assumption. Therefore, to address Andries et al.'s [48] recommendation that cost-effectiveness is an important component when assessing the maturity of EO technologies for providing data for the indicators of sustainable development, we performed a hypothetical cost-accuracy analysis on different literature approaches for SOC estimation on a 4-ha field using a combination of EO methods. We observed that the integration of data from satellite and airborne sensors with data from ground-based measurements represented a reasonable, current approach for prediction for SOC stocks at the field scale with an $R^2 = 0.81$ at a cost range of EUR 750 to EUR 1000. Relying on free satellite data (e.g., Sentinel-2) as a proxy for estimating SOC at the national and regional scale is much lower cost (equivalent to ~EUR 0.35 for the same area) but produces lower repeatability, with $R^2 = 0.32–0.44$. (Again, we note that R^2 provides only part of the information necessary to understand 'accuracy'.)

The importance of EO technologies as a reliable source of data for monitoring and reporting carbon stocks at a global scale within the context of the UN SDGs have been explored by the Global Soil Partnership (GSP) and its Intergovernmental Technical Panel on Soils (ITPS), which launched the Global Soil Organic Carbon GSOCMap [7]. The GSOCMap helps monitor and report on the carbon stock aspect of SDG indicator 15.3.1 ("Proportion of land that is degraded over the total land area, with three sub-indicators capturing trends in land cover, land productivity, and carbon stocks").

At national scales, several countries are attempting to make efforts to respond with initiatives such as '4 per 1000'. In the UK, the 25 Year Environment Plan aims for all of the UK's soils to be managed sustainably by 2030. To achieve this, the ELM scheme aims to update guidance for farmers, invest in developing a monitoring system and metrics for soil health, and deliver research to better understand how soil health supports wider environmental goals [76].

The suggestions of many of the respondents in this research informed the design we propose in Figure 3 for an MRV framework for SOC which could underpin policies supporting sustainable soils management. As is clear from our respondents, an MRV system should be robust, consistent, transparent, and accurate. It needs to rely on appropriate data, quality assurance and robust evaluation of the accuracy of the data, and collaboration with farmer communities and trained staff. With respect to linking MRV to fair payments, there are also requirements for defining clear SOC metrics and consistency in the SOC measurement approaches. Hence, Figure 3 presents the main three aspects of the framework as follows:

- *Monitoring*—Direct measurements through in situ soil sampling using a universal protocol and indirect measurements via proxies (e.g., UAS, airborne and satellite data). The measurements and estimation should determine the state and change of the SOC level. Key criteria to be considered should be consistency in soil sampling, high variability of the SOC, cost accuracy analysis, clear SOC metrics (that would be nationally/internationally recognised), and the scalability of the approach.
- *Reporting*—Documents to inform all relevant stakeholders, including information on the methodologies, assumptions, and data.
- *Verification*—A series of procedures for checking and verifying the quality of monitoring and reporting. Verification could be achieved through physical field inspections, independent remote sensing analysis, and/or smartphone apps (similar to BGS MySoil app [77] or Soilmentor [78]). Evidence suggests that there is a large potential for increasing carbon storage in agricultural soils through changes in land use management and agricultural practices. The importance of this SOC storage and the potential for

land managers to receive incentives for its enhancement suggests that VHR satellite data could well provide an economically viable, cost-effective tool for the verification stage. According to the UK National Audit Office (2019), the Rural Payments Agency (RPA), which pays subsidies to farmers (part of the Common Agricultural Policy-Basic Payments scheme), has increased remote inspections by 75% using satellite data with an operational cost saving of up to GBP 2.1 million per year [79]. Similar evidence has been shown in Ireland, which suggests that the average cost of a physical inspection in 2010 was EUR 1800 compared with EUR 60–EUR 70 for a remote sensing check [80].

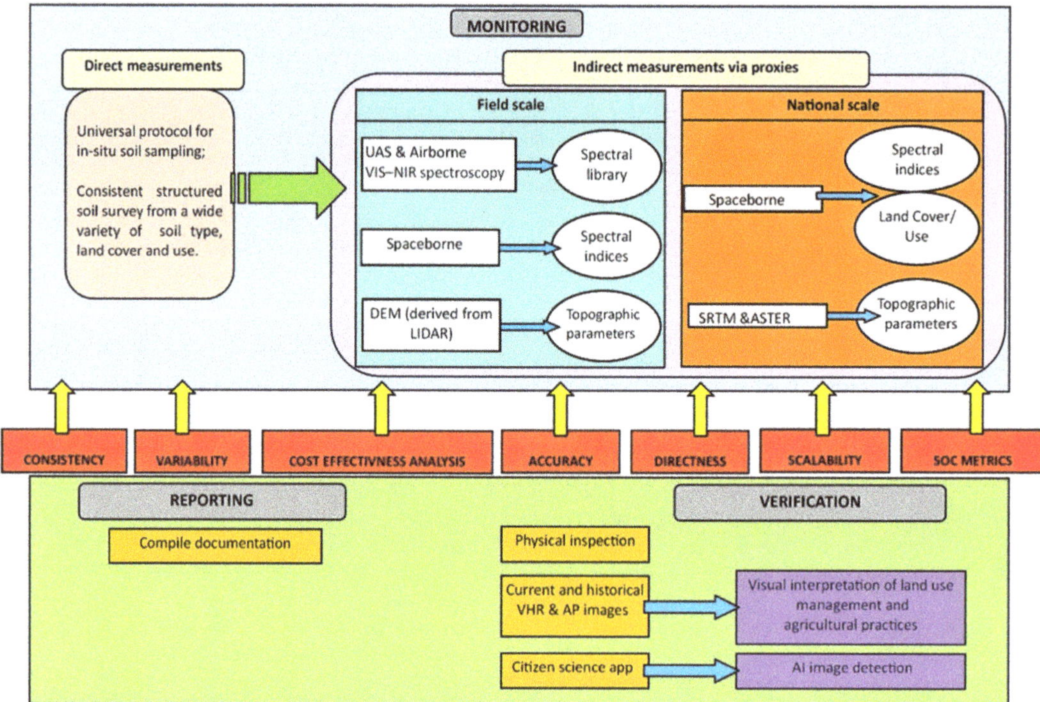

Figure 3. A framework for the measurement/monitoring, reporting, and verification (MRV) of SOC.

Overall, our research suggests that R&D to date has developed models at the field scale that, with appropriate in situ training data, can come close to the 'accuracy' expectations for policy when these are compared to results those obtained via traditional field assessment which employs sampling and laboratory analysis.

In effect, 'accuracy' within a land management context is about 'getting it right', with 'right' being what is gleaned from field measurement that can itself generate variable results based on methods of sampling and laboratory analysis that were employed. A desired 'accuracy' of 80% expressed by the respondents for EO methods appears well within reach at present, albeit for more expensive analyses. These are encouraging findings and suggest that only a small gap remains to be bridged between utility for policy and the accuracy/reliability available from EO-based SOC assessments at field scale. At regional and national scales, the accuracy of SOC assessments will likely remain limited for some time due to SOC determinations at these scales by EO-based methods being constrained by reliance upon free and lower spectral and spatial resolution EO imagery.

5. Conclusions

This research has led to the following conclusions:

- Satellite- and UAS-derived data provide unique capabilities for addressing current challenges in the MRV of SOC, at all scales;
- Well-established training datasets along with advances in machine learning are likely to increase the overall accuracy of SOC prediction models, especially when complemented by soil spectral libraries based on soil sampling;
- A more robust analysis of uncertainties associated with both EO derived data and in-situ measurements are required;
- Integration of UASs should be considered at farm scale as being the most cost-accurate approach, whereas free satellite data would be more appropriate to be used at the national scale but with lower accuracy;
- We propose a framework for the MRV of SOC, which includes direct measurements via a combination of in situ soil sampling and indirect measurements through EO.

Author Contributions: Conceptualization, A.A., S.M., R.J.M. and J.L.; methodology, A.A., S.M., R.J.M., E.R.W. and J.L.; software, A.A.; validation, A.A., S.M., R.J.M., E.R.W., B.M. and J.L.; formal analysis, A.A.; investigation, A.A.; resources, A.A., S.M., R.J.M., B.M. and J.L.; data curation, A.A. and B.M.; writing—original draft preparation, A.A.; writing—review and editing, A.A., S.M., R.J.M., E.R.W., B.M. and J.L., visualization, A.A.; supervision, S.M., R.J.M., E.R.W. and J.L.; project administration, R.J.M.; funding acquisition, S.M., R.J.M. and E.R.W. All authors have read and agreed to the published version of the manuscript.

Funding: This research was funded by the Natural Environment Research Council (NERC) SCENARIO Doctoral Training Partnership, Grant/Award NE/L002566/1, CASE award partner the National Physical Laboratory (NPL) and by UKRI/Research England SPF funding via the University of Surrey. The APC was funded by the University of Surrey.

Institutional Review Board Statement: Not applicable.

Informed Consent Statement: Not applicable.

Data Availability Statement: Not applicable.

Acknowledgments: This research was supported by the SCENARIO Doctoral Training Partnership of the UK Natural Environment Research Council (NERC), with PhD funding for the first author. We thank the National Physical Laboratory (NPL) for their contribution and co-funding as the CASE award partner. We also would like to thank the University of Surrey SPF UKRI/Research England support for AA for her secondment "Earth Observation for Advanced Monitoring, Reporting and Verification for Environmental Land Management (ELM) policy in the Surrey Hills Area of Outstanding Natural Beauty (SH-AONB)" to conduct this research with the SH-AONB Defra Test and Trial. The contribution of all expert interview respondents to the research is gratefully acknowledged.

Conflicts of Interest: The authors declare no conflict of interest.

References

1. FAO. *Measuring and Modelling Soil Carbon Stocks and Stock Changes in Livestock Production Systems—Guidelines for Assessment. Version 1—Advanced Copy*; FAO: Rome, Italy, 2019; p. 152.
2. Jobbagy, E.G.; Jackson, R.B. The Vertical Distribution of Soil Organic Carbon and Its Relation to Climate and Vegetation. *Ecol. Appl.* **2000**, *10*, 423–436. [CrossRef]
3. Batjes, N.H. Total carbon and nitrogen in the soils of the world. *Eur. J. Soil Sci.* **1996**, *47*, 151–163. [CrossRef]
4. Lal, R. Soil Carbon Sequestration Impacts on Global Climate Change and Food Security. *Science* **2004**, *304*, 1623–1627. [CrossRef] [PubMed]
5. ITPS. *Status of the World's Soil Resources (SWSR)—Main Report*; Food and Agriculture Organization of the United Nations and Intergovernmental Technical Panel on Soils: Rome, Italy, 2015.
6. House of Commons Environmental Audit Committee. Soil Health: Government Response to the Committee's First Report of Session 2016–2017. Available online: http://www.publications.parliament.uk/pa/cm201617/cmselect/cmenvaud/650/650.pdf (accessed on 12 August 2021).
7. GSOCMap. Available online: http://54.229.242.119/GSOCmap/ (accessed on 13 August 2021).

8. Lal, R.; Bouma, J.; Brevik, E.; Dawson, L.; Field, D.J.; Glaser, B.; Hatano, R.; Hartemink, A.E.; Kosaki, T.; Lascelles, B.; et al. Soils and sustainable development goals of the United Nations: An International Union of Soil Sciences perspective. *Geoderma Reg.* **2021**, *25*, e00398. [CrossRef]

9. Government, H. A Green Future: Our 25 Year Plan to Improve the Environment. Available online: https://assets.publishing.service.gov.uk/government/uploads/system/uploads/attachment_data/file/693158/25-year-environment-plan.pdf (accessed on 12 August 2021).

10. Smith, P.; Chapman, S.J.; Scott, W.A.; Black, H.I.J.; Wattenbach, M.; Milne, R.; Campbell, C.D.; Lilly, A.; Ostle, N.; Levy, P.E.; et al. Climate change cannot be entirely responsible for soil carbon loss observed in England and Wales, 1978–2003. *Glob. Chang. Biol.* **2007**, *13*, 2605–2609. [CrossRef]

11. Lal, R. Digging deeper: A holistic perspective of factors affecting soil organic carbon sequestration in agroecosystems. *Glob. Chang. Biol.* **2018**, *24*, 3285–3301. [CrossRef] [PubMed]

12. Sykes, A.J.; Macleod, M.; Eory, V.; Rees, R.M.; Payen, F.; Myrgiotis, V.; Williams, M.; Sohi, S.; Hillier, J.; Moran, D.; et al. Characterising the biophysical, economic and social impacts of soil carbon sequestration as a greenhouse gas removal technology. *Glob. Chang. Biol.* **2020**, *26*, 1085–1108. [CrossRef] [PubMed]

13. Zomer, R.J.; Bossio, D.A.; Sommer, R.; Verchot, L.V. Global Sequestration Potential of Increased Organic Carbon in Cropland Soils. *Sci. Rep.* **2017**, *7*, 15554. [CrossRef] [PubMed]

14. Bellamy, P.; Loveland, P.J.; Bradley, R.I.; Lark, R.; Kirk, G. Carbon losses from all soils across England and Wales 1978–2003. *Nat. Cell Biol.* **2005**, *437*, 245–248. [CrossRef] [PubMed]

15. Carey, P.D.; Wallis, S.; Emmett, B.A.; Maskell, L.C.; Murphy, J.; Norton, L.R.; Simpson, I.C.; Smart, S.M. *Countryside Survey: UK Headline Messages from 2007*; NERC/Centre for Ecology & Hydrology CEH Project Number: C03259; NERC: Wallingford, UK, 2008; p. 30.

16. Emmett, B.A.; Reynolds, B.; Chamberlain, P.M.; Rowe, E.; Spurgeon, D.; Brittain, S.A.; Frogbrook, Z.; Hughes, S.; Lawlor, A.J.; Poskitt, J.; et al. *Countryside Survey: Soils Report from 2007*; Technical Report No. 9/07 NERC/Centre for Ecology & Hydrology CEH Project Number: C03259; NERC: Wallingford, UK, 2010; p. 192.

17. Norton, L.; Maskell, L.; Smart, S.; Dunbar, M.; Emmett, B.; Carey, P.; Williams, P.; Crowe, A.; Chandler, K.; Scott, W.; et al. Measuring stock and change in the GB countryside for policy—Key findings and developments from the Countryside Survey 2007 field survey. *J. Environ. Manag.* **2012**, *113*, 117–127. [CrossRef]

18. Reynolds, B.; Chamberlain, P.; Poskitt, J.; Woods, C.; Scott, W.; Rowe, E.; Robinson, D.; Frogbrook, Z.; Keith, A.; Henrys, P.; et al. Countryside Survey: National "Soil Change" 1978–2007 for Topsoils in Great Britain-Acidity, Carbon, and Total Nitrogen Status. *Vadose Zone J.* **2013**, *12*, 1–15. [CrossRef]

19. Kirk, G.J.D.; Bellamy, P.H. Analysis of changes in organic carbon in mineral soils across England and Wales using a simple single-pool model. *Eur. J. Soil Sci.* **2010**, *61*, 406–411. [CrossRef]

20. FAO. Available online: http://www.fao.org/geonetwork/srv/en/metadata.show?id=14116 (accessed on 13 August 2021).

21. IUSS. Available online: https://www.iuss.org/soils-4-u/soil-data-and-information/ (accessed on 12 August 2021).

22. ESDB. Available online: https://esdac.jrc.ec.europa.eu/resource-type/european-soil-database-soil-properties (accessed on 13 August 2021).

23. LUCAS. Available online: http://eusoils.jrc.ec.europa.eu/library/esdac/index.html (accessed on 13 August 2021).

24. NSI. Available online: http://www.landis.org.uk/data/nsi.cfm (accessed on 13 August 2021).

25. CS. Available online: http://mapapps2.bgs.ac.uk/ukso/home.html (accessed on 13 August 2021).

26. UKSoil_Observatory. Available online: http://www.ukso.org/static-maps/community-soil-property-observations.html (accessed on 13 August 2021).

27. Gehl, R.J.; Rice, C.W. Emerging technologies for in situ measurement of soil carbon. *Clim. Chang.* **2006**, *80*, 43–54. [CrossRef]

28. Smith, P.; Soussana, J.; Angers, D.; Schipper, L.; Chenu, C.; Rasse, D.P.; Batjes, N.H.; van Egmond, F.; McNeill, S.; Kuhnert, M.; et al. How to measure, report and verify soil carbon change to realize the potential of soil carbon sequestration for atmospheric greenhouse gas removal. *Glob. Chang. Biol.* **2020**, *26*, 219–241. [CrossRef]

29. Garten, C.T., Jr.; Wullschleger, S.D. *Soil Carbon Inventories under a Bioenergy Crop (Switchgrass): Measurement Limitations*; Wiley Online Library: New York, NY, USA, 1999; pp. 47–2425.

30. Vanguelova, E.; Bonifacio, E.; De Vos, B.; Hoosbeek, M.; Berger, T.; Vesterdal, L.; Armolaitis, K.; Celi, L.; Dinca, L.; Kjønaas, O. Sources of errors and uncertainties in the assessment of forest soil carbon stocks at different scales—Review and recommendations. *Environ. Monit. Assess.* **2016**, *188*, 1–24. [CrossRef] [PubMed]

31. Minasny, B.; Malone, B.P.; McBratney, A.; Angers, D.A.; Arrouays, D.; Chambers, A.; Chaplot, V.; Chen, Z.-S.; Cheng, K.; Das, B.S.; et al. Soil carbon 4 per mille. *Geoderma* **2017**, *292*, 59–86. [CrossRef]

32. Neumann, M.; Smith, P. Carbon uptake by European agricultural land is variable, and in many regions could be increased: Evidence from remote sensing, yield statistics and models of potential productivity. *Sci. Total. Environ.* **2018**, *643*, 902–911. [CrossRef] [PubMed]

33. Chen, C.; Park, T.; Wang, X.; Piao, S.; Xu, B.; Chaturvedi, R.K.; Fuchs, R.; Brovkin, V.; Ciais, P.; Fensholt, R.; et al. China and India lead in greening of the world through land-use management. *Nat. Sustain.* **2019**, *2*, 122–129. [CrossRef]

34. Smith, P. Land use change and soil organic carbon dynamics. *Nutr. Cycl. Agroecosystems* **2008**, *81*, 169–178. [CrossRef]

35. Sims, N.C.; England, J.; Newnham, G.; Alexander, S.; Green, C.; Minelli, S.; Held, A. Developing good practice guidance for estimating land degradation in the context of the United Nations Sustainable Development Goals. *Environ. Sci. Policy* **2019**, *92*, 349–355. [CrossRef]
36. Conant, R.T.; Ogle, S.; Paul, E.A.; Paustian, K. Measuring and monitoring soil organic carbon stocks in agricultural lands for climate mitigation. *Front. Ecol. Environ.* **2010**, *9*, 169–173. [CrossRef]
37. Chen, F.; Kissel, D.E.; West, L.T.; Adkins, W. Field-Scale Mapping of Surface Soil Organic Carbon Using Remotely Sensed Imagery. *Soil Sci. Soc. Am. J.* **2000**, *64*, 746–753. [CrossRef]
38. Stevens, A.; Udelhoven, T.; Denis, A.; Tychon, B.; Lioy, R.; Hoffmann, L.; van Wesemael, B. Measuring soil organic carbon in croplands at regional scale using airborne imaging spectroscopy. *Geoderma* **2010**, *158*, 32–45. [CrossRef]
39. Nocita, M.; Stevens, A.; Noon, C.; van Wesemael, B. Prediction of soil organic carbon for different levels of soil moisture using Vis-NIR spectroscopy. *Geoderma* **2013**, *199*, 37–42. [CrossRef]
40. Gholizadeh, A.; Borůvka, L.; Saberioon, M.; Vašát, R. Visible, Near-Infrared, and Mid-Infrared Spectroscopy Applications for Soil Assessment with Emphasis on Soil Organic Matter Content and Quality: State-of-the-Art and Key Issues. *Appl. Spectrosc.* **2013**, *67*, 1349–1362. [CrossRef] [PubMed]
41. Peng, X.; Shi, T.; Song, A.; Chen, Y.; Gao, W. Estimating Soil Organic Carbon Using VIS/NIR Spectroscopy with SVMR and SPA Methods. *Remote Sens.* **2014**, *6*, 2699–2717. [CrossRef]
42. Kuang, B.; Tekin, Y.; Mouazen, A.M. Comparison between artificial neural network and partial least squares for on-line visible and near infrared spectroscopy measurement of soil organic carbon, pH and clay content. *Soil Tillage Res.* **2015**, *146*, 243–252. [CrossRef]
43. Vaudour, E.; Gilliot, J.M.; Bel, L.; Lefevre, J.; Chehdi, K. Regional prediction of soil organic carbon content over temperate croplands using visible near-infrared airborne hyperspectral imagery and synchronous field spectra. *Int. J. Appl. Earth Obs. Geoinf.* **2016**, *49*, 24–38. [CrossRef]
44. Morellos, A.; Pantazi, X.-E.; Moshou, D.; Alexandridis, T.; Whetton, R.; Tziotzios, G.; Wiebensohn, J.; Bill, R.; Mouazen, A.M. Machine learning based prediction of soil total nitrogen, organic carbon and moisture content by using VIS-NIR spectroscopy. *Biosyst. Eng.* **2016**, *152*, 104–116. [CrossRef]
45. Liu, Y.; Shi, Z.; Zhang, G.; Chen, Y.; Li, S.; Hong, Y.; Shi, T.; Wang, J.; Liu, Y. Application of Spectrally Derived Soil Type as Ancillary Data to Improve the Estimation of Soil Organic Carbon by Using the Chinese Soil Vis-NIR Spectral Library. *Remote Sens.* **2018**, *10*, 1747. [CrossRef]
46. Mosleh, Z.; Salehi, M.H.; Jafari, A.; Borujeni, I.E.; Mehnatkesh, A. The effectiveness of digital soil mapping to predict soil properties over low-relief areas. *Environ. Monit. Assess.* **2016**, *188*, 195. [CrossRef] [PubMed]
47. Zhou, T.; Geng, Y.; Chen, J.; Pan, J.; Haase, D.; Lausch, A. High-resolution digital mapping of soil organic carbon and soil total nitrogen using DEM derivatives, Sentinel-1 and Sentinel-2 data based on machine learning algorithms. *Sci. Total. Environ.* **2020**, *729*, 138244. [CrossRef] [PubMed]
48. Andries, A.; Morse, S.; Murphy, R.J.; Lynch, J.; Woolliams, E.R. Seeing Sustainability from Space: Using Earth Observation Data to Populate the UN Sustainable Development Goal Indicators. *Sustainability* **2019**, *11*, 5062. [CrossRef]
49. Aldana-Jague, E.; Heckrath, G.; Macdonald, A.; van Wesemael, B.; Van Oost, K. UAS-based soil carbon mapping using VIS-NIR (480–1000 nm) multi-spectral imaging: Potential and limitations. *Geoderma* **2016**, *275*, 55–66. [CrossRef]
50. Peng, Y.; Xiong, X.; Adhikari, K.; Knadel, M.; Grunwald, S.; Greve, M.H. Modeling Soil Organic Carbon at Regional Scale by Combining Multi-Spectral Images with Laboratory Spectra. *PLoS ONE* **2015**, *10*, e0142295. [CrossRef]
51. Castaldi, F.; Palombo, A.; Santini, F.; Pascucci, S.; Pignatti, S.; Casa, R. Evaluation of the potential of the current and forthcoming multispectral and hyperspectral imagers to estimate soil texture and organic carbon. *Remote Sens. Environ.* **2016**, *179*, 54–65. [CrossRef]
52. Mondal, A.; Khare, D.; Kundu, S.; Mondal, S.; Mukherjee, S.; Mukhopadhyay, A. Spatial soil organic carbon (SOC) prediction by regression kriging using remote sensing data. *Egypt. J. Remote Sens. Space Sci.* **2017**, *20*, 61–70. [CrossRef]
53. Gholizadeh, A.; Žižala, D.; Saberioon, M.; Borůvka, L. Soil organic carbon and texture retrieving and mapping using proximal, airborne and Sentinel-2 spectral imaging. *Remote Sens. Environ.* **2018**, *218*, 89–103. [CrossRef]
54. Schillaci, C.; Lombardo, L.; Saia, S.; Fantappiè, M.; Maerker, M.; Acutis, M. Modelling the topsoil carbon stock of agricultural lands with the Stochastic Gradient Treeboost in a semi-arid Mediterranean region. *Geoderma* **2017**, *286*, 35–45. [CrossRef]
55. Pouladi, N.; Møller, A.B.; Tabatabai, S.; Greve, M.H. Mapping soil organic matter contents at field level with Cubist, Random Forest and kriging. *Geoderma* **2019**, *342*, 85–92. [CrossRef]
56. Clerici, N.; Rubiano, K.; Abd-Elrahman, A.; Hoestettler, J.M.P.; Escobedo, F.J. Estimating Aboveground Biomass and Carbon Stocks in Periurban Andean Secondary Forests Using Very High Resolution Imagery. *Forests* **2016**, *7*, 138. [CrossRef]
57. Manns, H.R.; Berg, A.A.; Colliander, A. Soil organic carbon as a factor in passive microwave retrievals of soil water content over agricultural croplands. *J. Hydrol.* **2015**, *528*, 643–651. [CrossRef]
58. Bousbih, S.; Zribi, M.; Pelletier, C.; Gorrab, A.; Lili-Chabaane, Z.; Baghdadi, N.; Ben Aissa, N.; Mougenot, B. Soil Texture Estimation Using Radar and Optical Data from Sentinel-1 and Sentinel-2. *Remote Sens.* **2019**, *11*, 1520. [CrossRef]
59. Li, X.; Mccarty, G.W.; Du, L.; Lee, S. Use of Topographic Models for Mapping Soil Properties and Processes. *Soil Syst.* **2020**, *4*, 32. [CrossRef]

60. Rasel, S.; Groen, T.; Hussin, Y.; Diti, I.J. Proxies for soil organic carbon derived from remote sensing. *Int. J. Appl. Earth Obs. Geoinf.* **2017**, *59*, 157–166. [CrossRef]

61. Gomez, C.; Rossel, R.V.; McBratney, A. Soil organic carbon prediction by hyperspectral remote sensing and field vis-NIR spectroscopy: An Australian case study. *Geoderma* **2008**, *146*, 403–411. [CrossRef]

62. Crucil, G.; Castaldi, F.; Aldana-Jague, E.; Van Wesemael, B.; Macdonald, A.; Van Oost, K. Assessing the Performance of UAS-Compatible Multispectral and Hyperspectral Sensors for Soil Organic Carbon Prediction. *Sustainability* **2019**, *11*, 1889. [CrossRef]

63. Castaldi, F.; Chabrillat, S.; Van Wesemael, B. Sampling Strategies for Soil Property Mapping Using Multispectral Sentinel-2 and Hyperspectral EnMAP Satellite Data. *Remote Sens.* **2019**, *11*, 309. [CrossRef]

64. Castaldi, F.; Hueni, A.; Chabrillat, S.; Ward, K.; Buttafuoco, G.; Bomans, B.; Vreys, K.; Brell, M.; van Wesemael, B. Evaluating the capability of the Sentinel 2 data for soil organic carbon prediction in croplands. *ISPRS J. Photogramm. Remote Sens.* **2019**, *147*, 267–282. [CrossRef]

65. Žížala, D.; Minařík, R.; Zádorová, T. Soil Organic Carbon Mapping Using Multispectral Remote Sensing Data: Prediction Ability of Data with Different Spatial and Spectral Resolutions. *Remote Sens.* **2019**, *11*, 2947. [CrossRef]

66. Lacoste, M.; Minasny, B.; McBratney, A.; Michot, D.; Viaud, V.; Walter, C. High resolution 3D mapping of soil organic carbon in a heterogeneous agricultural landscape. *Geoderma* **2014**, *213*, 296–311. [CrossRef]

67. Padarian, J.; Minasny, B.; McBratney, A. Using deep learning to predict soil properties from regional spectral data. *Geoderma Reg.* **2019**, *16*, e00198. [CrossRef]

68. Wang, B.; Waters, C.; Orgill, S.; Cowie, A.; Clark, A.; Liu, D.L.; Simpson, M.; McGowen, I.; Sides, T. Estimating soil organic carbon stocks using different modelling techniques in the semi-arid rangelands of eastern Australia. *Ecol. Indic.* **2018**, *88*, 425–438. [CrossRef]

69. Angelopoulou, T.; Tziolas, N.; Balafoutis, A.; Zalidis, G.; Bochtis, D. Remote Sensing Techniques for Soil Organic Carbon Estimation: A Review. *Remote Sens.* **2019**, *11*, 676. [CrossRef]

70. Xiong, X.; Grunwald, S.; Myers, D.B.; Kim, J.; Harris, W.G.; Comerford, N.B. Holistic environmental soil-landscape modeling of soil organic carbon. *Environ. Model. Softw.* **2014**, *57*, 202–215. [CrossRef]

71. Xu, Y.; Smith, S.E.; Grunwald, S.; Abd-Elrahman, A.; Wani, S.P. Incorporation of satellite remote sensing pan-sharpened imagery into digital soil prediction and mapping models to characterize soil property variability in small agricultural fields. *ISPRS J. Photogramm. Remote Sens.* **2017**, *123*, 1–19. [CrossRef]

72. Brus, D.; Kempen, B.; Heuvelink, G. Sampling for validation of digital soil maps. *Eur. J. Soil Sci.* **2011**, *62*, 394–407. [CrossRef]

73. Vaudour, E.; Gomez, C.; Fouad, Y.; Lagacherie, P. Sentinel-2 image capacities to predict common topsoil properties of temperate and Mediterranean agroecosystems. *Remote Sens. Environ.* **2019**, *223*, 21–33. [CrossRef]

74. Dvorakova, K.; Shi, P.; Limbourg, Q.; Van Wesemael, B. Soil Organic Carbon Mapping from Remote Sensing: The Effect of Crop Residues. *Remote Sens.* **2020**, *12*, 1913. [CrossRef]

75. Roper, W.R.; Robarge, W.P.; Osmond, D.L.; Heitman, J.L. Comparing Four Methods of Measuring Soil Organic Matter in North Carolina Soils. *Soil Sci. Soc. Am. J.* **2019**, *83*, 466–474. [CrossRef]

76. DEFRA. *Environmental Land Management—Policy Discussion Document, Analysis of Responses*; Blue Marble Research: Radstock, UK, 2021; p. 135.

77. BGS MySoil App. Available online: https://www.bgs.ac.uk/technologies/apps/mysoil-app (accessed on 17 August 2021).

78. Soilmentor. Available online: https://soils.vidacycle.com (accessed on 17 August 2021).

79. Sadlier, G.; Flytkjær, R.; Sabri, S.; Robin, N. *Value of Satellite-Derived Earth Observation Capabilities to the UK Government Today and by 2020*; London Economics: London, UK, 2018.

80. Allen, M. Contextual Overview of the Use of Remote Sensing Data within CAP Eligibility Inspection and Control. Available online: http://www.niassembly.gov.uk/globalassets/documents/raise/publications/2015/dard/3115.pdf (accessed on 17 August 2021).

 *sustainability*

Article

Assessing Education from Space: Using Satellite Earth Observation to Quantify Overcrowding in Primary Schools in Rural Areas of Nigeria

Ana Andries [1,*], **Stephen Morse** [1], **Richard J. Murphy** [1], **Jim Lynch** [1] **and Emma R. Woolliams** [2]

[1] Centre for Environment and Sustainability, Faculty of Engineering and Physical Sciences, University of Surrey, Guildford GU2 7XH, UK; s.morse@surrey.ac.uk (S.M.); rj.murphy@surrey.ac.uk (R.J.M.); j.lynch@surrey.ac.uk (J.L.)

[2] Climate and Earth Observation Group, National Physical Laboratory, Teddington TW11 0LW, UK; emma.woolliams@npl.co.uk

* Correspondence: a.andries@surrey.ac.uk

Abstract: Nigeria is a country with a rapidly growing youthful population and the availability of good quality education for all is a key priority in the sustainable development of the country. An important element of this is the need to improve access to high-quality primary education in rural areas. A key indicator for progress on this is the provision of adequate classroom space for the more than 20 million learners in Nigerian public schools because overpopulated classrooms are known to have a strong negative impact on the performance of both pupils and their teachers. However, it can be challenging to rapidly monitor this indicator for the over 60 thousand primary schools, especially in rural areas. In this research, we used satellite Earth Observation (EO) and Nigerian government data to determine the size of available teaching spaces and evaluate the degree of overcrowding in a sample of 1900 randomly selected rural primary schools across 19 Nigerian states spanning all regions of the country. Our analysis shows that 81.4% of the schools examined were overcrowded according to the minimum standard threshold for school size of at least 1.2 m^2 of classroom space per pupil defined by the Federal Government of Nigeria. Such overcrowding can be expected to have a negative impact on educational performance, on achieving universal basic education and UN Sustainable Development Goal (SDG) 4 (Quality Education), and it can lead to poverty. While measuring floor area can be performed manually on site, collecting, and reporting such data for the number of rural primary schools in a large and populous country such as Nigeria is a serious, time-consuming administrative task with considerable potential for errors and data gaps. Satellite EO data are readily available including for remote areas, are reproducible and are easy to update over time. This paper provides a proof-of-concept example of how such EO data can contribute to addressing this socio-economic dimension of the SDGs framework.

Keywords: earth observation; UN sustainable development goals; education; socio-economic; overcrowded schools

Citation: Andries, A.; Morse, S.; Murphy, R.J.; Lynch, J.; Woolliams, E.R. Assessing Education from Space: Using Satellite Earth Observation to Quantify Overcrowding in Primary Schools in Rural Areas of Nigeria. *Sustainability* **2022**, *14*, 1408. https://doi.org/10.3390/su14031408

Academic Editor: Jamal Jokar Arsanjani

Received: 3 December 2021
Accepted: 20 January 2022
Published: 26 January 2022

Publisher's Note: MDPI stays neutral with regard to jurisdictional claims in published maps and institutional affiliations.

1. Introduction

Nigeria is the most populous country in Africa, and seventh in the world. It also has one of the largest populations of youth in the world [1]. From an estimated 42.5 million people at the time of independence in 1960, Nigeria's population has grown to around 195 million in 2018 [2]. Although Nigeria became the largest economy on the African continent in 2014 [3], the country still faces many serious issues such as violent rebellion and terrorism, endemic corruption, low life expectancy, inadequacies in public health systems, income inequalities, and high illiteracy rates [4,5].

The educational system in Nigeria is based on the Universal Basic Education (UBE) programme that was launched in 1999 aiming to provide free, universal, and compulsory basic education for children aged 6 to 15 years old. UBE covers six years of primary school and three years of junior secondary education. This can be followed by optional three years of senior secondary education and four years of tertiary education [6]. It needs to be noted that primary education is the only level of education that is available in urban and rural areas throughout the developed and developing world and is the largest subsector of any education system and so offers a unique opportunity to contribute to the transformation of societies [7,8].

Nigeria's education system struggles with the challenge of a persistent lack of adequate facilities. There is evidence that UBE is challenged by multiple issues such as insufficient classroom space in relation to high pupil enrolment, inadequate furniture and no functional chalkboards, lack of maintenance of building infrastructure and the lack of teachers [9,10]. All these lead to overcrowded classrooms and limit the quality of educational attainment [11–13]. While the UBEC's report does not specify a legal minimum space requirement for classroom dimensions, it does provide provisions and guidance on space norms which include a minimum standard learning space of 1.2 m^2/ pupil in rural primary schools (1.4 m^2/pupil for semi-urban and urban primary schools) [6]. Currently, among the suite of standards listed in the UBEC report, the key indicator that the Nigerian government uses for measuring quality education and equity is the Pupil—Teacher Ratio (PTR), with an ideal value set at 35:1 for primary schools [14]. Values higher than this equates to overcrowding in schools. However, while it is often claimed to be a key indicator, we found limited literature showing the PTR at the state and Local Government Areas (LGA) level in Nigeria, reflecting a weak and often highly politicised statistical system [15,16]. Many issues for national statistics offices in developing countries such as Nigeria often include a lack of timely data of suitable quality, a simple lack of data, limited independence of statistical information, unstable budgets, and misaligned incentives. These issues encourage the production of inaccurate data, the domination of national priorities by various sponsors, and limited access to and usability of the traditional data [17–20].

Despite these data challenges, researchers have conducted surveys in different states of Nigeria. For example, Opanuga et al. [14] noted that 81% of the 133 public primary schools in Ogun State have a PTR over 35:1. Moreover, a survey conducted by Ndem et al. [21] in Cross River State schools found an occurrence of high PTRs, such as 49:1 in primary and 62:1 at the junior secondary level. A most remarkable example of high PTR linked with overcrowding is noted in Sherry's 2008 [22]: *"all the schools I have seen are hugely overcrowded. In one record case, in a rural school, I saw a class of over 200 pupils of ages ranging from 11 to 21 with only one teacher to attend to them."* ([22], pp. 39–40). A five-year study that sought to support equitable access to education and improve the learning outcomes from basic education systems entitled 'Education Data, Research and Evaluation in Nigeria (EDOREN)' found that: *"consideration needs to be given to alternative ways of assessing classroom overcrowding, to complement pupil–teacher ratio rates, as the latter does not necessarily give an accurate indication of the numbers on the ground and can give the impression that classes are of manageable size when in reality they are not"* [9]. Overcrowded classrooms are well-known to be detrimental to educational outcomes [14,23] and have been reported in many studies as having a negative impact on adult and youth literacy [24,25]. Respondents of a survey conducted by Olaleye et al. [26] concluded that the shortage of building infrastructure of adequate quality was a major cause of overcrowded classrooms in Nigeria and Ikoya and Onoyase [27] presented a comprehensive national survey of primary school infrastructure that found 53% of the schools surveyed lacked fundamental structures. In addition, the assessment of basic education facilities in Kano, Jigawa, and Kaduna States by the Education Sector Support Programme in Nigeria (ESSPIN) concluded that around 75% of school infrastructure was *"very poor"* [28], while in Adamawa State 67% of public primary school classrooms were deemed to be in *"poor condition"* [13]. In a federal system such as in Nigeria, where taxes are raised at national and state levels, there can be disparities in wealth between states and this can influence the resources they have available to allocate to education [9]. Disparities in the resourcing of education between states can, in turn, lead to differences in education outcomes (e.g., literacy and numeracy) and differences in educational outcomes can in turn influence socio-economic indicators such as poverty; thus, there can be a negative reinforcement of inequality between states [29].

The present research aimed to evaluate the utility of data derived from satellite Earth Observation (EO) data as a direct measurement of classroom size and to determine the amount of classroom space allocated to pupils in rural primary schools in Nigeria. Although classroom areas could be individually measured by school staff, EO provides an opportunity for an independent, and, through image recognition machine learning algorithms, rapid assessment applicable to the whole country. Because Nigerian rural primary schools are built to a common pattern, they are easy to detect from satellite imagery. Each school is located near a road, with a playing field in front and a line of rectangular buildings (classrooms) behind and typically running parallel with the road. Thus, unlike surveys and measurements on site, EO satellite data provide the potential for a rapid, inexpensive, and accurate assessment [30].

In this study, we provide a first proof-of-concept assessment of the use of EO data for measuring school building footprints (area m^2) that could help governments and non-governmental organisations (NGOs) quickly identify schools with overcrowded classrooms. Classroom areas (m^2) were measured for 1900 rural primary schools across 19 Nigerian states and in combination with available enrolment data, a determination was made of the area per pupil for each school. Primary schools in rural areas were chosen for the research because i) they are an important component of delivering Nigeria's UBE ambitions and ii) most rural primary schools are single-storey buildings facilitating accurate measurements from satellite imagery.

Having estimated the area per pupil as an indicator of resourcing per pupil, the research then sought to use the data to explore whether there are links with existing data on educational outcomes such as literacy and numeracy. The latter has often been noted in studies based in the developed world [31], but literature is scarce to support such an association for the developing world, especially for Sub-Saharan Africa [26,32].

2. Materials and Methods

2.1. Study Area

Policies aimed at providing free universal primary education for all children in Nigeria pre-date independence from Britain in 1960. In the 1950s the colonial government recognised that secondary and tertiary education should be prioritised to provide the required number of teachers to achieve universal primary education (UPE). The attainment of UPE gathered pace during the 1960s but was piecemeal as separate states implemented their policies [33]. However, in 1976 the military government launched a major effort to implement free UPE across the entire country based on a significant school building programme and the recruitment and training of teachers [33]. Despite the prioritisation of UPE by successive governments (both civilian and military), the realisation of UPE remains a challenge and, since 1976, there have been many initiatives designed to address bottlenecks and constraints within the system. Later the UPE programme has morphed into UBE.

Nigeria has a federal system of governance, and the country comprises 36 states, each with its own governor and state assembly, plus the Federal Capital Territory (FCT) which houses the capital city of Abuja. Within the states, there are 768 Local Government Areas (LGA) and six Area Councils within the FCT, totalling 774 [34]. Responsibility for educational institutions is shared between different bodies at the federal, state, and local government levels, and a suite of indicators have been developed to help assess the quality of UBE received by pupils and their attainment. Various standards for basic education were urged by the Universal Basic Education Commission (UBEC) in 2000 as part of Nigeria's efforts to achieve the second Millennium Development Goal (MDG) of universal primary education [6]. For convenience, these states and the FCT are often classified in terms of six geopolitical zones primarily based on location, but which would also broadly encompass the major ethnic groups in the country. The South-West geopolitical zone, for example, largely comprises the Yoruba ethnic group while the South-East largely comprises the Igbo ethnic group [35,36]. But given the country comprises hundreds of ethnic groups a geopolitical zone will not be homogenous [36,37]. The same point applies to factors such as religious belief. In broad terms, the country comprises an Islamic north and Christian south, but geopolitical zones such as the South-West and North-Central, in particular, will be mixed [36]. The country also has an economic axis that runs from south to north in terms of wealth per capita; the southern zones tend to be richer than the northern ones [38] and this results in southern states having more resources available for investment in education. The authors hypothesised that these differences in wealth between states would, in turn, result in differences in area per pupil.

For this study, 19 states (Figure 1) were selected to span all six geopolitical regions of the country. These were:

1. North Central—Kogi, Benue, Kwara, Nasarawa;
2. North Eastern—Bauchi, Taraba, Gombe;
3. North Western—Sokoto, Kaduna, Zamfira;
4. South—Edo, Delta, Cross River;
5. South Eastern—Enugu, Abia, Anambra;
6. South Western—Ondo, Oyo, Osun.

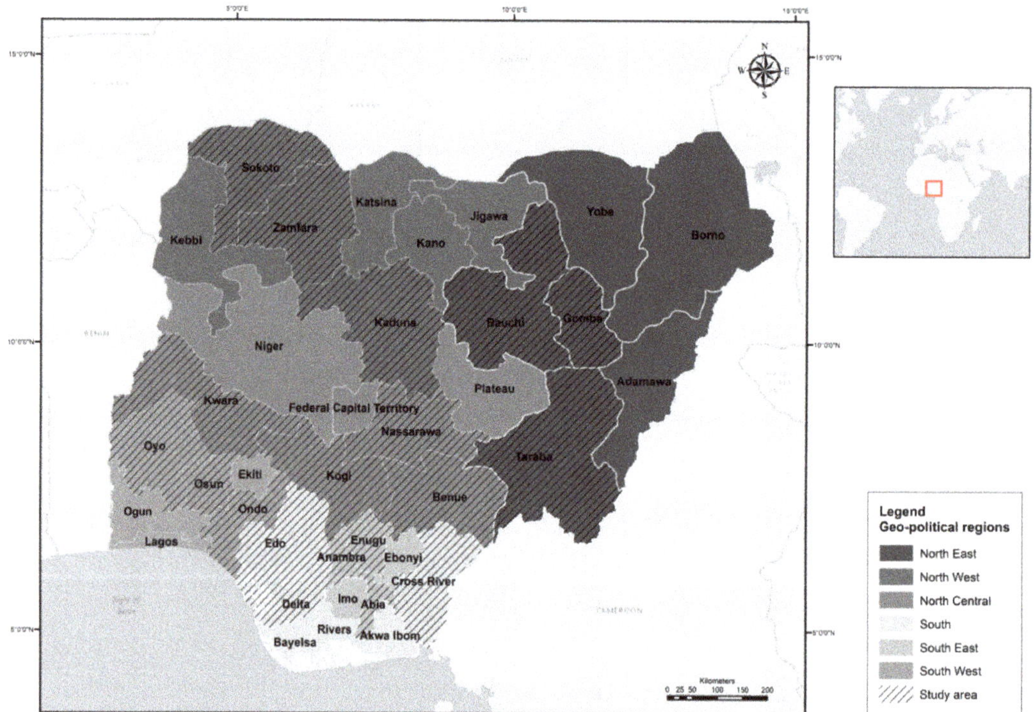

Figure 1. Nigeria geopolitical regions and study area.

2.2. Data Sources

Nigerian government data resources for rural primary schools [39] and satellite imageries from Google Earth Pro were used as the main data sources for the research. Firstly, the location of all primary schools was obtained via the Education Facilities in Nigeria (EFN) dataset [40] which includes school location (latitude and longitude), school type (primary, secondary, etc), school name, number of children registered, number of toilets, date of survey (survey period between 2009 and 2014), and number of teachers. The database comprises 98,667 primary schools across Nigeria and its goal was to build Nigeria's first nation-wide inventory of education facilities, to make the data collected available to planners, government officials, and the public, to be used to make strategic decisions for planning relevant interventions and to help achieve the MDGs.

The Google Earth Pro platform uses historical satellite and aerial imagery, at different spatial resolutions, which collect each image at a specific date and time. Most images used for this study came from satellites of very high resolution, and the date of the images was chosen to be close to the data of the pupil number survey (listed in [39]). For best results in measuring the footprints of school buildings, a top-down view of the images has been used as recommended in [39].

Given the lack of published and official data on the area (m^2)/pupil or pupil density in primary schools in Nigeria, but also to support validation of Google Earth measurements and later the calculation of the teaching area per pupil, classroom buildings in a subset of schools were physically measured. A professional town planner team was recruited to conduct on-site measurements of school buildings in Ogun State. Due to the COVID-19 travel restrictions and practical issues, this state was one of the few where travelling was allowed at the time of the research but still there were constraints in terms of accessing the school interior. Although this state was not part of our original selection, the fact that all rural schools were built in a nationwide characteristic pattern and in a typical morphology, we expect similar results in other parts of the country.

The team measured the exterior dimensions of the building for 21 primary schools from rural areas (listed in Table S1), and these were compared to the estimates made via satellite images (Table S2).

A series of national survey data was used to establish a causal relationship between the space per pupil and the educational outcomes of literacy and numeracy rates, as these variables have often been noted in the literature to have a significant negative association with pupil density.

The youth literacy and numeracy percentages (children age 5–16 able to read) by the state were taken from the Nigeria Education Data Survey [41] for the 19 studied states (Table 1). This survey was designed to provide information about the ability of children aged 5–16 years old and adults in a sample of 30,000 households to read and be numerate.

Table 1. Literacy and numeracy percentage at the state level, Source: [41].

State	Percentage of Children Ages 5–16 Able to Read	Percentage of Children Ages 5–16 Who are Numerate
Bauchi	8	18
Benue	33	59
Edo	76	79
Kaduna	46	62
Kogi	52	71
Kwara	53	61
Nasarawa	29	57
Ondo	78	92
Oyo	68	84
Sokoto	9	14
Enugu	51	81
Delta	65	81
Osun	83	92
Gombe	32	35
Zamfara	21	24
Taraba	21	41
Cross River	54	76
Anambra	84	69
Abia	83	92

As a further level of exploration, we sought to check whether the area/pupil indicator is linked with data on a variety of socio-economic measures of poverty available at the state level in Nigeria. For the latter, we used the following widely used indices for measuring poverty [42]: the poverty headcount ratio at US$3.20, consumption poverty headcount, Multidimensional Poverty Index (MPI) headcount, and relative poverty. The data have been collected under different surveys and methodologies and calculated at the state level in Nigeria (Table 2). The poverty rates of these indicators are given in Table 3 for each state in the present research.

Table 2. Poverty indices-Data description and source.

Index	Description	Data Source
Poverty headcount ratio at US $3.20 (2011 PPP) (% of population, 2013)	Data are based on primary household surveys obtained from Nigeria statistical agencies and the World Bank. The indicator is calculating the percentage of the population living on less than US $3.20 a day in 2011 international purchasing power parity (PPPs). A detailed description of this poverty indicator is presented by Ferreira et al. [42].	[43]
Consumption Poverty Headcount (2013)	Data on consumption are collected by the General Household Survey which asks the households about broad categories of consumed items of food, health care, schools. The indicator is obtained by aggregating information on food consumption and non-food consumption.	
MPI Headcount (2013)	MPI uses 10 indicators to measure poverty in three dimensions: education, health and living standard in which the intensity of poverty denotes the proportion of weighted indicators in which they are deprived. A person who is deprived in 90% of the weighted indicators has a greater intensity of deprivation than someone deprived in 40% of the weighted indicators. The proportion of the population that is multidimensionally poor is the incidence of poverty or MPI headcount ratio. This index was calculated using 2013 data from Demographic Health Surveys. The consumption poverty and MPI headcount indicators are both largely used to measure poverty, but the data are collected under two different surveys and methods, thus the poor according to the MPI does not always correspond to the poor measured according to consumption poverty.	[44,45]
Relative poverty (2010)	Relative poverty measurement is defined by the living standards of the majority and separates the poor from the non-poor. The threshold at which relative poverty is defined varies from one country to another, thus households with expenditure in Nigeria greater than two-thirds of the total household per capita expenditure are considered non-poor whereas those below it is poor.	[46]

2.3. Analysis

An overview of the analysis approach used in this paper is shown in Figure 2. The analysis had the main aim of exploring the utility of a method for evaluating the overcrowding of an individual school from satellite EO imagery of the school buildings and national statistical office data on pupil enrolments. The question being asked here was whether it is feasible to use EO-derived data to assess building footprint area and thereby use that measure as part of the area per pupil indicator? This process includes an analysis of the challenges involved in the measurement process. The EO-based measurements were used to assess area per pupil and based on the target employed in Nigeria of at least 1.2 m^2 being needed for a pupil it was possible to assess the degree of overcrowding (i.e., the proportion of schools having < 1.2 m^2/pupil). Firstly, we queried and extracted the EFN information relating to school type (e.g., primary), school management (e.g., public), schools' location, name, date of survey, and the number of pupils registered. Then, the width and length of each school building were measured using very high resolution (VHR) satellite images and the external area (m^2) was determined. The assessment of building footprint was directly checked against results obtained via ground-truthing (using a sample of schools in Ogun State; details below) and UBEC information that 15% of a school's built area should be attributed for administration [6]. This process allowed the authors to assess whether their measurements via EO had potential inaccuracies due to factors such as correct identification of buildings used for teaching rather than for other uses such as storage and also school buildings having a large veranda. However, even in the latter case, it is common for schools in Nigeria to use verandas as teaching spaces. Furthermore, we explored the results of the corrected teaching area to test for a possible association between educational attainment (youth literacy and numeracy rates) and poverty indices.

Table 3. Poverty indices by state.

State	Poverty Headcount Ratio Based on a Poverty Line of US $3.20 (2011 PPP) (% of Population, 2013) (Values 0 = Poorest and 1 = Non-Poor)	Consumption Poverty Headcount (% of Total Population)	MPI Headcount (% of Total Population)	Relative Poverty (% of Total Population)
Abia	0.65	17.8	8.8	63.4
Anambra	0.72	16	5	68
Bauchi	0.96	46.9	58.3	83.7
Benue	0.87	44	28	74.1
Cross River	0.71	51	14.6	59.7
Delta	0.65	13.6	10.7	70.1
Edo	0.73	17.4	8	72.5
Enugu	0.76	45.8	12.3	72.1
Gombe	0.93	29.2	47.1	79.8
Kaduna	0.85	41	31.1	73
Kogi	0.84	22.4	11.3	73.5
Kwara	0.87	34.4	9.9	74.3
Nassarawa	0.93	33.6	25.1	71.7
Ondo	0.74	15.6	12.7	57
Osun	0.61	21.4	4.3	47.5
Oyo	0.71	34.3	15.5	60.7
Sokoto	0.96	25.9	54.8	86.4
Taraba	0.84	51.8	44.8	76.3
Zamfara	0.90	49.2	60.5	80.2

2.4. Evaluating the Teaching Area

The EFN datasets were queried by school type, to obtain data on 60,000 public primary schools across the country. Having the approximate coordinates of the school location, these were overlaid on the Google Earth satellite images and 1900 schools (100 schools per state) from rural areas were selected using a random selection, with at least 2 km between schools, from rural areas using buffer measure tool in Google Earth, spanning all the geopolitical zones.

2.4.1. Google Earth Schools' Measurements

The total area (m^2) of the school building was obtained by measuring the length and width of each building within the selected school, using historical images from the same year as the EFN survey (from 2011–2016). The measurements were performed by the same operator manually measuring the length and width of each building within the school using the ruler tool in Google Earth and storing these measurements in an Excel spreadsheet. Where images from the same year were not available, the closest year for when images were available was selected. Identifying the school buildings was straightforward, as almost all sample schools followed a common pattern as expected from the rapid school building programme that took place during the mid-1970s. Figure 3 shows four examples of school configurations and locations that present characteristic patterns and a typical morphology, such as a schoolyard with bare soil or mowed grass, rectangular-shaped buildings, having up to 5 building units in a row, L or U building layout, and being in a peripheral location in the village adjacent to the main road (dirt road in most cases). Thus, teaching blocks (perhaps comprising more than one classroom) can be readily identified as being the larger buildings surrounding the playing field. While a typical school classroom block will have a veranda, the pressure on space is such that these are also often used for teaching. For operational use at the national scale, an image-processing classification algorithm supported by machine learning techniques could be alternatively used to perform this step and there are likely to be benefits in acquiring operational VHR data and using a more sophisticated processing platform.

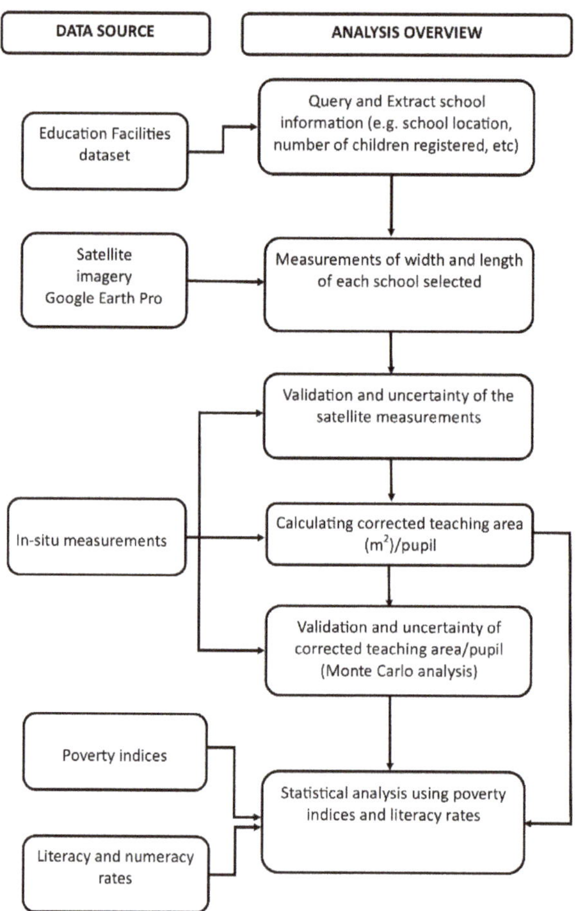

Figure 2. Data source and analysis overview.

2.4.2. Validation and Uncertainty Analysis

In any analysis method, and particularly when an approach is new, it is important to understand the uncertainty associated with the approach and to validate the results. This provides the information for users to judge the fitness for purpose of the data and the inferences drawn from them. Here, we wanted to evaluate a quantitative uncertainty associated with the area per pupil estimates and to validate the satellite measurements against on site measurements.

To validate the measurements taken from satellite images of Google Earth, we were aware of two main sources of uncertainties, one is that satellite images would not provide information about the building functionality (classroom, laboratory, office, veranda, etc.). The second concerns the repeatability of the measurement process in Google Earth.

The Guide to the Expression of Uncertainty in Measurement (GUM) provides a standard method for evaluating and propagating uncertainties [47] using two methods: the Law of Propagation of Uncertainties and Monte Carlo Analysis [48]. Both were employed in this analysis.

To evaluate the uncertainty, a sub-sample of 21 rural schools (with a total of 55 buildings) were selected for ground-truth measurement in Ogun State (South-Western Zone) and UBEC standards for administrative areas [6]. The team of planners provided for each school the width and length of each building and veranda (if present), along with a building plan as a pdf format document and a drone image (one example is presented in Figure 4). They also provided the number of buildings, number of building floors (always one for primary schools in our sample), details about building conditions, number of offices, and lavatory facilities within the school (see Table S1). From the on-site measurements, it was apparent that buildings with at least 6.5 m width had a veranda and toilets with a width of less than 4 m, and length less than 7 m are detached as small buildings (as seen in Figure 4— the building without roof) (see Table S2). As the planner team was not allowed to take internal measurements within buildings, the administrative area is unknown. UBEC [6] recommends that 15% of the total building area for each school should be attributed for administrative purposes (e.g., office, storage).

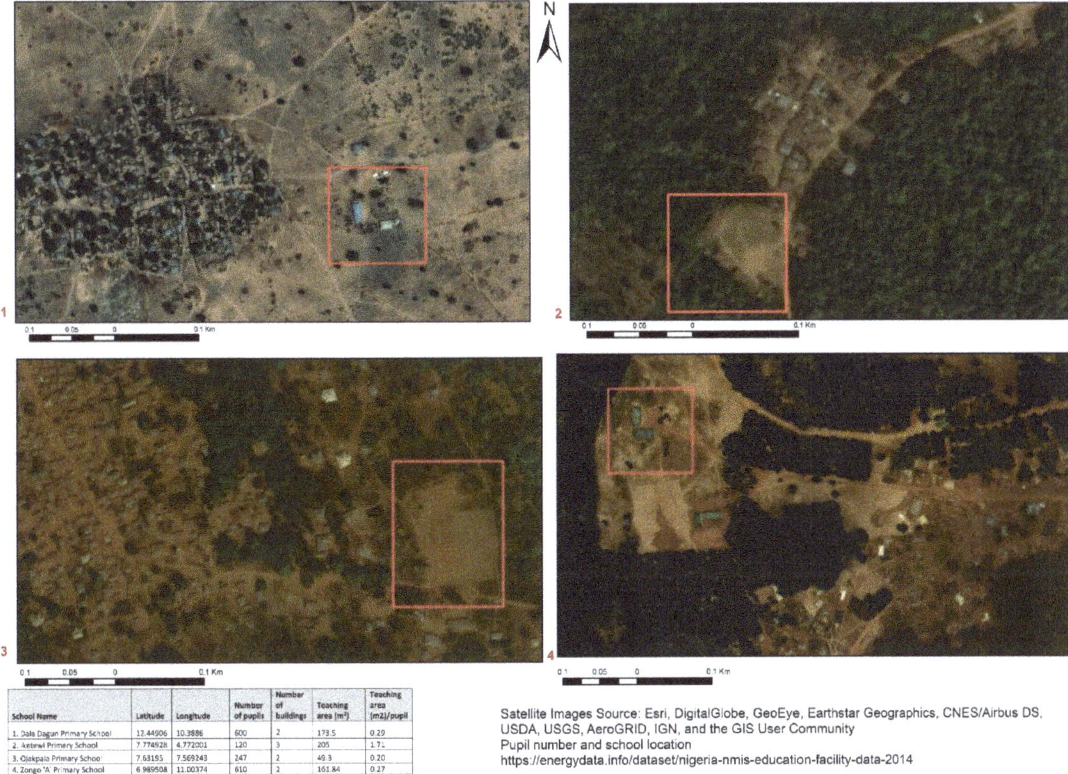

School Name	Latitude	Longitude	Number of pupils	Number of buildings	Teaching area (m²)	Teaching area (m2)/pupil
1. Dala Dagun Primary School	12.44906	10.3886	600	2	173.5	0.29
2. ketewi Primary School	7.774928	4.772001	120	3	205	1.71
3. Ojekpala Primary School	7.63195	7.569243	247	2	49.3	0.20
4. Zongo 'A' Primary School	6.989508	11.00374	610	2	161.84	0.27

Figure 3. Examples of the characteristic spatial patterns and locations of Nigerian rural primary schools. Note: Satellite images used are those closest to the date of the education facility survey [39] and some school structures might have changed since these images.

To estimate the uncertainty associated with the measured building sizes, we considered two components that we defined as "reproducibility uncertainty" (accuracy of the Google Earth measurements compared to on-site measurements) and "repeatability uncertainty" (repeatability of multiple measurements in Google Earth).

It should be noted that Google imagery does not have the resolution needed to provide sharp demarcations for the buildings. Thus, it was not possible to know precisely where the edge of the building was in the image. Hence it is important to establish an uncertainty associated with the Google-based assessments of area and inferences.

(a)

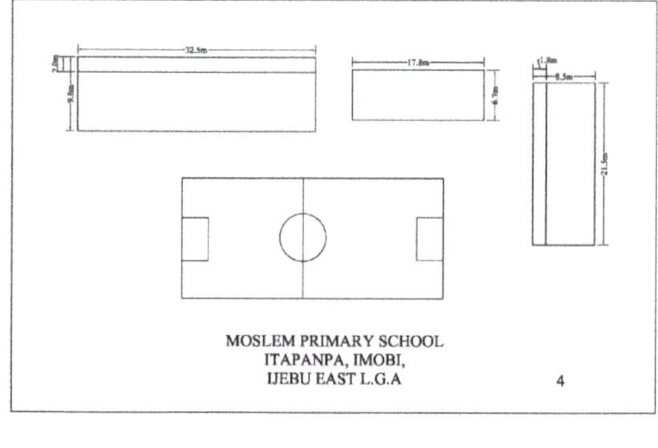

(b)

Figure 4. On site measurements of Moslem primary school Itapanpa, Imobi, Ijebu east LGA, Ogun State (**a**) drone image, (**b**) school plan and building on site measurements.

2.5. Calculating Corrected Teaching Area Per Pupil

2.5.1. Model Development

As described above, the initial measurements of buildings using Google Earth might contain inaccuracies associated with building functionality since there are aspects that are opaque to satellite EO, such as the size of administrative areas (e.g., offices), storage areas, and verandas. The on-site measurements revealed that small buildings with less than 4 m width and 7 m length are non-teaching areas (e.g., toilet blocks or storage areas) and hence they were removed from further analysis. For this analysis, and despite the fact that schools often use verandas as teaching areas, we removed a veranda area for larger buildings. We also removed our best estimate of administrative areas.

The calculation of "teaching area per pupil" thus followed Equations (1) and (2).

$$A'_{T,j} = \left[\sum_{i=1}^{n} L_i(W_i - W_{V,i}) \right] \times \left(1 - S_{\text{off}} B_j\right) \tag{1}$$

where,

$A'_{T,j}$ is the corrected teaching area of the jth school (in m^2),

i is an index representing the individual school buildings (in total there are n buildings),

L_i is the measured (from satellite imagery) external length of the ith building, in metres,

W_i is the measured (from satellite imagery) external width of the ith building, in metres,

$W_{V,i}$ is the assumed width of the veranda for school building i.

Based on the analysis of the 21 schools measured on-site, it was assumed that the veranda width would be 2 m for buildings wider than 10 m and 1.6 m for buildings from 6.5 m to 10 m wide. Smaller buildings less than 6.5 m had no veranda. Mathematically expressed:

$$W_{V,i} = \begin{cases} 2 \text{ m}, & W_i > 10 \text{ m} \\ 1.6 \text{ m}, & 10 \text{ m} > W_i > 6.5 \text{ m} \\ 0 \text{ m}, & W_i < 6.5 \text{ m} \end{cases} \tag{2}$$

To account for the space of administrative areas, an extra term was included: $(1 - S_{\text{off}})B_j$ is a term that reduces the teaching area of the school if the buildings are big enough to have office and storage space. If the school has more than three buildings, then it was assumed that the office space is 15% of the total school area after removing the veranda.

$S_{\text{off}} = 0.15$ is the proportion of the building area taken up by offices.

$$B_j = \begin{cases} 1 & n \geq 3 \\ 0 & n = 1,2 \end{cases} \tag{3}$$

is a Boolean that takes the value 1 if there are 3 or more buildings in school j and 0 otherwise. Therefore, the teaching area per pupil (α') is given by

$$\alpha' = \frac{A'_T}{N_P} \tag{4}$$

where,

A'_T is the corrected teaching area,

N_P is the number of pupils in a measured school.

2.5.2. Estimation of the Uncertainty of the Google Earth Measurements of the Buildings

To estimate the repeatability uncertainty, each of the 55 buildings in the 21 reference schools was measured 10 times (by the same operator) in Google Earth. The standard deviations of the 10 measurements are shown in Figure 5a and show the spread (28%) expected from a standard deviation calculated from just ten measurements. Those values also show no pattern as a function of actual length or width and therefore we consider the mean value (0.3 m) to be the uncertainty associated with random effects in a single Google Earth Pro measurement (the 1900 schools in the main set were each only measured once).

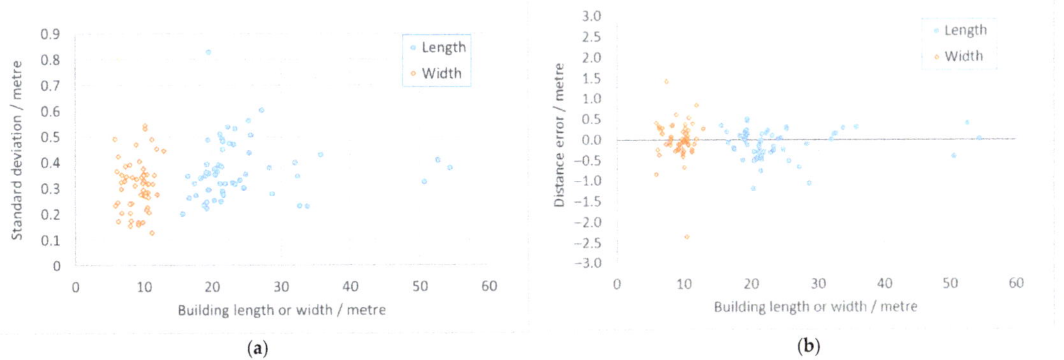

Figure 5. (**a**) Absolute standard deviation (in metres) of the ten independent measurements using Google Earth Pro of the length and width of the 55 buildings in the 21 reference set schools. Shown as a function of building length or width. (**b**) Distance error (calculated from the mean of the ten Google Earth Pro measurements minus the on site distance) for the 55 buildings in the 21 reference schools.

To estimate the reproducibility uncertainty, we took the average value of the 10 measurements of the length and width from the Google Earth Pro measurements and subtracted the on-site measured length or width from this. The differences obtained are given in Figure 5b. If the difference could be entirely explained by the random repeatability effects, we would expect the uncertainty associated with the mean of the ten measurements to be equal to the uncertainty associated with a single measurement (0.3 m) divided by $\sqrt{10}$ (i.e., 0.095 m). We see from Figure 5b that the actual spread is closer to 0.4 m (the standard deviation of the points is 0.43 m).

The increased spread of 0.4 m, in Figure 5b is symmetrical around the 0 axis; there is no obvious systematic bias between on site measurements and Google Earth Pro, but there is a randomly distributed difference between the two that cannot be accounted for by the random spread in the Google Earth Pro measurements alone.

From these two analyses, we can determine that the uncertainty associated with a single (rather than the mean of ten) measurement in Google Earth Pro contains a repeatability component of 0.3 m and a reproducibility component of 0.4 m. The uncertainty associated with a single building's length and/or width is obtained by combining these two quantities according to the GUM's law of propagation of uncertainties, and is, therefore:

$$u(L_i) = \sqrt{u_{repeat}^2 + u_{reproduce}^2} = \sqrt{0.3^2 + 0.4^2} = 0.5 \text{ m} \tag{5}$$

2.5.3. Uncertainty Analysis for Teaching Area per Pupil

To establish the uncertainty associated with the teaching area per pupil for a single school, Monte Carlo analysis was performed, using the uncertainty distributions described in Table 4. For this, a Python-coded algorithm that calculates Equations (1) and (2) was placed within a "for loop" and run 50 times (see Table S3). Within each loop, the different parameters were varied, for example by adding to the length and width of each building a random error from a Gaussian distribution with a standard deviation of 0.5 m, or by treating the office proportion, S_{off} as a quantity taken from a uniform random distribution between 0.10 and 0.20. The different errors were treated entirely independently —that is a separate random number was generated for every length and width measurement of every building in every school and for every Monte Carlo iteration.

Table 4. Uncertainties associated with the terms in Equations (1) and (2), and the probability distribution used to create the error for the Monte Carlo simulation.

Equations (1) and (2) term	Probability Distribution the Monte Carlo Error Is Drawn From	Where This Came From
N_p, number of pupils	0	It is assumed that the number of pupils is known from enrolment statistics without uncertainty.
Set of L_i, W_i for this school: lengths and widths of the external buildings measured in Google Earth	A Gaussian (normal) distribution centred on the original measurement, with a standard deviation of 0.5 m. Note each length and width has a different random error drawn from this distribution.	The analysis is described in Section 2.5.2 and Figure 5a,b. Statistically determined.
$W_{V,i} = \begin{cases} 2 \text{ m}, & W_i > 10 \text{ m} \\ 1.6 \text{ m}, & 10 \text{ m} > W_i > 6.5 \text{ m} \\ 0 \text{ m}, & W_i < 6.5 \text{ m} \end{cases}$	No uncertainty is associated with the step points (6.5 m and 10 m). Veranda width is described by a Gaussian (normal) distribution centred on the calculated width (2 m or 1.6 m), with a standard deviation of 0.3 m.	The on situ data showed this variety in the veranda widths (see Section 2.5.1).
$S_{off} = 0.15$, the proportion of the buildings taken up by offices (for a school big enough) is 15%	Office proportion has taken as a uniform distribution from $S_{off} = 0.10$ to $S_{off} = 0.20$. That is each school that is big enough for an office is assigned an office proportion randomly from this interval with an equally likely probability of any value in this interval	The authors do not have any strong justification for this range and have made a "best guess" based on the UBEC report [6] requirement of 15% area for a school, and allowing for a "reasonable" range of values around this.
$B_j = \begin{cases} 1 & n \geq 3 \\ 0 & n = 1, 2 \end{cases}$ Boolean criterion to decide whether or not to subtract office space.	No uncertainty is assumed.	Arguably, other criteria could be used to decide whether or not to select office space, but this was not analysed in the Monte Carlo simulation.
Form of equation.	No uncertainty is assumed.	Arguably, the form of Equation (1) could be different–for example, the office area could be removed before subtracting a veranda. But for this analysis, alternative forms were not considered.

Monte Carlo simulations are a method of uncertainty analysis described in the GUM [48]. The standard deviation of the results of the 50 individual Monte Carlo iterations provides an uncertainty estimate for the results obtained without perturbing the data. For Monte Carlo simulation to provide a reliable estimate of the uncertainty, an estimate of the uncertainty associated with the individual input parameters is required in order to define the probability distribution from which the random errors are calculated. Table 4 lists the uncertainties that we assumed.

2.6. Summary Statistics

2.6.1. Corrected Teaching Area Per Pupil

Basic summary statistics are calculated for the data set, both coming from the original data set and the Monte Carlo outputs. The original data set results are presented as the results of the analysis and the standard deviation of the Monte Carlo outputs are used to evaluate the uncertainty.

2.6.2. Statistical Analysis using Socio-Economic Indicators

It has been well-reported in the literature that pupil density (or area/pupil) is linked to outcomes such as literacy and numeracy [14,23–25]. Therefore, we attempted to test the validity of the values obtained from Equation (2) and evaluate the relationship between teaching area/pupil and percentage of children (age 5–16) able to read and be numerate at the state level using Welch's ANOVA.

Welch's ANOVA is a test of multiple comparisons of means (a modified one-way ANOVA) that is appropriate to use when there are unequal sample sizes and heterogeneity variance. Non-parametric methods such as Kruskal Wallis can be also used but Welch's ANOVA fits better especially with heterogeneous large datasets [49]. Welch's ANOVA was performed in Excel (extension Sigma XL), so we tested the hypothesis as follows:

- Null Hypothesis (H0): all five groups means are the same
- Alternative hypothesis (Ha): at least one mean is different

First, we used the teaching area (m^2)/pupil data measured for 1900 schools and the literacy and numeracy rates related to the states where the schools are located. Then, we grouped the states into 5 classes for the literacy and numeracy rates in Tables 5 and 6. To demonstrate Welch's ANOVA, we used the literacy and numerate groups in relation to the mean teaching area (m^2)/ pupil.

Table 5. Groups by Youth Literacy rates (%).

Youth Literacy (%)	Group 1 0–25%	Group 2 26–50%	Group 3 51–64%	Group 4 65–80%	Group 5 81–100%
Studied States	Bauchi Taraba Sokoto Zamfara	Benue Nasarawa Gombe Kaduna	Kogi Kwara Cross River Enugu	Edo Delta Ondo Oyo	Abia Anambra Osun

Table 6. Groups by Numeracy rates (%).

Youth Numeracy (%)	Group 1 0–25%	Group 2 26–60%	Group 3 61–75%	Group 4 76–81%	Group 5 82–100%
Studied States	Sokoto Bauchi Zamfara	Gombe Taraba Nasarawa Benue	Kwara Kaduna Anambra Kogi	Cross River Edo Enugu Delta	Oyo Ondo Osun Abia

3. Results

3.1. Values and Associated Uncertainties for Each School

To establish the uncertainty associated with the teaching area per pupil calculated according to Equation (2), a Python program was written to calculate Equations (1) and (2)

50 times as a Monte Carlo simulation (Table S3). As an example of the output, Figure 6 shows the 50 Monte Carlo runs on the teaching area/pupil and the original data analysis (Equations (1) and (2)) (illustrated in black diamond) on the first 50 schools in the dataset. For the vast majority of schools, the originally measured value is close to the centre of the Monte Carlo distribution. However, there are cases when it comes closer to the top or bottom, due to those schools where at least one building is close in width to the boundary conditions for having a veranda or not. Such cases create a bias between the Monte Carlo result set and the original data. Moreover, Tables S4 and S5, Figures S1–S3 include a further investigation on both biases (difference between the average of the Monte Carlo output and the originally determined value) and the standard deviation of the Monte Carlo output.

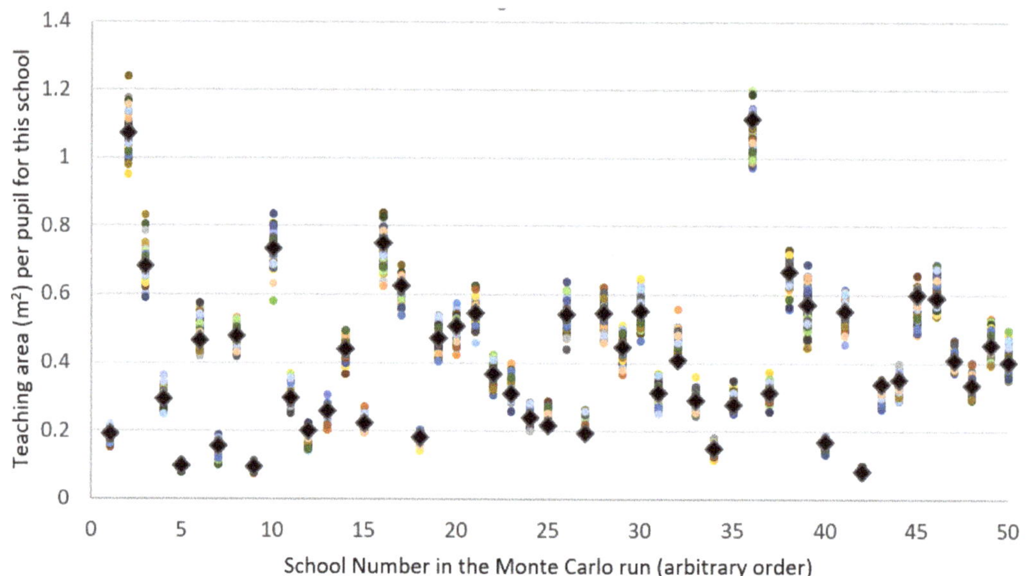

Figure 6. Output of the Monte Carlo analysis (circles) and original data analysis (black diamond) for each of the 50 runs for the first 50 schools in the study.

3.2. Basic Summary Statistics

From the Monte Carlo analysis, we determined that, typically, for a single school with the teaching area per pupil calculated according to Equations (1) and (2), the uncertainty associated with that school "teaching area per pupil" was 10% of the value (see Supplementary Material Table S5, Figures S2 and S3).

A histogram of the area per pupil calculated for the 1900 individual schools is given in Figure 7. This histogram was calculated an additional 50 times, each time using the results of a separate Monte Carlo run for the 1900 schools. The error bars in Figure 7 are calculated as the standard deviation of the histograms calculated for the 50 Monte Carlo runs results.

We calculated the summary statistics for the set of 1900 schools (Table 7) and found that 81.4% of the measured schools do not have a minimum required teaching space for children (a school is overcrowded if the teaching area per pupil is less than 1.2 m^2). Note that the standard uncertainties for the statistical values in Table 7 are considerably less than the 10% standard uncertainty associated with a single school. This is because the uncertainties are random from school to school and are reduced in effect by the very large number of schools considered.

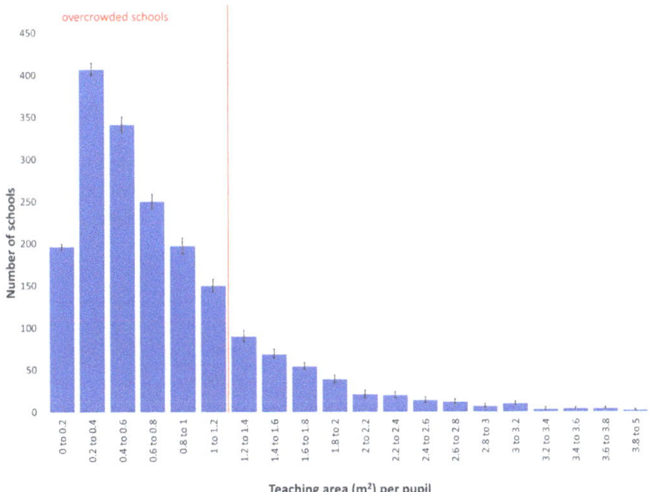

Figure 7. The number of schools with each corrected teaching area (m^2) per pupil for the full set of 1900 schools considered. Main bars are based on the calculation of Equations (1) and (2), error bars represent the standard uncertainty associated with this calculated from the Monte Carlo outputs.

Table 7. Summary statistics for the 1900 schools in the test data set. Overcrowded is defined as having a teaching area per pupil less than 1.2 m^2.

Statistics	Value	Standard Uncertainty Associatedwith this Value
Number of schools measured	1900	
Total of buildings measured	4490	
Mean teaching area per pupil	0.782 m^2	0.002 m^2
Median teaching area per pupil	0.601 m^2	0.004 m^2
Proportion of schools overcrowded	81.4%	0.2%

3.3. Teaching Area/Pupil in Relation to Socio-Economic Indicators

3.3.1. Literacy and Numeracy Rates

Welch's ANOVA was applied to understand the relationship between the mean area m^2/ pupil and education performance (literacy and numeracy rates) analyses. Therefore, Table 8 presents the sample size (number of schools), the mean of area m^2/ pupil, standard error and standard deviation of each literacy and numeracy group (also presented above in Tables 5 and 6, Section 2.6.2), as well as the results of Welch's ANOVA. The results of applying Welch's ANOVA's show that there are statistically significant ($p < 0.0001$) differences in teaching area/pupil between the five literacy and numeracy groups, indicating that lower teaching means area m^2/pupil (overcrowded) are associated with low literacy and numeracy rates (% children 5–12 able to read and numerate), and conversely when more space is allocated to the pupil gradually increased, there is better performance in schools.

3.3.2. Poverty Indices

Figure 8 illustrates the percentage of schools measured that are overcrowded or meet the minimum size required, overlaid on the poverty indices: (a) poverty headcount ratio at \$3.20 (% of total population, state level), (b) consumption poverty headcount (by state), (c) MPI headcount (by state), (d) poverty relative (% of total population, state-level). The poverty rates described by the four indicators have similar trends, Northeastern and Northwestern States are the poorest states, while Southwestern and South-Southern states show the lowest poverty rates, with slight differences in Consumption poverty (b) probably because our estimates were performed in the rural areas and the consumption rates are normally higher in urban zones. Overall, schools with lower teaching areas/pupils (<1.2 m^2) are associated with the populations of that state being poorer. States such as Sokoto, Zamfara, Bauchi, Gombe, Kaduna, Oyo, Benue, Taraba, Nasarawa, and Kwara, where over 80% of measured schools per state are overcrowded, poverty is also at the highest level. Conversely, the states with lower poverty rates (e.g., Anambra, Enugu, Delta, Cross River, Abia, Ondo, Osun, and Cross River) generally have less overcrowded schools.

Table 8. Descriptive statistics and results of Welch's ANOVA for literacy and numeracy groups.

	Literacy Groups (as per Table 5)					Numeracy Groups (as per Table 6)				
	1	**2**	**3**	**4**	**5**	**1**	**2**	**3**	**4**	**5**
Statistics	0–25%	26–50%	51–64%	65–80%	81–100%	0–25%	26–60%	61–75%	76–81%	82–100%
Sample size (N)	398	400	391	406	302	298	400	391	400	408
Mean (area/pupil)	0.43	0.50	0.86	1.04	1.14	0.43	0.51	0.65	1.11	1.11
Standard Error (SE)	0.01	0.01	0.03	0.03	0.04	0.02	0.02	0.03	0.03	0.04
Standard Deviation (SD)	0.28	0.37	0.65	0.64	0.75	0.28	0.36	0.53	0.69	0.72

<table>
<tr><td colspan="5" align="center">Welch's F Ratio = 144.10 (p < 0.0001)
df = 4879.14</td><td colspan="5" align="center">Welch's F Ratio = 139.86 (p < 0.0001)
df = 4936.59</td></tr>
</table>

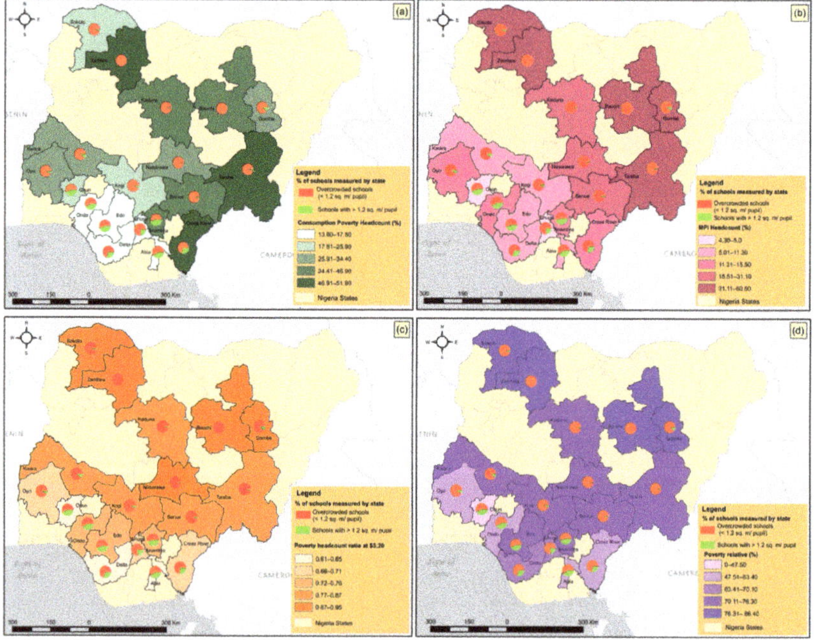

Figure 8. Poverty indices and % of schools overcrowded in the studied states, (**a**) consumption poverty headcount (%), (**b**) MPI Headcount (%), (**c**) Poverty headcount ratio at $3.20, (**d**) Poverty relative (%).

4. Discussion

This paper presents a novel application of EO satellite data in measuring the teaching floor area of a sample of 1900 rural public primary school buildings across 19 Nigerian States. We relate these measurements to nationally reported pupil enrolment data, thus determining how many of these schools would be deemed overcrowded. This approach identified that 81.4% ($\pm$0.2%) of the schools measured appeared to be overcrowded. In order to illustrate the potential value of our results, we performed further exploration of the distribution of the overcrowded schools and their interlinkage with literacy and numeracy rates, and poverty indices.

While measuring floor area could be performed manually on-site *by school staff or others,* the collection and reporting of such data for the number of rural primary schools in a large and populous country such as Nigeria is a substantial, expensive, and time-consuming administrative task, with potential for miscalculation and data gaps. On the other hand, EO data are readily available, address issues of accessibility in remote areas, are easily operated (convenient and free use of Google Earth), and are easy to update over time as schools add more classroom buildings. For example, from the historical satellite imagery incorporated in Google Earth, we were able to observe from establishing our sample of 1900 schools, that 113 of the schools had been extended and another 130 schools had been demolished between 2011 and 2019. Therefore, we suggest that EO data can provide a reliable, accurate, and convenient means for assessing classroom areas at the national scale and this has the potential to be automated via Artificial Intelligent/machine learning approaches (see Yazdani et al., [50] for identifying rural schools in Liberia). The advantages of this approach are probably most likely to be realised in the developing world where issues of accessibility to rural schools are especially challenging. Indeed, we consider the ability to rapidly and remotely evaluate overcrowding in the rural primary schools presented in this study, can help government agencies and NGOs in recognising priorities and to target attention and investment. Likewise, the method can be used in other countries where the spatial pattern of the school buildings and the number of students enrolled in school is understood.

As noted, EO data are convenient and easily available, but some limitations exist. We based our analysis on the total area of the school buildings and assumed that this was primarily dedicated to classrooms. Classrooms are understood to be the major use of the space, but satellite images cannot distinguish other usages within the building (e.g., administration uses, storerooms, etc.). Therefore, for calculation of classroom space per pupil, small buildings (width < 4 m and length < 7 m) were removed from the analysis for schools with more than 2 buildings assuming these are lavatory facilities. Veranda space was also extracted from classroom space based on the building width measurements, detailed exclusion criteria, and uncertainty analysis. All assumptions were established using the trends observed from the on-site measurements. The total remaining school building area obtained for classroom space was further reduced by 15% as per UBEC recommendations for administration uses [6].

Secondly, the EO images cannot determine the quality of the classroom space such as the internal condition of the building, availability of desks, availability of equipment, blackboards, etc., or lack of teachers (e.g., 'ghost' teachers—a type of fraud that often occurs in developing countries) [51]. Thirdly, the approach is not straightforward for schools that have more than one storey (mostly in urban areas). In Nigeria, the majority of rural schools were built to a common single-storey design, but urban schools often have multiple storeys. It may be possible to estimate height using shadow length taken at a particular time of the day, but even so, the assumptions become more complicated.

In our previous work [52,53], we discussed the importance of providing validation information and estimating uncertainty associated with indicators calculated from EO data sets. Despite the pandemic time, we obtained the on-site school measurements of 21 schools from one state and consider that to be a representative sample of common single-storey designed buildings used across rural areas in the country. Hence, the results of this study have been validated by both on-site measurements and statistical analysis using socio-economic indicators.

Overcrowded schools result from increases in student enrollment that are unmatched by school resources. School overcrowding and poverty are two distinct social issues, but literature shows that they are related and take part in a cause-effect relationship (shown in Figure 9). Insufficient classroom area can cause indirectly poverty, as children drop out of primary school without achieving minimal academic performance [54,55] and is also reflected in literacy and numeracy results [14,23–25], and vice versa when poverty (state not having enough money to spend in education) causes a lack of school resources [56–58]. Our study has shown that EO approaches can be important, efficient, enabling tools revealing school overcrowding in this case and helping authorities to make progress on a number of the UN SDGs. Figure 9 provides a conceptual model showing where EO can play a key role in making progress on the critical socio-economic relationships that affect the achievement of both SDG4 for quality education and promotion of lifelong learning and SDG 1 on poverty alleviation.

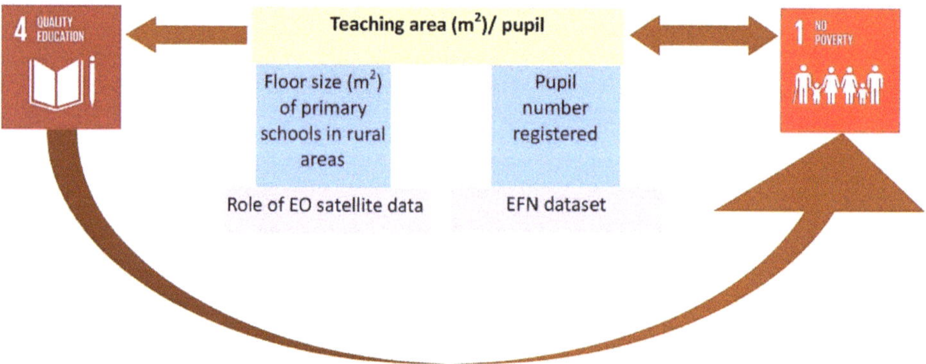

Figure 9. Cause effect relationship of overcrowded schools and social indices.

This study supports the need for an increased awareness of the value of satellite EO approaches for identifying both the specific example of overcrowded classrooms and, more generally, in supporting the socio-economic SDGs (as well as the environmentally focussed ones). Given the challenges involved in implementing the SDGs then the adaptation of a number of indicators to enable them to make use of readily available EO data presents a valuable opportunity to calibrate progress efficiently and economically.

5. Conclusions

The following conclusions can be drawn from this research:

- Overcrowded classrooms with less than 1.2 m²/pupil in rural primary schools in Nigeria were readily identified using satellite EO tools in combination with available school enrolment data.
- Results show that 84.4% (±0.2%) of schools measured are overcrowded and these are reflected in the education attainment (using literacy and numeracy rates) and poverty levels. The use of satellite images offers cost and time-efficient data to support improvements to education in Nigeria and elsewhere, particularly for those schools with one floor, and a simple measurement model and Monte Carlo Analysis can provide uncertainty to those satellite estimations that can be used to assess the fitness-for-purpose of the satellite data.
- Assessing pupil density using satellite EO can provide important information to help progress towards the UN SDGs for quality education and lifelong learning (SDG 4), equal access to opportunities (SDG 10), and reduce poverty (SDG 1). In wider terms, this study has also highlighted how EO-derived information can offer effective and

complementary support for sustainable development, including for indicators that are more closely aligned with social dimensions.

Supplementary Materials: The following supporting information can be downloaded at: https://www.mdpi.com/article/10.3390/su14031408/s1, Table S1. Information about the on-site school measurements, Table S2. On site measurements compared to satellite images measurements, Table S3. Python codes are used to calculate school overcrowding and the associated uncertainty Table S4. Further explanation of differences between original calculations and the Monte Carlo analysis, Table S5. Monte Carlo Analysis for individual schools. Figure S1. School number 6 (blue dots) and 36 (grey dots) show the 50 Monte Carlo (MC) runs and the original measurements (brown and green lines) on the teaching area (m^2)/pupil, Figure S2. In blue is the standard deviation of the school teaching area per pupil as a function of school teaching area per pupil. Negative values are given (calculated as -1 times the standard deviation) as well, to show the full spread. In orange the bias, calculated as the difference between the mean of the Monte Carlo output and the teaching area per pupil calculated from the original dataset, Figure S3. In blue is the standard deviation of the school teaching area per pupil as a function of school teaching area per pupil. Negative values are given (calculated as -1 times the standard deviation) as well, to show the full spread. In orange, the bias, calculated as the difference between the mean of the Monte Carlo output and the teaching area per pupil calculated from the original dataset.

Author Contributions: Conceptualization, A.A., E.R.W., S.M., R.J.M. and J.L.; methodology, A.A., S.M., E.R.W.; software, A.A., E.R.W.; validation, A.A., S.M., R.J.M., E.R.W. and J.L.; formal analysis, A.A., E.R.W.; investigation, A.A., S.M., R.J.M., E.R.W.; resources, A.A., S.M., R.J.M. and J.L.; data curation, A.A., E.R.W.; writing—original draft preparation, A.A., E.R.W.; writing—review and editing, A.A., S.M., R.J.M., E.R.W. and J.L., visualization, A.A., E.R.W.; supervision, S.M., R.J.M., E.R.W. and J.L.; project administration, R.J.M.; funding acquisition, S.M., R.J.M. and E.R.W. All authors have read and agreed to the published version of the manuscript.

Funding: This research was funded by the Natural Environment Research Council (NERC) SCENARIO Doctoral Training Partnership, Grant/Award NE/L002566/1, CASE award partner the National Physical Laboratory (NPL) and by UKRI/Research England SPF funding via the University of Surrey. ERW's time on this research was funded by the UK Government's Department for Business, Energy and Industrial Strategy (BEIS) through the UK's National Measurement System programmes. The APC was funded by the University of Surrey.

Institutional Review Board Statement: Not applicable.

Informed Consent Statement: Not applicable.

Data Availability Statement: Not applicable.

Acknowledgments: This research was supported by the SCENARIO Doctoral Training Partnership of the UK Natural Environment Research Council (NERC), with PhD funding for the first author. We thank the National Physical Laboratory (NPL) for their contribution and co-funding as the CASE award partner. We also would like to thank the Nigerian team of town planners who made the on-site measurements possible despite challenging conditions brought on by COVID.

Conflicts of Interest: The authors declare no conflict of interest.

References

1. The World Bank in Nigeria. Available online: https://www.worldbank.org/en/country/nigeria/overview (accessed on 25 September 2021).
2. UN. Nigeria General Information. Available online: http://data.un.org/en/iso/ng.html (accessed on 30 August 2021).
3. Terwase, I.T.; Abdul-Talib, A.-N.; Zengeni, K.T. Nigeria, Africa's largest economy: International business perspective. *Int. J. Manag. Sci.* **2014**, *3*, 534–543. [CrossRef]
4. Ajiye, S. Achievements of millennium development goals in Nigeria: A critical examination. *Int. Aff. Glob. Strategy* **2014**, *25*, 24–36.
5. UNDP. *National Human Development Report 2018—Timor-Leste*; United Nations: San Francisco, CA, USA, 2018.
6. UBEC. Minumum standard for basic education in Nigeria. 2000. Available online: http://wbgfiles.worldbank.org/documents/hdn/ed/saber/supporting_doc/AFR/Nigeria/TCH/Minimum%20Standards%20for%20Basic%20Education.pdf (accessed on 30 August 2021).

7. Boland, T. The importance of being literate: Reading development in primary school and its consequences for the school career in secondary education. *Eur. J. Psychol. Educ.* **1993**, *8*, 289–305. [CrossRef]
8. Lockheed, M.E.; Verspoor, A.M. *Improving Primary Education in Developing Countries*; Oxford University Press for World Bank: Oxford, UK, 1991.
9. Humphreys, S.; Crawfurd, L. Review of the literature on basic education in Nigeria: Issues of access, quality, equity and impact. *Educ. Data Res. Eval. Niger.* **2014**, *1*, 1–189. [CrossRef]
10. Okoroma, N. Educational policies and problems of implementation in Nigeria. *Aust. J. Adult Learn.* **2006**, *46*, 243–263.
11. Chege, F.; Zakariya, J.O.; Okojie, C.; Aregbeyen, O. *Girls' Education Project (GEP) Evaluation Report*; UNICEF: Abuja, Nigeria, 2008.
12. Okojie, C. *Formative Evaluation of the United Nations Girls' Education Initiative: Nigeria Report*; UNGEI: New York, NY, USA, 2012.
13. Dunne, M.; Humphreys, S.; Dauda, M.; Kaibo, J.; Garuba, A. Adamawa State primary education research: Access, quality and outcomes, with specific reference to gender. In *Centre for International Education*; University of Sussex: Brighton, UK, 2013.
14. Opanuga, A.A.; Okagbue, H.I.; Oguntunde, P.E.; Bishop, S.A.; Ogundile, O.P. Learning Analytics: Issues on the Pupil-Teacher Ratio in Public Primary Schools in Nigeria. *Int. J. Emerg. Technol. Learn.* **2019**, *14*, 180–199. [CrossRef]
15. Glassman, A.; Ezeh, A. Delivering on the Data Revolution in Sub-Saharan Africa. 2014. Available online: https://www.cgdev. org/publication/delivering-data-revolution-sub-saharan-africa-0 (accessed on 30 August 2021).
16. In the News: The Nigerian Census. Available online: https://www.prb.org/resources/in-the-news-the-nigerian-census/ (accessed on 25 September 2021).
17. Demombynes, G.; Sandefur, J. Costing a Data Revolution. 2014. Available online: https://www.cgdev.org/publication/costing-data-revolution (accessed on 25 September 2021).
18. Organisation for Economic Co-operation and Development. *Overview: What Will It Take for Data to Enable Development?* OECD: Paris, France, 2017. [CrossRef]
19. SciDevNet. Africa's 'Sluggish Data Collection Needs a Revolution'. Available online: https://www.scidev.net/global/news/ africa-s-sluggish-data-collection-needs-a-revolution/ (accessed on 18 August 2021).
20. Bhattacharya, D.; Khan, T.; Rezbana, U.; Mostaque, L. *Moving Forward with the SDGs: Implementation Challenges in Developing Countries*; Center for Policy Dialogue (CPD): Rehman Sobhan, 2016.
21. Ndem, A.; Uwem, E.; Udah, E. The universalization of basic education in Nigeria: The Cross River state experience. *Wudpecker J. Pub. Admin* **2013**, *1*, 7–19.
22. Sherry, H. Teachers' Voice: A Policy Research Report on Teachers' Motivation and Perceptions of Their Profession in Nigeria, London, Voluntary Service Overseas. Available online: https://www.vsointernational.org/sites/default/files/nigeria_teachers_voice_tcm76-22699.pdf (accessed on 29 August 2021).
23. Boyi, A. Education and Sustainable National Development in Nigeria: Challenges and Way Forward. *Int. Lett. Soc. Humanist. Sci.* **2012**, *14*, 65–72. [CrossRef]
24. Fowler, W.J.; Walberg, H.J. School Size, Characteristics, and Outcomes. *Educ. Eval. Policy Anal.* **1991**, *13*, 189–202. [CrossRef]
25. Chikadibia Domike, G.; Ogar Odey, E. An Evaluation of the Major Implementation Problems of Primary School Curriculum in Cross River State, Nigeria. *Am. J. Educ. Res.* **2014**, *2*, 397–401. [CrossRef]
26. Olaleye, F.; Ajayi, A.; Oyebola, O.; Ajayi, O. Impact of overcrowded classroom on academic performance of students in selected public secondary schools in Surelere local government of Lagos state, Nigeria. *Int. J. High. Educ. Res.* **2017**, *7*, 110–132.
27. Ikoya, P.O.; Onoyase, D. Universal basic education in Nigeria: Availability of schools' infrastructure for effective program implementation. *Educ. Stud.* **2008**, *34*, 11–24. [CrossRef]
28. ESSPIN. School Infrastructure and Maintenance Report. Available online: https://www.esspin.org/reports/download/67-file-1266228690-school_infrastr.pdf (accessed on 29 August 2021).
29. Oxfam. Inequality in Nigeria—Exploring the Drivers. Available online: https://www-cdn.oxfam.org/s3fs-public/file_attachments/cr-inequality-in-nigeria-170517-en.pdf (accessed on 15 November 2021).
30. Harrington, S.; Teitelman, J.; Rummel, E.; Morse, B.; Chen, P.; Eisentraut, D.; McDonough, D. Validating google earth pro as a scientific utility for use in accident reconstruction. *SAE Int. J. Transp. Saf.* **2017**, *5*, 135–167. [CrossRef]
31. Isingoma, P. *Overcrowded Classrooms and Learners' Assessment in Primary Schools of Kamwenge District*; Uganda University of South Africa: Pretoria, South Africa, 2013.
32. Lawal, I. Overcrowding As Metaphor for Declining Educational Quality. Available online: https://guardian.ng/features/ overcrowding-as-metaphor-for-declining-educational-quality/ (accessed on 15 November 2021).
33. Oni, J. Universality of primary education in Nigeria: Trends and issues. *Int. J. Afr. Afr.-Am. Stud.* **2008**, *7*, 23–31.
34. The Local Government System in Nigeria—Country Profile. 2020. Available online: http://www.clgf.org.uk/default/assets/ File/Country_profiles/Nigeria.pdf (accessed on 25 September 2021).
35. Ayomola, O.; Oketokun, O. Ethnicity, Separatists Agitations and National Stability in Nigeria: A Study of the South East Geo-Political Zone (1999–2020). *KIU J. Soc. Sci.* **2021**, *7*, 35–43.
36. Adegbami, A.; Uche, C.I. Ethnicity and ethnic politics: An impediment to political development in Nigeria. *Public Adm. Res.* **2015**, *4*, 59. [CrossRef]
37. Chikwado, N.K. An Overview of Ethnicity as a Leading Cause of Conflict in Nigeria. *Int. Digit. Organ. Sci. Res.* **2021**, *7*, 167–171.
38. Aderounmu, B.; Azuh, D.; Onanuga, O.; Oluwatomisin, O.; Ebenezer, B.; Azuh, A. Poverty drivers and Nigeria's development: Implications for policy intervention. *Cogent Arts Humanit.* **2021**, *8*, 1927495. [CrossRef]

39. Educational Facilities in Nigeria. Available online: https://africaopendata.org/dataset/educational-facilities-in-nigeria (accessed on 15 September 2021).
40. Measure Distances and Areas in Google Earth. Available online: https://support.google.com/earth/answer/9010337?co=GENIE.Platform%3DDesktop&hl=en (accessed on 15 September 2021).
41. NPC. DHS EdData Survey. 2010. Available online: https://catalog.ihsn.org/catalog/3344/related-materials (accessed on 15 September 2021).
42. Ferreira, F.H.; Chen, S.; Dabalen, A.; Dikhanov, Y.; Hamadeh, N.; Jolliffe, D.; Narayan, A.; Prydz, E.B.; Revenga, A.; Sangraula, P. A global count of the extreme poor in 2012: Data issues, methodology and initial results. *J. Econ. Inequal.* **2016**, *14*, 141–172. [CrossRef]
43. Nigeria—Poverty Map (Admin 1–2012). Available online: https://data.apps.fao.org/catalog/dataset/poverty-map-admin-1-2012/resource/dcbb7b30-6ec2-44a4-8e7d-cf2c681103d6?view_id=e60b6917-8ccf-4d65-9bc0-206314bdcd14 (accessed on 15 September 2021).
44. Gething, P.; Molini, V. Developing an Updated Poverty Map for Nigeria. Available online: https://data.apps.fao.org/catalog/dataset/54753faa-7f30-4841-ba8d-9b3685183f37/resource/beea75bc-afbf-41d3-9bdf-cd068acffb6b/download/developing-an-updated-poverty-map-for-nigeria.pdf (accessed on 15 September 2021).
45. WorldBank. Federal Republic of Nigeria Poverty Work Program Poverty Reduction in Nigeria in the Last Decade. Available online: https://documents1.worldbank.org/curated/en/103491483646246005/pdf/ACS19141-REVISED-PUBLIC-Pov-assessment-final.pdf (accessed on 15 September 2021).
46. (NBS), N.B.o.S. The Nigeria Poverty PROFILE 2010 Report. Available online: https://reliefweb.int/sites/reliefweb.int/files/resources/b410c26c2921c18a6839baebc9b1428fa98fa36a.pdf (accessed on 15 September 2021).
47. Metrology, J.C.F.G.I. Evaluation of measurement data—guide to the expression of uncertainty in measurement. *JCGM* **2008**, *100*, 1–116.
48. Shimaoka, K.; Kinoshita, M.; Fujii, K.; Tosaka, T. Evaluation of measurement data-supplement 1 to the guide to expression of uncertainty in measurement-propagation of distributions using a monte carlo method. *JCGM* **2008**, *101*, 1–82.
49. Hangcheng, L. Comparing W Comparing Welch's ANOVA, a Kruskal-W A, a Kruskal-Wallis Test and Trallis Test and Traditional Aditional ANOVA in Case of Heterogeneity of Variance, Virginia Commonwealth University. 2015. Available online: https://scholarscompass.vcu.edu/cgi/viewcontent.cgi?article=5026&context=etd (accessed on 15 September 2021).
50. Yazdani, M.; Nguyen, M.H.; Block, J.; Crawl, D.; Zurutuza, N.; Kim, D.; Hanson, G.; Altintas, I. Scalable Detection of Rural Schools in Africa Using Convolutional Neural Networks and Satellite Imagery. In Proceedings of the 2018 IEEE/ACM International Conference on Utility and Cloud Computing Companion (UCC Companion), Zurich, Switzerland, 17–20 December 2018; pp. 142–147.
51. Ibeh, A.I.; Uzegbu, A.J. The Effects of Corruption on the Implementation of Curriculum at the Secondary School Level in Nigeria: The Way Forward. *Niger. J. Curric. Stud.* **2021**, *27*, 120–133.
52. Andries, A.; Morse, S.; Murphy, R.; Lynch, J.; Woolliams, E.; Fonweban, J. Translation of Earth observation data into sustainable development indicators: An analytical framework. *Sustain. Dev.* **2018**, *27*, 366–376. [CrossRef]
53. Andries, A.; Morse, S.; Murphy, R.J.; Lynch, J.; Woolliams, E.R. Seeing Sustainability from Space: Using Earth Observation Data to Populate the UN Sustainable Development Goal Indicators. *Sustainability* **2019**, *11*, 5062. [CrossRef]
54. Opotow, S. Rationalities and levels of analysis in complex social issues: The examples of school overcrowding and poverty. *Soc. Justice Res.* **2006**, *19*, 135–150. [CrossRef]
55. Oriahi, C.; Aitufe, A. Education for the Eradication of Poverty. *Curr. Res. J. Soc. Sci.* **2010**, *2*, 306–310.
56. Amzat, I.H. The effect of poverty on education in Nigeria: Obstacles and Solutions. *OIDA Int. J. Sustain. Dev.* **2010**, *1*, 55–72.
57. Knight, J.; Shi, L.; Quheng, D. Education and the Poverty Trap in Rural China: Setting the Trap. *Oxf. Dev. Stud.* **2009**, *37*, 311–332. [CrossRef]
58. Van der Berg, S. Poverty and Education, International Academy of Education. Available online: https://citeseerx.ist.psu.edu/viewdoc/download?doi=10.1.1.464.9607&rep=rep1&type=pdf (accessed on 25 September 2021).

 *sustainability*

Article

Implications for Tracking SDG Indicator Metrics with Gridded Population Data

Cascade Tuholske [1,*], Andrea E. Gaughan [2,3], Alessandro Sorichetta [3], Alex de Sherbinin [1], Agathe Bucherie [4], Carolynne Hultquist [1], Forrest Stevens [2], Andrew Kruczkiewicz [4,5], Charles Huyck [6] and Greg Yetman [1]

[1] Center for International Earth Science Information Network, The Earth Institute, Columbia University, Palisades, NY 10964, USA; asherbin@ciesin.columbia.edu (A.d.S.); c.hultquist@columbia.edu (C.H.); gyetman@ciesin.columbia.edu (G.Y.)

[2] Department of Geography and Geosciences, University of Louisville, Louisville, KY 40292, USA; ae.gaughan@louisville.edu (A.E.G.); forrest.stevens@louisville.edu (F.S.)

[3] WorldPop, School of Geography and Environmental Science, University of Southampton, Southampton SO17 1BJ, UK; A.Sorichetta@soton.ac.uk

[4] International Research Institute for Climate and Society, The Earth Institute, Columbia University, Palisades, NY 10964, USA; agathe@iri.columbia.edu (A.B.); andrewk@iri.columbia.edu (A.K.)

[5] Red Cross Red Crescent Climate Centre, 2502 KC The Hague, The Netherlands

[6] ImageCat, Inc., Long Beach, CA 90802, USA; ckh@imagecatinc.com

* Correspondence: cascade@ciesin.columbia.edu

Citation: Tuholske, C.; Gaughan, A.E.; Sorichetta, A.; de Sherbinin, A.; Bucherie, A.; Hultquist, C.; Stevens, F.; Kruczkiewicz, A.; Huyck, C.; Yetman, G. Implications for Tracking SDG Indicator Metrics with Gridded Population Data. *Sustainability* **2021**, *13*, 7329. https://doi.org/10.3390/su13137329

Academic Editors: Stephen Morse, Richard Murphy and Ana Andries

Received: 12 May 2021
Accepted: 21 June 2021
Published: 30 June 2021

Publisher's Note: MDPI stays neutral with regard to jurisdictional claims in published maps and institutional affiliations.

Abstract: Achieving the seventeen United Nations Sustainable Development Goals (SDGs) requires accurate, consistent, and accessible population data. Yet many low- and middle-income countries lack reliable or recent census data at the sufficiently fine spatial scales needed to monitor SDG progress. While the increasing abundance of Earth observation-derived gridded population products provides analysis-ready population estimates, end users lack clear use criteria to track SDGs indicators. In fact, recent comparisons of gridded population products identify wide variation across gridded population products. Here we present three case studies to illuminate how gridded population datasets compare in measuring and monitoring SDGs to advance the "fitness for use" guidance. Our focus is on SDG 11.5, which aims to reduce the number of people impacted by disasters. We use five gridded population datasets to measure and map hazard exposure for three case studies: the 2015 earthquake in Nepal; Cyclone Idai in Mozambique, Malawi, and Zimbabwe (MMZ) in 2019; and flash flood susceptibility in Ecuador. First, we map and quantify geographic patterns of agreement/disagreement across gridded population products for Nepal, MMZ, and Ecuador, including delineating urban and rural populations estimates. Second, we quantify the populations exposed to each hazard. Across hazards and geographic contexts, there were marked differences in population estimates across the gridded population datasets. As such, it is key that researchers, practitioners, and end users utilize multiple gridded population datasets—an ensemble approach—to capture uncertainty and/or provide range estimates when using gridded population products to track SDG indicators. To this end, we made available code and globally comprehensive datasets that allows for the intercomparison of gridded population products.

Keywords: Sustainable Development Goals; hazards; Earth observations; remote sensing; demography; urbanization; gridded population

1. Introduction

The United Nations Sustainable Development Goals (SDGs) aim to end global poverty by 2030 and ensure a sustainable future [1]. To accomplish this, the SDGs outline a set of seventeen interlinked and shared objectives to improve economic and health outcomes in low- and middle-income countries (LMICs) while simultaneously reducing environmental degradation and tackling climate change for all countries [2,3]. The SDGs were designed to overcome the measurement challenges of the Millennium Development Goals [4] by outlining a clear set of indicators to track progress. Accordingly, global partnerships—such as the Sustainable Development Solutions Network [5] and the Global Partnership for Sustainable Development Data [6]—were established to provide countries with best practices to monitor SDG indicators and to support decision making to achieve the SDGs [7]. These partnerships increasingly advocate for countries to leverage the considerable amount of Earth observations (EO) data to track SDG indicators. In particular, analysis-ready EO data present a systematic, affordable, and longitudinal pathway to track SDG indicators that are specifically tractable for decision makers [8–12]. However, researchers, practitioners, and decision makers collectively lack guidance on how to best utilize wide-ranging and context-specific EO data to monitor SDG indicators.

This lack of guidance creates challenges for the SDGs' "leave no one behind" agenda. Not only does tracking many SDG indicators require accurate and accessible information on where people live, but achieving the SDGs entails providing services to people. Indeed, 73 SDG indicators require population data to track them [5]. Yet, due to the lack of reliable and consistent census data for many countries [13–15], we do not have information regarding where people live at sufficiently fine spatial scales to measure, monitor, and map changes in the distribution of populations relevant to many of the SDGs [13]. As such, EO-derived gridded population products present a vital source of population information to track the SDGs that continue to gain attention [5,16–19]. Each gridded population product offers spatially explicit representations of population distributions in a comparable, consistent manner that can be well suited to monitor SDG indicators across disparate geographies and time points. We chose to focus on gridded population data for two reasons. First, the gridded population products are receiving increasing use by a wide range of researchers, practitioners, and decision makers. Second, the methodologies and EO input data used to develop gridded population products continue to advance rapidly.

Given the wide range of gridded population products available, new "fitness for use" guidelines outline the tradeoffs and benefits the various gridded products offer for monitoring many SDG indicators [5,20]. However, the findings from recent studies [21–25] that have compared how gridded populations products allocate population illuminate the need for comparison of these datasets in the context of measuring and monitoring the SDGs. For example, a recent comparison of gridded population products against government-derived gridded census data in Sweden found wide variation in the accuracy of gridded population datasets to measure pixel-level population estimates [21]. Similarly, another recent study identified wide disagreement in both the location of and population estimates for urban settlements in Africa across gridded population products [25].

Broadly, the variation across products is explained by three reasons (for a recent review, see [20]). First, each gridded population product relies on different types of EO-based ancillary input data and production methods. Second, gridded population products differ in spatial resolution, projections, and temporal coverage. Third, for all but one product (Table 1), the process of creating the gridded distribution of population involves a disaggregation of census or administrative unit area counts into individual cells by relying on various EO-based (e.g., land cover type, settlement presence, nighttime light intensity) and geospatial covariates (e.g., buildings, roads, elevation) that vary in characteristics, quality, and accuracy [20]. Simply put, each gridded population product will likely tell a different story. In countries where reliable or recent fine-resolution census data are not available—precisely the context in which EO-enhanced gridded population datasets were designed to be employed—few independent fine-resolution micro censuses exist against which to assess the accuracy of gridded population products. Furthermore, aside from World Population Estimate 2016 (WPE-16), all current global gridded population products lack uncertainty estimates [5,20]. Until robust validation studies are available or uncertain estimates are produced, there remains a need to assess how different population data sets may influence the methods and results employed to monitor and evaluate progress towards SDGs.

Here we present three case studies to illuminate how gridded population datasets compare in measuring and monitoring SDGs and to advance the "fitness for use" guidance. Our focus is on SDG Target 11.5:

"By 2030, significantly reduce the number of deaths and the number of people affected and substantially decrease the direct economic losses relative to global gross domestic product caused by disasters, including water-related disasters, with a focus on protecting the poor and people in vulnerable situations."

In the context of SDG Target 11.5, we compare how five gridded population data products (Table 1) measure the population exposed to the 2015 earthquake in Nepal and Cyclone Idai in Mozambique, Malawi, and Zimbabwe (henceforth referred as MMZ) in 2019, as well as populations susceptible to flash floods in Ecuador. By focusing on a range of geographic and country-specific contexts across several hazards, our results provide insights into the ways in which the construction of different gridded population products across geographies affects the resulting calculation of the potentially affected populations.

Furthermore, we explore how gridded population products can be applied to other global frameworks, such as towards the Sendai Framework for Disaster Risk Reduction Targets [26]. Specifically, to achieve the first four targets (a–d), disaster risk monitoring requires accurate estimates of impacts on people and/or property [27]. Our work contributes to the Sendai Framework Global Target B: "Substantially reduce the number of affected people globally by 2030, aiming to lower the average global figure per 100,000 between 2020–2030 compared with 2005–2015." Beyond the importance of gridded data for calculating these indicators, however, we note that accurate population estimates are also vital to emergency management and humanitarian agencies in the post-disaster response phase, when assessments of the number of people affected directly translates to supplies and disaster finance being prioritized (or deprioritized) across spatial units of interest [28]. Thus, the work presented here contributes to an understanding of the potential for operational uses of various gridded population products.

We chose five gridded populations products (Table 1) due to their availability at the time of analysis and their global coverage. Our analysis focuses on the comparison of these different gridded products and why estimates are similar or dissimilar in different socio-economic and hazard contexts. Given the lack of "ground truth" micro census population estimates for the regions compared, we do not assess the accuracy of the gridded population products themselves. However, the analysis provided does inform end users of the potential pros and cons of using these datasets in the context of measuring SDG 11.5.

We have two interrelated objectives. First, we map and quantify geographic patterns of agreement/disagreement across gridded population products for Nepal, MMZ, and Ecuador, including delineating urban from rural populations estimates. Several methodologies have been used to compare products [21–25]. For the initial step, we identify the number of rasters that agree if a given pixel is inhabited or not. Next, we assess pixel-level variation across the five gridded population products by plotting the minimum pixel values against pixel ranges to identify outliers and showcase the contrast in pixel-level measurements. Then, for each gridded product, we examine transects through the primary urban centers impacted by the hazards to both visually and quantitatively demonstrate the variability in population estimates by product across the urban-rural continuum [29].

The second objective aims to situate our first objective within the context of measuring, monitoring, and mapping SDG 11.5. For each dataset, we estimate the total number of people, stratified by urban and rural populations, exposed to each hazard. For Nepal, we compare estimates across seismic intensity levels during the 2015 Earthquake. With Cyclone Idai, we compare estimates of population inundated by water detected by Sentinel-1 EO platform and exposed to wind speed zones. Lastly, in Ecuador, we quantify and map populations living in zones across levels of susceptibility to flash floods. We emphasize that our results do not quantify error or validate population estimates across gridded population products. We note that the urban land cover designation we employ (Section 2.1) is for comparative purposes only. Our analysis does not assess how gridded population products measure urban populations and urban boundaries (for further detail see [25,29]). All our code and data used in this analysis is open source and freely available for other scholars and practitioners to develop their own use cases. This includes global raster datasets that allow for the intercomparison of gridded population products.

2. Materials and Methods

2.1. Dataset Descriptions

2.1.1. Population Data

We focus on three geographies of interest—Nepal, the region of Mozambique, Malawi, and Zimbabwe (MMZ), and Ecuador (Figure 1)—to explore how the gridded population products measure populations related to SDG 11.5 across a range of geographic contexts. Nepal, a relatively small country, is landlocked between China's Tibet Autonomous Region and India, and is very mountainous. As hazards do not respect political boundaries, we present MMZ to measure exposure in a cross-border use case. Indeed, population exposure to Cyclone Idai spanned from Mozambique's low-elevation coastal zones, to Malawian settlements near Lake Malawi, to the relatively high-elevation settlements in Zimbabwe. Ecuador presents both a mountainous and a coastal geography to examine hazards, as well as higher levels of economic development compared to the other two geographies of interest. We intentionally chose countries and study areas that coincided with hazard events that matched the gridded population product dates and span LMIC contexts.

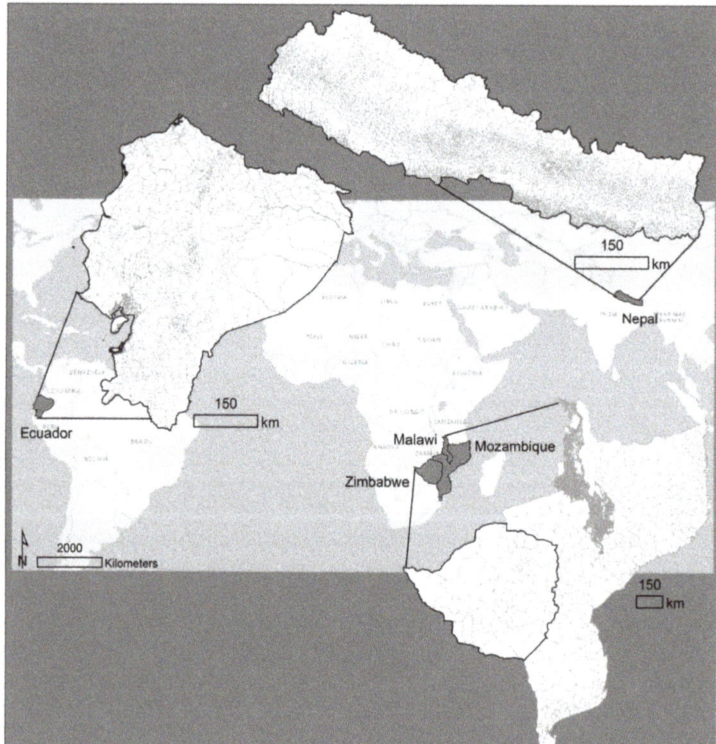

Figure 1. Map of three geographies of interest with administrative input units for GWP-15 overlaid.

The five gridded population products we use are (Table 1): World Population Estimates 2016 [30]; the Global Human Settlement Layer Population 2015 (GHS-15) [31]; Gridded

Population of the World version 4 2015 (GPW-15) [32]; LandScan 2015 (LS-15) [33]; and WorldPop 2016 (WP-16) [34]. For a complete description of how each gridded population product is produced, see [20,25], as well as the PopGrid Data Collaborative [35].

Aside from GPW-15, the gridded population datasets used in this study rely on relationships between EO data and human settlement patterns to disaggregate the finest available administrative unit level population data into pixels (Table 1). A higher administrative level corresponds to a finer resolution administrative unit. Additionally, only LS-15 focuses on daytime, or ambient population, whereas the other products aim to capture nighttime residential population [35]. Last, aside from GHS-15, which disaggregates administrative unit-level population only within pixels identified as containing built settlements (constrained), all other products disaggregate administrative unit-level population over all land pixels globally (unconstrained). All products we use are at 1 km spatial resolution. We use GPW-15 as a baseline for comparison, as it is the underlying population data for both GHS-15 and WP-16. Finally, we include United Nations population estimates for each geography of interest for 2015 (Table 2) [36].

Table 1. Summary of near-global coverage gridded population datasets included in this study. For further information, see [20] and the PopGrid Data Collaborative [35].

Dataset	Producer	EO Data	Population	Constrained	Model Description	Citation
GPW-15: Gridded Population of the World v4.11, 2015	CIESIN, Columbia University	None	Residential	No	Equal allocation of population to cells within admin. units	[32]
GHS-15: Global Human Settlement Layer-POP, 2015	European Commission, Joint Research Centre (JRC)	Landsat	Residential	Yes	Binary dasymetric, proportional allocation to built-up areas extracted primarily from 30 m Landsat imagery	[31]
WP-16: WorldPop Global, Unconstrained, 2016	WorldPop, Univ. of Southampton	Landsat, DMSP-OLS, VIIRS, MODIS, MERIS	Residential	No	Random Forest model with 24 covariates and weighted dasymetric redistribution	[34,37]
LS-15: LandScan, 2015	Oak Ridge National Laboratory	Landsat, MODIS, DMSP-OLS	Ambient (24-h average)	No	Multivariable dasymetric model with 4 covariate types and weighted redistribution	[33,38]
WPE-16: ESRI World Population Estimate, 2016	Esri Inc.	Landsat	Residential	No	Dasymetric algorithm with 16 covariate weighting data sets	[30]

Table 2. Allocation of population and pixel-level agreement across five gridded population products for Nepal, MMZ, and Ecuador circa 2015. For each geography of interest the number and level of input administrative units is listed. Note urban, rural, and total populations are in millions.

Geography	Urban Pop	Rural Pop	Total Pop	Pct Urban	Urban Max	Rural Max	Urban Pixels	Rural Pixels	Uninhabited Pixels
				Nepal 3990 level 3 units					
WPE-16	2.66	30.71	33.37	7.97%	45,982	25,237	275	118,437	76,844 (39%)
GHS-15	3.3	25.16	28.46	11.6%	46,472	117,462	271	104,208	91,077 (47%)
GPW-15	2.88	27.84	30.72	9.38%	32,592	28,114	275	175,048	20,233 (10%)
LS-15	2.85	28.69	31.54	9.04%	57,668	44,892	275	145,639	49,642 (25%)
WP-16	3.64	28.6	32.24	11.29%	48,358	46,939	275	167,188	28,093 (14%)
UN-15	5.32	23.34	28.66	18.56%					
			MMZ—Mozambique 413 level 3 units, Malawi 12,647 level 3 units, Zimbabwe 92 level 2 units						
WPE-16	6.84	59.52	66.36	10.31%	26,168	17,138	1695	232,593	1,356,048 (85%)
GHS-15	8.34	52.13	60.47	13.79%	81,852	156,171	1810	176,110	1,412,416 (89%)
GPW-15	3.46	51.9	55.37	6.25%	26,555	17,190	2011	1,467,545	120,780 (8%)
LS-15	6.51	50.93	57.45	11.33%	61,126	40,592	1976	1,370,043	218,317 (14%)
WP-16	5.01	51.81	56.82	8.82%	26,995	25,233	2009	1,409,324	179,003 (11%)
UN-15	17.61	43.75	61.36	28.7%					
				Ecuador 1047 level 3 units					
WPE-16	7.97	10.28	18.26	43.65%	18,108	14,867	1457	41,897	247,071 (85%)
GHS-15	8.19	7.86	16.05	51.03%	31,851	43,017	1510	41,755	247,160 (85%)
GPW-15	2.06	13.83	15.89	12.96%	4172	4172	1664	235,809	52,952 (18%)
LS-15	7.58	8.24	15.82	47.91%	44,304	31,740	1645	192,868	95,912 (33%)
WP-16	5.35	10.87	16.23	32.96%	8782	8428	1663	224,407	64,355 (22%)
UN-15	10.24	5.91	16.14	63.44%					

2.1.2. Hazards Impacts & Data

2.1.2.1. Nepal Earthquake

On 25 April 2015, a 7.8-magnitude earthquake struck approximately 80 km northwest of Kathmandu, the country's capital, at 6:11 UCT [39]. Official estimates state that the earthquake killed more than 8000 people, injured 21,000 more people, and displaced at least 2 million people in total [40]. Some 600,000 homes were destroyed, with another quarter million damaged [41]. The government estimated that reconstruction costs would surpass $7 billion, or a third of Nepal's GDP in the prior fiscal year [42]. Data and information on the earthquake was obtained from the US Geological Survey (USGS) [39]. USGS ShakeMap shapefiles were used to estimate earthquake impacts by "Instrumental Intensity", which is a proxy for Modified Mercalli Intensity (a qualitative index that can not strictly be determined by instruments).

2.1.2.2. Cyclone Idai

Cyclone Idai made landfall near Beira, Mozambique, on 14 March 2019. The storm had sustained wind >120 km/h. By March 16, the storm had tracked across Southern Mozambique into Zimbabwe. Flooding was observed throughout Malawi, Mozambique, and Zimbabwe, directly impacting 1.85 million people across MMZ [43]. The immediate financial requirement of the response was estimated to be nearly $300 million [43].

To measure maximum flood extent, we use a 90 m raster available from the World Food Program that captures maximum flood extent as of 21 March 2019 [44]. The raster is derived from Sentinel-1 data obtained from 12 to 21 March 2019 and ARC Flood Extent Depiction Model (AFED) detecting non-persistent water (7–16 March and 20 March 2019). We resampled the 90 m flood raster to 1 km and reprojected it to match GPW-15. Data on wind speeds were downloaded from the Global Disaster Alert and Coordination System [45], with shapefiles delineating zones impacted by 60, 90, 120 km/h wind speed thresholds.

Ecuador Flash Flood Susceptibility

In Ecuador, as well as on a global scale, flash floods are one of the most deadly types of flood with distinct spatiotemporal physical and impact-related characteristics [46–49]. Early warning systems exist for floods in many countries; however, they are rarely linked to resilience programming that can decrease risk of a flash flood disaster. While a long time series of impact data for flash floods (and any type of floods) does not openly exist in Ecuador [50], financial estimates of flood impact in the country can be acquired in some instances, with a reported US$238 million in flood impact in 2012 [51].

To represent the susceptibility for flash flooding at the catchment scale in Ecuador, a new vector dataset is derived from geophysical and non-geophysical data [52]. The susceptibility layer was built using a principal component analysis (PCA) derived weighted mean [53] of geographic indicators known to drive the flash flood potential of a catchment, related to geomorphology, drainage systems, and surface characteristics [54–56]. Geographic indicators such as slope, curvature, stream order, area of contributing sources, density of drainage, land cover, and sand content [57–60] were attributed to each catchment, using the predefined level 12 watershed units of HydroSHEDS [61], developed by the Conservation Science Program of World Wildlife Fund (WWF). The resulting flash flood susceptibility composite layer is normalized and reclassified into an equal count discrete flash flood susceptibility index from 1 to 10, low to high susceptibility, respectively, and represents the relative ranking of Ecuador catchments according to their increased susceptibility to generate flash flooding in the case of heavy rain.

Urban/Rural Data

To identify urban versus rural population estimates across the five gridded population products, we use an urban-rural binary land cover classification derived from MODIS data—the MODIS global urban extent product (MGUP) [62]. This dataset is available from 2002 to 2018 at 500 m spatial resolution. We resample the 2015 MGUP data to 1 km and projected it to match GPW-15. We employ MGUP as a relatively independent estimation of where urban settlements exist, as other MODIS products are an input in three of the five gridded population products (Table 1). In addition, we recognize that MGUP is one among many datasets that delineate urban from rural land cover globally and that binary urban/rural categorizations have well-known limitations [29]. As such, the MGUP urban/rural designation we employ is, to a degree, arbitrarily defined with an intent more on trying to better understand underlying population distribution methods, not a statement on what population is urban and what is rural.

2.2. Raster Processing & Analysis

The five global gridded population rasters and the MGUP urban/rural land cover raster were spatially co-registered (see Supplemental Information for the detail) and clipped using the GADM level 0 administrative units for Nepal, MMZ, and Ecuador (excluding islands). Across all five gridded population datasets, for the three study areas we map uninhabited pixels, calculate maximum and minimum population (as well as the range (maximum–minimum) at the pixel level), and measure urban and rural population estimates according to the MGUP urban/rural classification (Table 2). To identify outliers and highlight the variation in pixel-level estimates, we plotted pairwise pixel minimum population estimates against pixel ranges (Figure 2). We then examine 7 km by 61 km transects through three urban areas—Katmandu for Nepal, Beira for MMZ, and Quito for Ecuador—to visually and quantitatively demonstrate how the products' population estimates compare at the pixel level across the urban-rural continuum.

To compare estimates of populations impacted by the three hazards under study, we sum the populations by hazard criteria for the five gridded population products, separating urban and rural estimates. For the 2015 Earthquake in Nepal, we sum the population exposed by the USGS Shakemap "Instrumental Intensity" contour polygons. For Cyclone Idai, we sum the population exposed by wind speed buffer and flooded area.

Finally, for the flash flood in Ecuador we sum the population exposed by the susceptibility index layer.

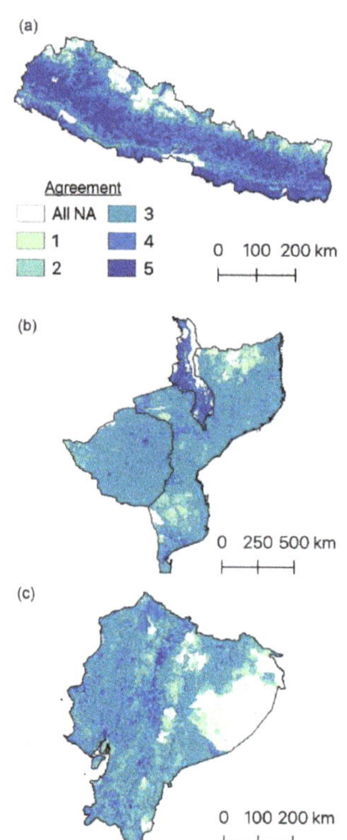

Figure 2. Spatial agreement of if a pixel is inhabited for five gridded population datasets for (**a**) Nepal, (**b**) MMZ, and (**c**) Ecuador. Values correspond to the number of rasters in agreement that a pixel is inhabited. White shows that all five gridded population datasets agree that a pixel is uninhabited. The higher agreement for Nepal (**a**) and Malawi (**b**) is a result of the higher number of administrative input units. Note that the spatial scale of each panel is different.

3. Results

3.1. Pixel-Level Comparisons

Four broad patterns emerged when comparing how the five gridded population datasets allocate populations in Nepal, MMZ, and Ecuador. First, we found widespread pixel-level variation in agreement across gridded population products of whether or not a given pixel is inhabited. Broadly, GHS-15 and WPE-16 identify a smaller proportion of inhabited pixels regardless of geography. We found that Nepal had the highest proportion of agreement, with all five gridded population products agreeing that 76% of pixels are either inhabited or uninhabited (Figure 2a). Only 67% for MMZ (Figure 2b) and 62% for Ecuador (Figure 2c) had full agreement by all five products. WP-16, LS-15, and GPW-15 tended to distribute population to a far greater number of pixels, unlike GHS-15 and WPE-16 (Table 2). For example, for MMZ, 85% of pixels in WPE-16 and 89% of pixels in GHS-15 were uninhabited. In contrast, WP-16, LS-15, and GPW-15 identified that only 11%, 14%, and 8% of pixels in MMZ are uninhabited, respectively.

Second, we documented extreme pixel-level population estimation disparities across gridded population products (Figure 3) and identified outliers. In pairwise comparison between the minimum pixel values with the range identified across all five gridded population products, for Nepal and MMZ, 27 rural pixels with minimums of 0–1000 people had ranges that exceed 50,000 people. Rural outliers in Ecuador do not have the same magnitude as Nepal or MMZ. Yet we still identified 8 rural pixels with minimum values of 0–1000 that have a range that exceeds 25,000 people in Ecuador. In the most extreme example, one pixel on the border of Nepal and India was estimated by GHS-15 to have nearly 120,000 residents (Figure 3a, Table 2). GPW-15 and WP-16 allocate 2392 and 730 people, respectively, to the same pixel, and the other two products identify fewer than 125 people. A visual inspection of high-resolution WorldView imagery from Google Earth reveals that the pixel mostly corresponds to a river with sand bars. For context, recently the United Nations Statistical Commission released standards identifying pixels within urban cores as having a population density of at least >1500 people per km^2 [63].

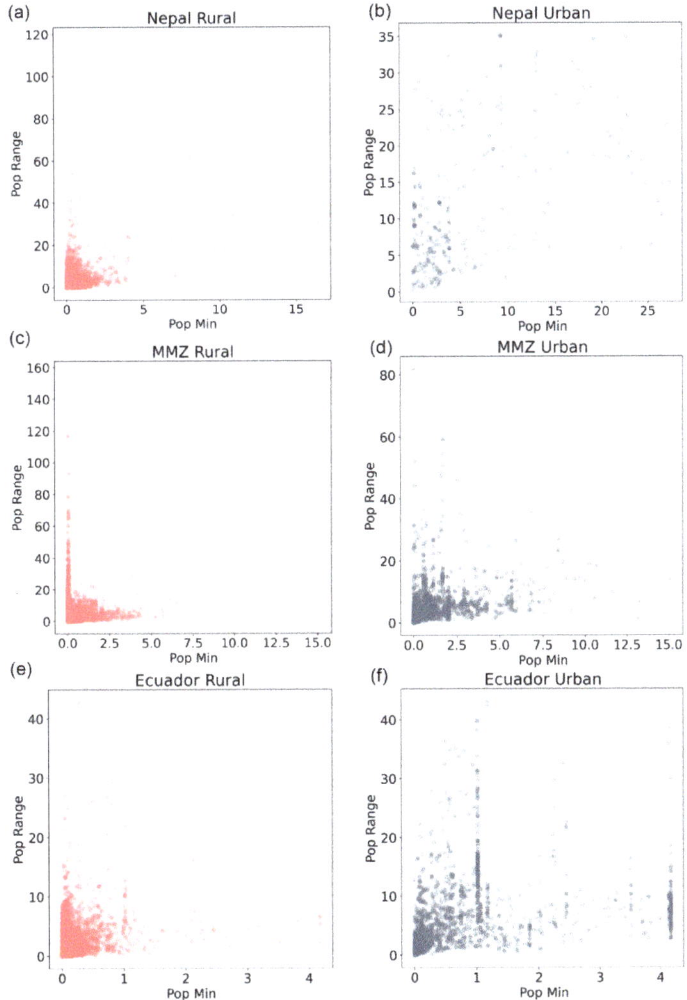

Figure 3. To identify outliers and pixel-level variation across gridded population products, the minimum population estimates (thousands) for each pixel are plotted against the population range (thousands), separated by rural and urban pixels, for Nepal (**a**,**b**), MMZ (**c**,**d**), and Ecuador (**e**,**f**). An example of an outlier is evident in panel (**a**) with one rural pixel having a range of more than 120,000 people.

Third, we found a clear pattern that WPE-16 total population estimates greatly exceeded the other four datasets. For instance, in MMZ, we found that WPE-16 exceeds the total population measured by the other gridded population datasets by 6–11 million people (Table 2), and exceeds UN population estimates for all three geographies of interest as well. As another example, while GHS-15 tended to prioritize allocating population to urban settlements compared to the other gridded population datasets, WPE-16 identified more urban residents in Nepal, by as much as 500,000 people, than the other four gridded population products.

Fourth, GHS-15 allocated a greater share of the total population to urban areas (as defined by the MGUP dataset) than the other four products for all three regions under study. For example, in Ecuador, GHS-15 estimated that 51% of the total population is urban. LS-15 was ranked second for Ecuador, allocating 47.9% of the total population to urban areas, followed by WPE-16 with 43.67% of the total population estimated as urban. In MMZ, GHS-15 again led in terms of the share of total population allocated to urban areas, again followed by LS-15. In Nepal, WP-16 and GPW-15, respectively, followed GHS-15 in terms of the share of total populations allocated to urban areas. Nonetheless, all five gridded population products underestimated the total urban population for all three geographies of interest when compared to 2018 United Nations World Urbanization Prospect estimates [36]. Indeed, even GHS-15 estimated nearly 50% fewer urban residents in MMZ than official UN counts.

Because the MGUP rural-urban binary designation does not capture how population density varies across that rural-urban continuum [29], Figure 4 illustrates how each gridded population product allocates population moving away from major urban centers. For the Kathmandu transect (Figure 4a), there is far closer agreement in populations among the products compared to Beira (Figure 4b) and Quito (Figure 4c). GPW-15, LS-15, and WP-16 capture less dense populations to the west of Beira (Figure 4b), as well as to the east of Quito (Figure 4c). In contrast, WPE-15 and GHS-15 do not capture these rural populations near Beira and Quito. This finding reinforces the preference of GHS-15 to allocate population to urban pixels.

Figure 5 presents the comparison of population estimates exposed to seismic intensity across the five gridded population products, stratified by MGUP-identified urban and rural settlements in Nepal. For the total population exposed to an intensity greater than seven, the difference between the highest and lowest populations estimated to be exposed by the products was more than 1 million people. WP-15 estimated the maximum number of people exposed at 9.85 million people, whereas GHS-15 finds 8.64 million people. The products furthermore measured a wide range in both the total number and the proportion of urban populations exposed to an intensity > 7. On the low end, WPE-16 categorized 22% (2.11 million people) of the population exposed to an intensity greater than 7 as urban, whereas, on the high end, WP-16 indentifed 33% (3.33 million people) of the total population exposed to an intensity greater than 7 as urban. As such, the elevated number of urban residents exposed according to WP-16 paralleled the previous finding that WP-16 identified more urban residents in Nepal compared to the other gridded population products (Table 2).

For intensities less than 7 (Figure 5), we found a broad range of population estimates across the gridded population products. For example, for an intensity between 5 and 6, WPE-16 measured more than 7.73 million people exposed, yet GHS-15 found only 6.46 million people exposed. Generally, the gridded products were largely in agreement for these lower intensities that the vast majority of people impacted lived in rural areas, although LS-15 still identified nearly 500,000 urban residents exposed to an intensity between 5 and 6, and 200,000 urban residents exposed to an intensity between 4 and 5.

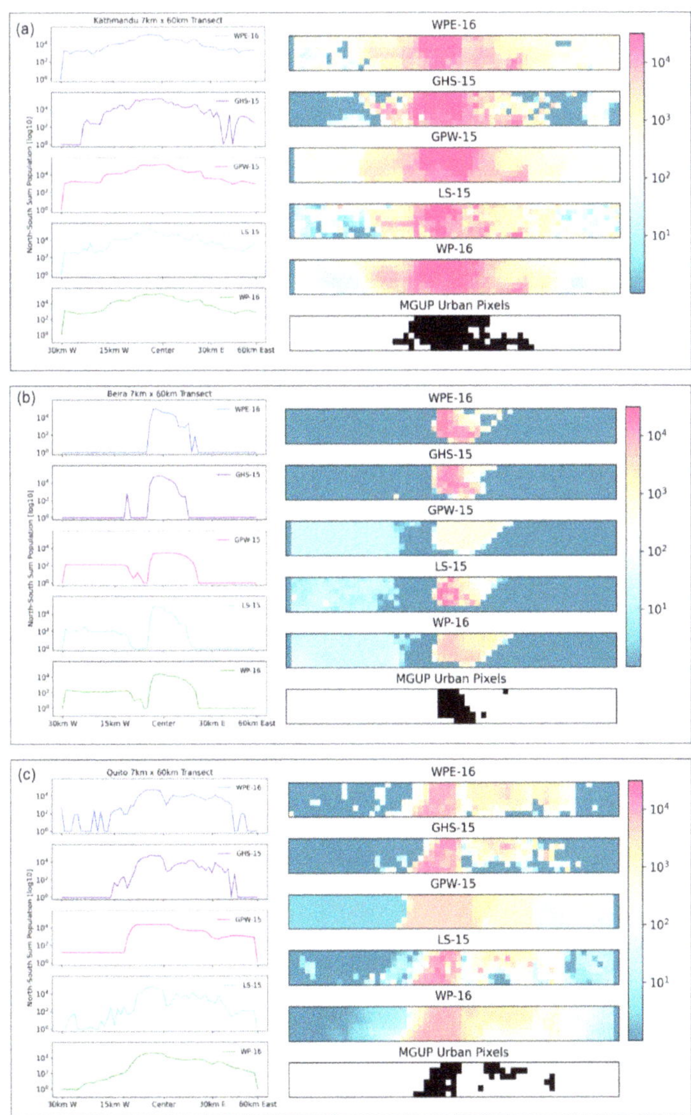

Figure 4. Transects (7 km by 61 km) of population estimates for each gridded population product centered on urban cores for (**a**) Kathmandu, (**b**) Beiria, Mozambique, and (**c**) Quito. MGUP urban/rural delineation is indicated for each pixel in the transect in the bottom of the plot of each panel. Note that the population sums presented in the line plots are the summation of 7 north-south pixels along the 61 km west-east transect. The color bar corresponds to log10 pixel-level population counts for the transects shown.

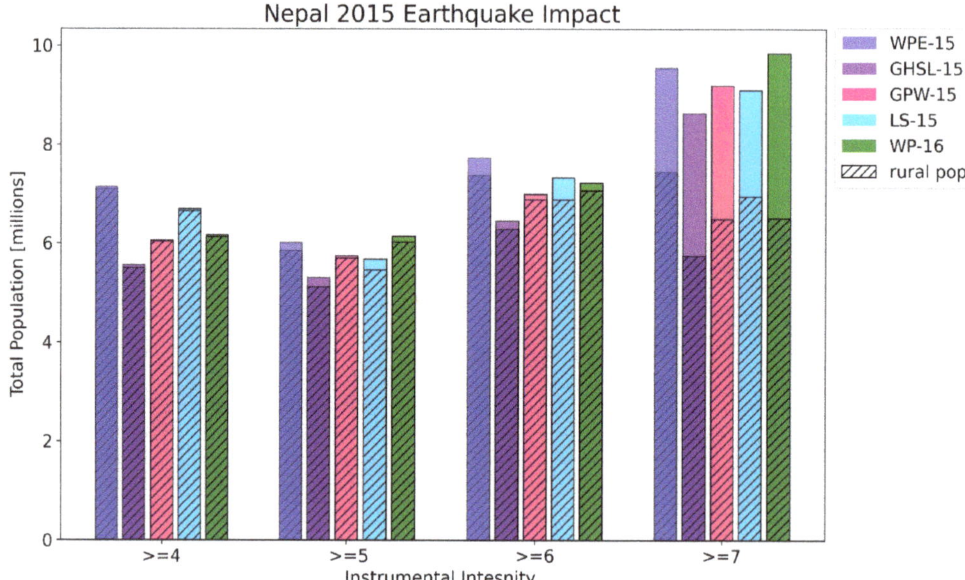

Figure 5. The total population impacted by the 2015 earthquake that struck Nepal estimates by "instrumental intensity" for five gridded population datasets. Hatching on bars signifies rural populations, with unhatched portions corresponding to MGUP-identified urban populations.

3.1.1. Cyclone Idai Exposure in MMZ

Generally, for Cyclone Idai, estimates of populations exposed to high wind speeds and living in flood-inundated areas in MMZ varied across the five gridded population data products, both in total agreement and divided by MGUP-identified urban and rural areas (Figure 6). Wind speeds of 60 km/h, though the least severe of wind categories, had the greatest variation. For instance, WPE-16 measured 7.41 million people (80% rural), the most impacted by wind speeds of 60 km/h. GPW-15 identified only 7.01 million people (88% rural) exposed. Estimates of populations exposed to wind speeds of 120 km/h and flood-inundated areas, the most damaging hazards, also showed substantial variation. Again, WPE-16 ranked first, with 2.39 million people (83% rural) exposed to wind speeds of 120 km/h. The other four gridded products had similar estimates of the total population exposed to wind speeds of 120 km/h, which ranged from 1.89 to 1.95 million people, though GHS-15 and LS-15 identify a greater proportion of urban populations impacted compared to GPW-15 and WP-16. Similarly, estimates of populations living in flood-inundated areas ranged from WPE-16 identifying 817,000 people (88% rural) to GPW-15 identifying 1.28 million people (99% rural). Only for wind speeds of 90 km/h were the products relatively consistent—rural populations were almost exclusively exposed, with high-end estimates of 1.56 million people by WPE-16 and a low-end estimate of 1.46 million people by WP-16.

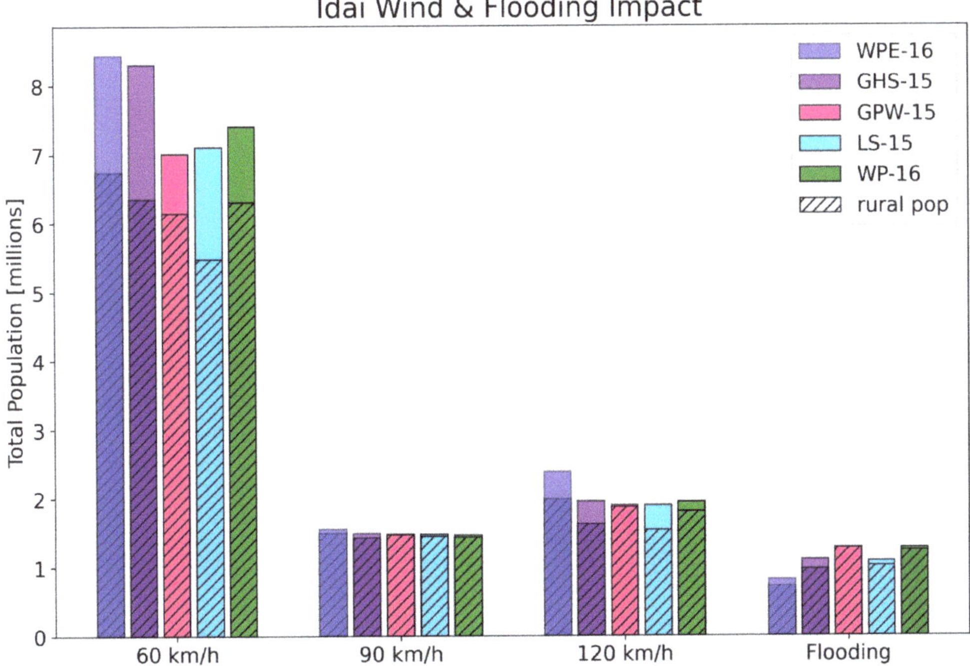

Figure 6. The total population impacted by wind speeds and flooding from Cyclone Idai across MMZ estimated by five gridded population datasets. Hatching on bars signifies rural populations, with unhatched portions corresponding to MGUP-identified urban populations.

3.1.2. Flash Flood Susceptibility in Ecuador

We found two main trends in the comparative analysis of population datasets and the estimation of the rural and urban share within each susceptibility decile (Figure 7). First, for most deciles, WPE-15 identified more population compared to the other four gridded population data products. For example, for the 10th decile, which was the most populated, WPE-15 estimated 2.96 million people, while on the low end, GPW-15 identified 2.47 million people. Second, across all susceptibility deciles GPW-15 estimated a greater share of rural population compared to the other four gridded population datasets. This is demonstrated by clear differences in rural/urban proportions in decile 3, whereby GPW-15 estimated almost all population as rural, compared to the under 50% estimation of rural population identified using the other products. GHS-15 and LS-15, on the other hand, allocated a greater share of population to urban areas. Again, using the 10th decile as an example, which indicates the areas with the highest likelihood of flash flood susceptibility, GHS-15 and LS-15 allocate 72% and 69% of the total population to urban areas, respectively. WPE-15 (62% urban) and WP-16 (56% urban) tend to fall between the preference of GHS-15 and LS-15 for urban areas and GPW-15's (32% urban) preference for rural areas.

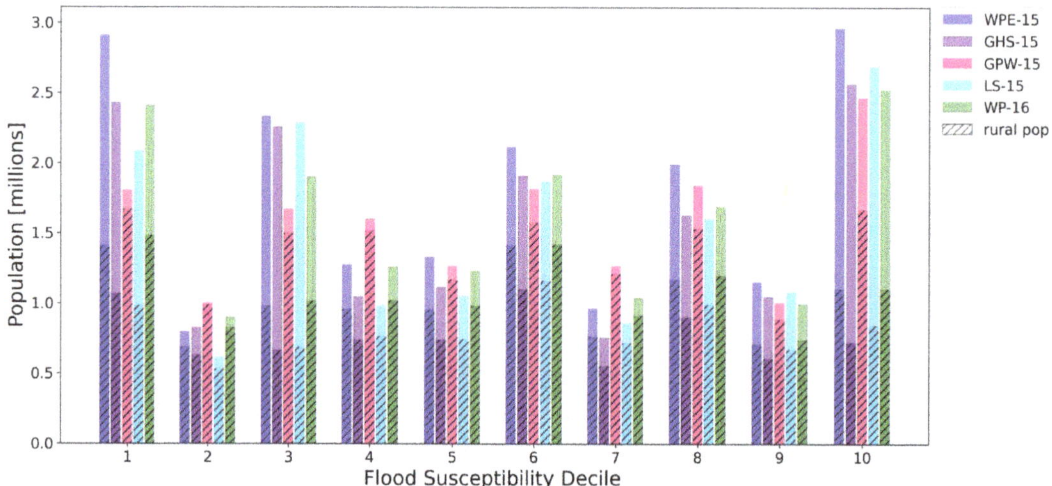

Figure 7. The total population susceptible to flash floods in Ecuador by susceptibility decile estimated derived from five gridded population datasets. Hatching on bars signifies rural populations, with unhatched portions corresponding to MGUP-identified urban populations.

4. Discussion

Gridded population data provide valuable population counts and densitity estimates for regions in the world where census data is lacking, coarse-scaled, or outdated [13]. The comparable and consistent means by which individual gridded data products are created ensures simple integration with other geospatial data products for use in measuring and monitoring various SDG indicators. However, using SDG 11.5 and a hazard context for three different geographies, we demonstrate that gridded population estimates can vary widely depending on the product of choice. Given the broad geographic contexts of our three case studies, our results suggest that gridded population products will similarly vary across many low- and middle-income countries (LMICs).

While variation in gridded population datasets has been documented by previous studies [21–24], many widely cited hazards studies (e.g., [64,65]), recent media narratives [66], and United Nations reports [67] continue to employ a single gridded population dataset without justification. In all of these cases, the authors neglect to acknowledge the wide variation in pixel-level population estimates. Indeed, we note that a recent review of estimates of population exposure to sea-level rise or living in low-elevation coastal zones [68] identified multiple global studies published since 2016 that use gridded population products. None of these studies employed multiple gridded population products in their risk estimates. Single-source population estimates, if framed in the context of hazard risk reduction related to SDG 11.5, are thus presented as facts to decision makers. This has broad implications for the allocation of scarce resources. If deployed in the immediate aftermath of a natural disaster, it could also affect humanitarian response and allocation of disaster relief.

Take two examples: first, we found that GHS-15 and LS-15 tend to allocate a greater share of population to the MODIS global urban extent product (MGUP)-designated urban areas compared to the other three products. While the MGUP rural-urban delineation is specific to the context of MODIS built-environment detection [62], and it is not the only criterion to identify urban settlement locations with EO data [25,29], this finding suggests that GHS-15 and LS-15 also prioritize allocating population to where the built environment is detected by Earth observation (EO)-derived spatial correlates. WP-16 also has been shown to prioritize built environment [69]; but our results indicate that WP-16 allocates population more evenly across land cover classes. Second, WPE-16 estimates a far greater population in Nepal, MMZ, and Ecuador compared to both UN estimates (Table 1) [1] and the other gridded population products. Should decision makers be presented with hazard risk reduction solutions or disaster preparedness scenarios based on GHS-15, they may implement policies that overly support urban populations. Communities not easily identified by the spectral signature of the EO-based data product would then be neglected. Similarly, relying only on WPE-16, in which population estimates are largely driven by leveraging the Landsat archive, may overestimate populations exposed or impacted by a hazard, leading to the over-allocation of resources, compared to policies developed with the other gridded population products. As such, efforts to monitor SDG indicators like SDG 11.5, which depend on detailed population data, can vary as a function of the gridded population product employed.

It is important to note that tracking any SDG indicator will also depend on the granularity of hazard data used in association with gridded population data. Indeed, in the context of SDG 11.5 the granularity of hazard data will affect estimates. For example, when we zoomed in on major urban centers from the three case studies presented (Figure 4, Figure 8), significant pixel-level population ranges are identified that may not actually affect the total population measured over larger geographic areas. The USGS earthquake instrument intensity data (Figure 8a) is at a much coarser granularity over Kathmandu than the EO-observed flood inundated area around Beira, Mozambique (Figure 8b), or the flash flood susceptibility layer for Quito (Figure 8c). Yet, unlike the more standardized analysis-ready hazard datasets we presented here, there is not a widely accepted set of criteria for deciding which gridded population dataset should be used to measure exposure to a given hazard.

While recent "fitness for use" guidelines provide key information for researchers, practitioners, and decision makers [5,20], our results suggest that the single use of a gridded population product should be avoided in tracking SDG indicators. These guidelines emphasize that spatial scale, reliability, and granularity of underlying census data, the population under study, and geography must be considered in data selection. Yet the variation across gridded population products we found for a range of hazards across geographic contexts signals that the single use of a product has realworld financial consequences. For instance, using our results from Cyclone Idai and per capita dollar basic disaster emergency costs of $112 per person [70], relief costs solely for those in flood inundated areas in MMZ range from US$92 million using the WPE-16 estimate of 817,000 people to US$143 million using GHS-15 estimate 1.28 million. Both estimates are below the official estimate of 1.85 million people impacted by Idai and the immediate financial estimate of US$300 million [43]. As such, we caution against the single use of a gridded population product. We reaffirm the need for the validation of existing products and urge future producers of gridded population products to provide error estimations.

Different gridded population modeling approaches and EO input data can result in varying population estimates per grid cell [20]. However, the finer the spatial resolution of input administrative units (Table 2) associated with population counts, the more similar the output population values per pixel will be across products regardless of the disaggregation approach. This is clear from Figure 4a, which shows much less variation in population estimates across the urban-rural gradient around Kathmandu. Generally speaking, finer administrative units tend to concentrate in higher population density areas. Administrative units that are larger, and coarser in spatial extent, tend to associate with lower population density areas, and are affected more by the different disaggregation approaches and their underlying assumptions. The result is greater variation in population estimates per pixel within the same administrative unit. We thus encourage national census agencies to release data with associated boundary files for the highest resolution units possible.

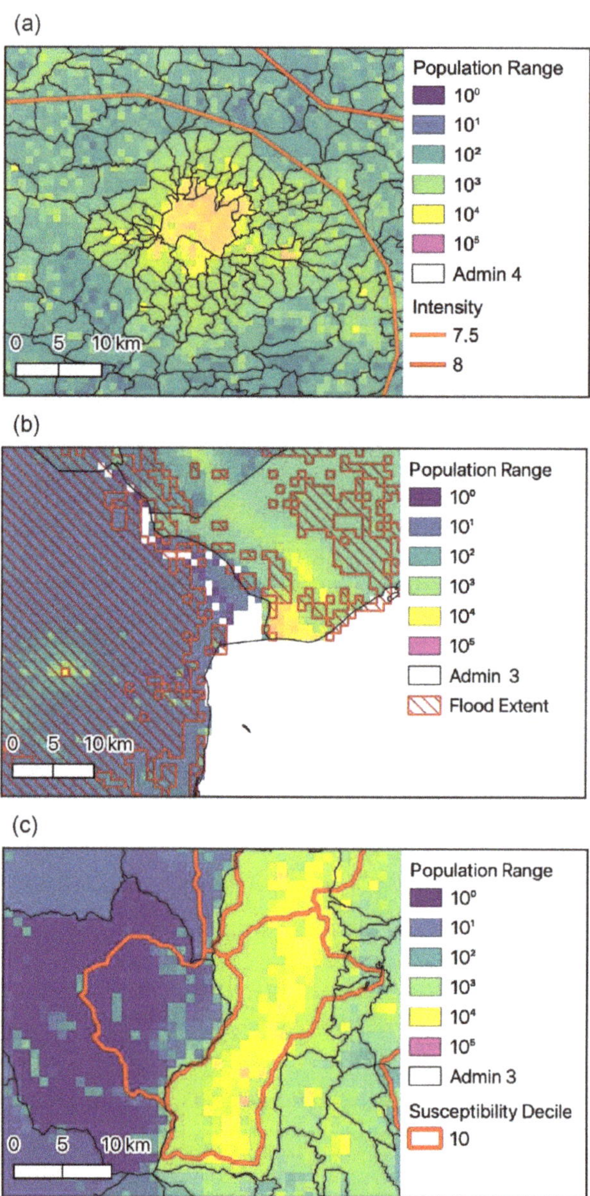

Figure 8. Pixel-level population range (maximum–minimum population estimated across gridded population products) overlaid with administrative units used for deriving gridded population datasets and the hazard under study. Panel (**a**) is situated over Kathmandu, with the earthquake instrument intensity contours shown. Panel (**b**) covers Quito with the highest flash flood sustainability area outlined in red. Panel (**c**) focuses on Beira, Mozambique and the mouth of the Pungwe River. Red hatching is the 90 m flood layer from Cyclone Idai resampled to 1 km.

Our results further highlight differences between constrained and unconstrained gridded populations products. Constrained approaches disaggregate population counts linked to administrative units only within pixels identified as "settled." In doing so, there may be a tendency to overestimate the number of people "distributed" within high-density (urban) settings. Our results indicate that this is the case with GHS-15. In contrast, the unconstrained approach will disaggregate population counts into any pixel identified as "land," which may overestimate the number of people allocated within low-density, i.e., more rural, settings. In the case of GPW-15, the overestimation in low-density (rural) settings is expected to be even more pronounced than in the other unconstrained products given that the disaggregated population totals are evenly redistributed within all pixels identified as "land" and there is no additional ancillary data used in the model.

While we recommend using multiple products in hazard analysis, if selecting a single gridded population product, it is important to identify not only if the gridded population product is constrained in some way, but also how it is constrained. Underlying data sources of constrained products may have validation or uncertainty estimates that may vary by region or time point. For input settlement products that represent the constrained "built" areas in which population counts are disaggregated, there is a general expectation that those gridded products will have a more accurate representation of population distribution. However, the final gridded population product will be greatly impacted by the presence of omission and commission errors in the input settlement dataset, with small/isolated rural settlements potentially being more difficult to detect and certain types of land cover (e.g., rock outcrops or sandy soils) potentially being misclassified as settlement areas [71].

On the other hand, unconstrained approaches use a dasymetric approach to disaggregate input administrative population values within all pixels identified as "land." This will introduce error in the final gridded output for those areas that are actually uninhabited. There will be some trade-off with a minimization of the presence of omission and commission errors in whatever input settlement data is used in the modeling due to the influence of other ancillary covariates representing factors correlated with population density and presence. There is error in all products, but some basic understanding of how the gridded data products are produced can help identify which product makes the most sense for a given application. Reed et al. (2018) [72] demonstrates the relative robustness of the unconstrained WorldPop dataset compared to the constrained HRSL dataset. The similarities in error metrics for these products emphasizes the importance of considering omission/commission errors in the input built datasets and subsequent allocation of population values in the final product.

Finally, it should be noted that most of these gridded population datasets represent a residential population, or where people are most likely to be when at home. LandScan is an exception and represents the ambient population, or the average population distribution over 24 h. That type of data product is potentially useful for assessing exposure as it represents not only the static census-based residence information but also the ambient nature of population movement over a 24 h period.

5. Conclusions

Vector-based administrative-level population data often fails to disaggregate population at the spatial scales requisite to identify where people actually live on the planet. This type of data can fail to provide useful information for the delivery of services required to achieve the SDGs. As such, our findings reinforce the many advantages of using gridded population products to track SDG indicators. This is especially important in the context of measuring, monitoring, and mapping SDG 11.5. Indeed, reducing exposure to hazards requires accurate population estimates, and for many LMICs, Earth observation-derived gridded population products are the best available data. Likewise, gridded population products can provide crucial information in post-disaster contexts as well. The case studies we showcase here reinforce the broad utility of these products and advance our understanding of "fitness for use" for both SDG 11.5 and the 73 SDG indicators that require accurate, comparable, and timely high-resolution population estimates.

Nonetheless, we highlight that for some geographical regions (and/or hazards), population estimates will vary depending on the choice of gridded population product.

Despite the variation we identify across gridded population datasets, we emphasize that uncertainty is not per se a limitation in employing gridded population datasets to track the SDGs. Without externally derived validation data or producer error metrics, it remains difficult to provide definitive recommendations in terms of what product to use, where and for what type of hazard. As such, we recommend further resources be dedicated to micro census data collection and encourage producers to quantify uncertainty in future gridded population products.

Most importantly, we recommend that researchers, practitioners, and decision makers acknowledge that inherent uncertainty when using these products. Along with leveraging the "fitness for use guidelines" [5,20], a key step to doing so is to perform a sensitivity analysis [23] and/or present a range of estimates using multiple gridded populations using an ensemble approach. To this end, we have provided the code used in the analysis and made available a global raster dataset that allows for the intercomparison of gridded population products. Furthermore, researchers and practitioners who develop tools for decision makers to track the SDGs should incorporate multiple gridded population datasets. Decision makers can thus develop policies and allocate resources informed by information that captures some of the inherent uncertainties of leveraging EO data to measure human populations across the planet. Indeed, single-use population estimates could open the door for bad actors to select the gridded population product that maximizes progress towards achieving an SDG. Lastly, we emphasize that the development of indicators, including the use of datasets like gridded population products, should be a collective effort between users and producers across decision-making levels [73].

Supplementary Materials: The following are available online at https://www.mdpi.com/article/10.3390/su13137329/s1.

Author Contributions: C.T., A.E.G. and A.d.S. developed the concept for the paper. C.T. wrote all code, performed all analysis, and produced all figures. C.T., A.E.G., A.S., A.d.S. drafted the initial text. A.B., C.H. (Carolynne Hultquist), and A.K. developed the Ecuador flood susceptibility layer. F.S. and G.Y. assisted with gridded population data selection, interpretation, and use. C.H. (Charles Huyck) consulted on 2015 Nepal earthquake data selection and interpretation. All authors contributed intellectually to the project throughout and all contributed to the final manuscript text. All authors have read and agreed to the published version of the manuscript.

Funding: C.T. was supported by the Earth Institute Postdoctoral Research Fellowship Program. A.B., C.H. (Carolynne Hultquist) and A.K. were supported by NASA Grant #80NSSC18K0342. A.S. was supported by funding from the Bill & Melinda Gates Foundation (OPP1134076 and 3(PG009268-05)). C.T., A.E.G., F.S., A.d.S. and G.Y. were supported by NASA Grant #80NSSC18K0328 for "Population and Infrastructure on Our Human Planet: Supporting Sustainable Development through Improved Spatial Data and Models for Human Settlements, Infrastructure and Population".

Institutional Review Board Statement: Not applicable.

Informed Consent Statement: Not applicable.

Data Availability Statement: All code and data are available at: https://github.com/cascadet/PopGridCompare (accessed on 20 January 2021).

Acknowledgments: We thank three anonamous reviewers for their valuable feedback.

Conflicts of Interest: The authors declare no conflict of interest.

References

1. UN-DESA. Sustainable Development Goals. Available online: https://sdgs.un.org/goals (accessed on 1 March 2021).
2. Sachs, J.D. From millennium development goals to sustainable development goals. *Lancet* **2012**, *379*, 2206–2211. [CrossRef]
3. Griggs, D.J.; Nilsson, M.; Stevance, A.; McCollum, D. (Eds.) *A Guide to SDG Interactions: From Science to Implementation*; International Council for Science (ICSU): Paris, France, 2017. Available online: https://council.science/wp-content/uploads/2017/05/SDGs-Guide-to-Interactions.pdf (accessed on 20 January 2021).
4. Attaran, A. An immeasurable crisis? A criticism of the millennium development goals and why they cannot be measured. *PLoS Med.* **2005**, *2*, e318. [CrossRef]

5. Thematic Research Network on Data and Statistics (TReNDS). Leaving No One Off the Map: A Guide for Gridded Population Data for Sustainable Development. 2020. Available online: https://static1.squarespace.com/static/5b4f63e14eddec374f416232/t/5eb2b65ec575060f0adb1feb/1588770424043/Leaving+no+one+off+the+map-4.pdf (accessed on 20 January 2021).

6. About the Global Partnership for Sustainable Development Data. Available online: https://www.data4sdgs.org/index.php/about-gpsdd (accessed on 1 March 2021).

7. Avtar, R.; Aggarwal, R.; Kharrazi, A.; Kumar, P.; Kurniawan, T.A. Utilizing geospatial information to implement SDGs and monitor their Progress. *Environ. Monit. Assess.* **2019**, *192*, 35. [CrossRef] [PubMed]

8. Kavvada, A.; Metternicht, G.; Kerblat, F.; Mudau, N.; Haldorson, M.; Laldaparsad, S.; Friedl, L.; Held, A.; Chuvieco, E. Towards delivering on the sustainable development goals using earth observations. *Remote Sens. Environ.* **2020**, *247*, 111930. [CrossRef]

9. Kussul, N.; Lavreniuk, M.; Kolotii, A.; Skakun, S.; Rakoid, O.; Shumilo, L. A workflow for sustainable development goals indicators assessment based on high-resolution satellite data. *Int. J. Digit. Earth* **2020**, *13*, 309–321. [CrossRef]

10. Estoque, R.C. A review of the sustainability concept and the state of SDG monitoring using remote sensing. *Remote Sens.* **2020**, *12*, 1770. [CrossRef]

11. Kavvada, A.; Held, A. Analysis-ready earth observation data and the united nations sustainable development goals. In Proceedings of the IGARSS 2018—2018 IEEE International Geoscience and Remote Sensing Symposium, Valencia, Spain, 22–27 July 2018; pp. 434–436.

12. De Lamo, X.; Simonson, W. Earth Observation for SDG: Compendium of Earth Observation Contributions to the SDG Targets and Indicators. ESA. May 2020. Available online: https://eo4society.esa.int/wp-content/uploads/2021/01/EO_Compendium-for-SDGs.pdf (accessed on 20 January 2021).

13. Wardrop, N.A.; Jochem, W.C.; Bird, T.J.; Chamberlain, H.R.; Clarke, D.; Kerr, D.; Bengtsson, L.; Juran, S.; Seaman, V.; Tatem, A.J. Spatially disaggregated population estimates in the absence of national population and housing census data. *Proc. Natl. Acad. Sci. USA* **2018**, *115*, 3529–3537. [CrossRef]

14. Juran, S.; Broer, P.N.; Klug, S.J.; Snow, R.C.; Okiro, E.; Ouma, P.O.; Snow, R.; Tatem, A.J.; Meara, J.G.; Alegana, V.A. Geospatial mapping of access to timely essential surgery in sub-Saharan Africa. *BMJ Glob. Health* **2018**, *3*, e000875. [CrossRef]

15. Dotse-Gborgbortsi, W.; Dwomoh, D.; Alegana, V.; Hill, A.; Tatem, A.J.; Wright, J. The influence of distance and quality on utilisation of birthing services at health facilities in Eastern Region, Ghana. *BMJ Glob. Health* **2019**, *4*, e002020. [CrossRef]

16. Palacios-Lopez, D.; Bachofer, F.; Esch, T.; Marconcini, M.; MacManus, K.; Sorichetta, A.; Zeidler, J.; Dech, S.; Tatem, A.; Reinartz, P. High-resolution gridded population datasets: Exploring the Capabilities of the world settlement footprint 2019 imperviousness layer for the African continent. *Remote Sens.* **2021**, *13*, 1142. [CrossRef]

17. Melchiorri, M.; Pesaresi, M.; Florczyk, A.J.; Corbane, C.; Kemper, T. Principles and applications of the global human settlement layer as baseline for the land use efficiency indicator—SDG 11.3.1. *ISPRS Int. J. Geo Inf.* **2019**, *8*, 96. [CrossRef]

18. Qiu, Y.; Zhao, X.; Fan, D.; Li, S. Geospatial disaggregation of population data in supporting SDG assessments: A case study from Deqing county, China. *ISPRS Int. J. Geo Inf.* **2019**, *8*, 356. [CrossRef]

19. Schiavina, M.; Melchiorri, M.; Corbane, C.; Florczyk, A.; Freire, S.; Pesaresi, M.; Kemper, T. Multi-scale estimation of land use efficiency (SDG 11.3.1) across 25 years using global open and free data. *Sustain. Sci. Pract. Policy* **2019**, *11*, 5674. [CrossRef]

20. Leyk, S.; Gaughan, A.E.; Adamo, S.B. The spatial allocation of population: A review of large-scale gridded population data products and their fitness for use. *Earth Syst. Monit.* **2019**. Available online: https://www.earth-syst-sci-data.net/11/1385/2019/essd-11-1385-2019-discussion.html (accessed on 20 January 2021). [CrossRef]

21. Archila Bustos, M.F.; Hall, O.; Niedomysl, T.; Ernstson, U. A pixel level evaluation of five multitemporal global gridded population datasets: A case study in Sweden, 1990–2015. *Popul. Environ.* **2020**, *42*, 255–277. [CrossRef]

22. Chen, R.; Yan, H.; Liu, F.; Du, W.; Yang, Y. Multiple global population datasets: Differences and spatial distribution characteristics. *ISPRS Int. J. Geo Inf.* **2020**, *9*, 637. [CrossRef]

23. Smith, A.; Bates, P.D.; Wing, O.; Sampson, C.; Quinn, N.; Neal, J. New estimates of flood exposure in developing countries using high-resolution population data. *Nat. Commun.* **2019**, *10*, 1814. [CrossRef] [PubMed]

24. Fries, B.F.; Guerra, C.A.; García, G.A.; Wu, S.L.; Smith, J.M.; Oyono, J.N.M.; Donfack, O.T.; Nfumu, J.O.O.; Hay, S.I.; Smith, D.L.; et al. Measuring the accuracy of gridded human population density surfaces: A case study in Bioko Island, Equatorial Guinea. *Cold Spring Harb. Lab.* **2020**. [CrossRef]

25. Tuholske, C.; Caylor, K.; Evans, T.; Avery, R. Variability in urban population distributions across Africa. *Environ. Res. Lett.* **2019**, *14*, 85009. [CrossRef]

26. Green, H.K.; Lysaght, O.; Saulnier, D.D.; Blanchard, K.; Humphrey, A.; Fakhruddin, B.; Murray, V. Challenges with disaster mortality data and measuring progress towards the implementation of the sendai framework. *Int. J. Disaster Risk Sci.* **2019**, *10*, 449–461. [CrossRef]

27. Mizutori, M. Reflections on the sendai framework for disaster risk reduction: Five years since its adoption. *Int. J. Disaster Risk Sci.* **2020**, *11*, 147–151. [CrossRef]

28. Dagys, K. The Effectiveness of Forecast-Based Humanitarian Assistance in Anticipation of Extreme Winters: Evidence from an Intervention for Vulnerable Herders in Mongolia. 2020. Available online: https://www.researchgate.net/profile/Kadirbyek_Dagys/publication/347885328_The_effectiveness_of_forecast-based_humanitarian_assistance_in_anticipation_of_extreme_winters_Evidence_from_an_intervention_for_vulnerable_herders_in_Mongolia/links/5fe570d0a6fdccdcb8fc06a2/The-effectiveness-of-forecast-based-humanitarian-assistance-in-anticipation-of-extreme-winters-Evidence-from-an-intervention-for-vulnerable-herders-in-Mongolia.pdf (accessed on 2 March 2021).

29. Cattaneo, A.; Nelson, A. Global mapping of urban–rural catchment areas reveals unequal access to services. *Proc. Natl. Acad. Sci. USA* **2021**. Available online: https://www.pnas.org/content/118/2/e2011990118.short (accessed on 1 June 2021). [CrossRef] [PubMed]

30. Frye, C.; Wright, D.J.; Nordstrand, E.; Terborgh, C.; Foust, J. Using classified and unclassified land cover data to estimate the footprint of human settlement. *Data Sci. J.* **2018**, *17*. Available online: https://datascience.codata.org/articles/10.5334/dsj-2018-020/?toggle_hypothesis=on (accessed on 20 January 2021). [CrossRef]

31. Florczyk, A.J.; Corbane, C.; Ehrlich, D.; Freire, S.; Kemper, T.; Maffenini, L.; Melchiorri, M.; Pesaresi, M.; Politis, P.; Schiavina, M.; et al. *GHSL Data Package 2019*; JRC Technical Report: Luxembourg, 2019. [CrossRef]

32. NASA Socioeconomic Data and Applications Center (SEDAC). *Gridded Population of the World, Version 4*; (GPWv4): Population Density Adjusted to Match 2015; NASA Socioeconomic Data and Applications Center (SEDAC): Palisades, NY, USA, 2018.

33. LandScan. LandScan Datasets, LandScanTM. Available online: https://landscan.ornl.gov/index.php/landscan-datasets (accessed on 24 February 2019).

34. Worldpop. WorldPop: Population. 18 November 2016. Available online: https://www.worldpop.org/doi/10.5258/SOTON/WP00004 (accessed on 24 February 2019).

35. POPGRID. Input Layers for Global Gridded Data Sets. 2019. Available online: https://www.popgrid.org/data-docs-table2 (accessed on 24 February 2019).

36. UN-DESA. World Urbanization Prospects 2018. Available online: https://esa.un.org/unpd/wup/ (accessed on 30 August 2018).

37. Stevens, F.R.; Gaughan, A.E.; Linard, C.; Tatem, A.J. Disaggregating census data for population mapping using random forests with remotely-sensed and ancillary data. *PLoS ONE* **2015**, *10*, e0107042. [CrossRef] [PubMed]

38. Dobson, J.E.; Bright, E.A.; Coleman, P.R.; Durfee, R.C.; Worley, B.A. LandScan: A global population database for estimating populations at risk. *Photogramm. Eng. Remote Sens.* **2000**, *66*, 849–857.

39. USGS. M 7.8–36 km E of Khudi, Nepal. In Earthquake Hazards Program. Available online: https://earthquake.usgs.gov/earthquakes/eventpage/us20002926/executive (accessed on 4 March 2021).

40. Hall, M.L.; Lee, A.C.K.; Cartwright, C.; Marahatta, S.; Karki, J.; Simkhada, P. The 2015 Nepal earthquake disaster: Lessons learned one year on. *Public Health* **2017**, *145*, 39–44. [CrossRef] [PubMed]

41. Reid, K. 2015 Nepal Earthquake: Facts, FAQs, and How to Help. World Vision. 2018. Available online: https://www.worldvision.org/disaster-relief-news-stories/2015-nepal-earthquake-facts (accessed on 4 March 2021).

42. Government of Nepal National Planning Commission, Kathmandu. Nepal Earthquake 2015 Post Disaster Needs Assessment Vol. B: Sector Reports. 2015. Available online: https://www.npc.gov.np/images/category/PDNA_volume_BFinalVersion.pdf (accessed on 20 January 2021).

43. Ocha, U.N. Humanitarian Response Plan 2018–2019 (Revised Following Cyclone Idai, March 2019); United Nations Office for the Coordination of Humanitarian Affairs Mozambique. Available online: https://reliefweb.int/sites/reliefweb.int/files/resources/ROSEA_20190325_MozambiqueFlashAppeal.pdf (accessed on 20 January 2021).

44. World Food Program. Mozambique Satellite Detected Waters, Cyclone Idai. 2019. Available online: https://data.humdata.org/dataset/mozambique-flood-detected-waters-cyclone-idai (accessed on 20 January 2021).

45. EC-JRC. Overall Red Tropical Cyclone Alert for IDAI-19 in Mozambique, Zimbabwe, Miscellaneous (French) Indian Ocean Islands from 09 March 2019 06:00 UTC to 15 March 2019 00:00 UTC; Global Disaster Alert and Coordination System. Available online: https://www.gdacs.org/resources.aspx?eventid=1000552&episodeid=24&eventtype=TC (accessed on 20 January 2021).

46. Ahern, M.; Kovats, R.S.; Wilkinson, P.; Few, R.; Matthies, F. Global health impacts of floods: Epidemiologic evidence. *Epidemiol. Rev.* **2005**, *27*, 36–46. [CrossRef] [PubMed]

47. Montz, B.E.; Gruntfest, E. Flash flood mitigation: Recommendations for research and applications. *Glob. Environ. Chang. Part B Environ. Hazards* **2002**, *4*, 15–22. [CrossRef]

48. Hallegatte, S.; Green, C.; Nicholls, R.J.; Corfee-Morlot, J. Future flood losses in major coastal cities. *Nat. Clim. Chang.* **2013**, *3*, 802–806. [CrossRef]

49. SENPLADES. Boletín Informativo Costos de las Pérdidas por las Inundaciones, mes de Julio. 2012. Available online: https://reliefweb.int/sites/reliefweb.int/files/resources/Costos%20de%20las%20perdidas%20por%20las%20inundaciones%202012.pdf (accessed on 20 January 2021).

50. Pinos, J.; Orellana, D.; Timbe, L. Assessment of microscale economic flood losses in urban and agricultural areas: Case study of the Santa Bárbara river, Ecuador. *Nat. Hazards* **2020**, *103*, 2323–2337. [CrossRef]

51. Informe a la Nación 2007–2017; Planifica Ecuador Secretaría Técnica; 2017. Report No.: Informe Nacional, 214. Available online: www.planificación.gob.ec (accessed on 20 January 2021).

52. Kruczkiewicz, A.; Bucherie, A.; Ayala, F.; Hultquist, C.; Vergara, H.; Mason, S.; Bazo, J.; de Sherbinin, A. Development of a flash flood confidence index from disaster reports and geophysical susceptibility. *Remote Sens.* **2021**, *13*, submitted.

53. Chao, Y.-S.; Wu, C.-J. Principal component-based weighted indices and a framework to evaluate indices: Results from the Medical Expenditure Panel Survey 1996 to 2011. *PLoS ONE* **2017**, *12*, e0183997. [CrossRef]

54. Janizadeh, S.; Avand, M.; Jaafari, A.; Van Phong, T.; Bayat, M.; Ahmadisharaf, E.; Prakash, I.; Pham, B.T.; Lee, S. Prediction success of machine learning methods for flash flood susceptibility mapping in the Tafresh watershed, Iran. *Sustain. Sci. Pract. Policy* **2019**, *11*, 5426. [CrossRef]

55. Pham, B.T.; Avand, M.; Janizadeh, S.; Van Phong, T.; Al-Ansari, N.; Ho, L.; Das, S.; Van Le, H.; Amini, A.; Bozchaloei, S.K.; et al. GIS based hybrid computational approaches for flash flood susceptibility assessment. *Water* **2020**, *12*, 683. [CrossRef]

56. Ullah, K.; Zhang, J. GIS-based flood hazard mapping using relative frequency ratio method: A case study of Panjkora river Basin, eastern Hindu Kush, Pakistan. *PLoS ONE* **2020**, *15*, e0229153. [CrossRef]

57. Azmeri Hadihardaja, I.K.; Vadiya, R. Identification of flash flood hazard zones in mountainous small watershed of Aceh Besar Regency, Aceh Province, Indonesia. *Egypt J. Remote Sens. Space Sci.* **2016**, *19*, 143–160.

58. Saharia, M.; Kirstetter, P.-E.; Vergara, H.; Gourley, J.J.; Hong, Y.; Giroud, M. Mapping flash flood severity in the United States. *J. Hydrometeorol.* **2017**, *18*, 397–411. [CrossRef]

59. Mahmood, S.; Rahman, A.-U. Flash flood susceptibility modelling using geomorphometric approach in the Ushairy basin, eastern Hindu Kush. *J. Earth Syst. Sci.* **2019**, *128*. [CrossRef]

60. Oruonye, E.D. Morphometry and flood in small drainage basin: Case study of Mayogwoi river basin in Jalingo, Taraba state Nigeria. *J. Geogr. Environ. Earth Sci. Int.* **2016**, *5*, 1–12. [CrossRef]

61. Lehner, B.; Verdin, K.; Jarvis, A. New global hydrography derived from spaceborne elevation data. *Eos* **2008**, *89*, 93. [CrossRef]

62. Huang, X.; Huang, J.; Wen, D.; Li, J. An updated MODIS global urban extent product (MGUP) from 2001 to 2018 based on an automated mapping approach. *Int. J. Appl. Earth Obs. Geoinf.* **2021**, *95*, 102255. [CrossRef]

63. Global Human Settlement—Global Definition of Cities, Urban and Rural Areas—European Commission. 6 July 2016. Available online: https://ghsl.jrc.ec.europa.eu/degurba.php (accessed on 30 March 2021).

64. Kulp, S.A.; Strauss, B.H. New elevation data triple estimates of global vulnerability to sea-level rise and coastal flooding. *Nat. Commun.* **2019**, *10*, 5752. [CrossRef]

65. Mora, C.; Dousset, B.; Caldwell, I.R.; Powell, F.E.; Geronimo, R.C.; Bielecki, C.R.; Counsell, C.W.W.; Dietrich, B.S.; Johnston, E.T.; Louis, L.; et al. Global risk of deadly heat. *Nat. Clim. Chang.* **2017**, *7*, 501. [CrossRef]

66. Lustgarten, A. The Great Climate Migration Has Begun; The New York Times. 16 December 2020. Available online: https://www.nytimes.com/interactive/2020/07/23/magazine/climate-migration.html (accessed on 6 March 2021).

67. Gu, D. Exposure and Vulnerability to Natural Disasters for World's Cities; UN DESA; 2019. Available online: https://www.un.org/en/development/desa/population/publications/pdf/technical/TP2019-4.pdf (accessed on 20 January 2021).

68. McMichael, C.; Dasgupta, S.; Ayeb-Karlsson, S.; Kelman, I. A review of estimating population exposure to sea-level rise and the relevance for migration. *Environ. Res. Lett.* **2020**, *15*, 123005. [CrossRef] [PubMed]

69. Nieves, J.J.; Stevens, F.R.; Gaughan, A.E.; Linard, C.; Sorichetta, A.; Hornby, G.; Patel, N.N.; Tatem, A.J. Examining the correlates and drivers of human population distributions across low- and middle-income countries. *J. R. Soc. Interface* **2017**, *14*. [CrossRef]

70. IFRC. The Cost of Doing Nothing: The Humanitarian Price of Climate Change and How It Can Be Avoided. 2019. Available online: https://media.ifrc.org/ifrc/wp-content/uploads/sites/5/2019/09/2019-IFRC-CODN-EN.pdf (accessed on 20 January 2021).

71. Stevens, F.R.; Gaughan, A.E.; Nieves, J.J.; King, A.; Sorichetta, A.; Linard, C.; Tatem, A.J. Comparisons of two global built area land cover datasets in methods to disaggregate human population in eleven countries from the global south. *Int. J. Digit. Earth* **2020**, *13*, 78–100. [CrossRef]

72. Reed, F.J.; Gaughan, A.E.; Stevens, F.R.; Yetman, G.; Sorichetta, A.; Tatem, A.J. Gridded population maps informed by different built settlement products. *Data* **2018**, *3*, 33. [CrossRef]

73. Bell, S.; Morse, S. Breaking through the glass ceiling: Who really cares about sustainability indicators? *Local Environ.* **2001**, *6*, 291–309. [CrossRef]

sustainability

Article

Assessing Urban Vulnerability to Flooding: A Framework to Measure Resilience Using Remote Sensing Approaches

Mercio Cerbaro [1,*], Stephen Morse [1], Richard Murphy [1], Sarah Middlemiss [2] and Dimitrios Michelakis [2]

[1] Centre for Environment & Sustainability, University of Surrey, Guildford GU2 7XH, UK; s.morse@surrey.ac.uk (S.M.); rj.murphy@surrey.ac.uk (R.M.)

[2] Ecometrica Limited, Orchard Brae House, 30 Queensferry Road, Edinburgh EH4 2HS, UK; sarah.middlemiss@ecometrica.com (S.M.); dimitrios.michelakis@ecometrica.com (D.M.)

* Correspondence: m.cerbaro@surrey.ac.uk

Abstract: Assessing and measuring urban vulnerability resilience is a challenging task if the right type of information is not readily available. In this context, remote sensing and Earth Observation (EO) approaches can help to monitor damages and local conditions before and after extreme weather events, such as flooding. Recently, the increasing availability of Google Street View (GSV) coverage offers additional potential ways to assess the vulnerability and resilience to such events. GSV is available at no cost, is easy to use, and is available for an increasing number of locations. This exploratory research focuses on the use of GSV and EO data to assess exposure, sensitivity, and adaptation to flooding in urban areas in the cities of Belem and Rio Branco in the Amazon region of Brazil. We present a Visual Indicator Framework for Resilience (VIFOR) to measure 45 indicators for these characteristics in 1 km² sample areas in poor and richer districts in the two cities. The aim was to assess critically the extent to which GSV-derived information could be reliable in measuring the proposed indicators and how this new methodology could be used to measure vulnerability and resilience where official census data and statistics are not readily available. Our results show that variation in vulnerability and resilience between the rich and poor areas in both cities could be demonstrated through calibration of the chosen indicators using GSV-derived data, suggesting that this is a useful, complementary and cost-effective addition to census data and/or recent high resolution EO data. Furthermore, the GSV-linked approach used here may assist users who lack the technical skills to process raw EO data into usable information. The ready availability of insights on the vulnerability and resilience of diverse urban areas by straightforward remote sensing methods such as those developed here with GSV can provide valuable evidence for decisions on critical infrastructure investments in areas with low capacity to cope with flooding.

Keywords: vulnerability; flooding; remote sensing; Earth Observation (EO); Google Street View (GSV); climate change

Citation: Cerbaro, M.; Morse, S.; Murphy, R.; Middlemiss, S.; Michelakis, D. Assessing Urban Vulnerability to Flooding: A Framework to Measure Resilience Using Remote Sensing Approaches. *Sustainability* **2022**, *14*, 2276. https://doi.org/10.3390/su14042276

Academic Editor: Michalis Diakakis

Received: 21 December 2021
Accepted: 11 February 2022
Published: 17 February 2022

Publisher's Note: MDPI stays neutral with regard to jurisdictional claims in published maps and institutional affiliations.

1. Introduction

Floods are one of the most common and severe hazards to disrupt people's livelihoods globally [1]. The effects of climate change and widespread flooding can exacerbate urban challenges and make it more difficult to tackle issues and help vulnerable communities in informal settlements [2]. The Intergovernmental Panel on Climate Change (IPCC) outlines that climate-related risks for natural and human systems are higher for global warming of 1.5 °C than at present, but risks depend on the magnitude and rate of warming, levels of development and vulnerability, and on the choices of adaptation [2]. Given that the world will further urbanize during the next decade, from 56.2% in 2020 to about 60.4% by 2030 [3], these vulnerabilities are likely to intensify. Climate change can cause events such as flooding with higher frequency, intensity, and variability, affecting urban areas where density of housing is high and widespread [4,5]. Given that urban areas are expanding, along with the proportion of a population living and working within them, there is a growing need for assessing their vulnerability to disasters such as flooding. Indeed, there have been various international initiatives to address issues of uneven development and vulnerabilities within urban areas, and the New Urban Agenda (NUA) of the United Nations Human Settlement Programme (UN-Habitat) adopted in Quito, Ecuador, in October 2016 is one example of a core commitment for a transformative agenda in urban areas [4].

Implementation of the New Urban Agenda and promoting actions on urban-related Sustainable Development Goals (SDGs), such as the development of land use policies for climate resilience and adaptation to climate change, housing and slum upgrading policies, and preparation of existing institutions for disasters will require significant mobilization of financial resources to improve infrastructure and services [4]. Making cities and human settlements inclusive, safe, resilient, and sustainable (SDG11), and enhancing urban resilience in cities with rapid population growth, informal settlements, unplanned public services, and extreme income inequality requires several coordinated efforts [4,5]. The World Cities Report 2020 reaffirms that unplanned urban living leaves people vulnerable, and the COVID-19 pandemic has exposed deep inequalities which suggest that tackling the virus is more challenging in urban areas [3].

Vulnerability involves an individual or group's exposure to, capacity to cope with, and potentiality to recover from crises [5–7], and there are several international programmes and frameworks designed to assess vulnerability to disasters. One example is the Disaster Recovery Framework (DRF) of the World Bank's Global Facility for Disaster Reduction and Recovery's (GDDRR), and another is the UN Sendai Framework for Disaster Risk Reduction 2015–2030 [5]. The Sendai framework includes a set of global targets and indicators, but these operate mainly at the national level policies rather than being focused on local strategies [6,7]. Another example of a specific framework to assess vulnerability is provided by the Notre Dame Global Adaptation Index (ND-GAIN); a free open-source index that shows a country's current vulnerability to climate disruptions [8]. The aim of the index is to support the private and public sector in prioritizing climate adaptation [9]. The details of the ND-GAIN methodology are not covered here, but it assesses the vulnerability of a country by considering six life-supporting sectors: food, water, health, ecosystem services, human habitat, and infrastructure [10,11]. The index is based on over 74 variables, which are used to create 45 indicators spanning critical environmental, economic, and social aspects designed to measure vulnerability and readiness of 192 UN countries [11]. The inclusion of social aspects towards assessing vulnerability is included in all these frameworks and is important [12].

According to the United Nations Office for Disaster Risk Reduction (UNDRR), resilience of vulnerable communities is associated with various factors, including poverty and inequality [12]. De Almeida et al. [13] demonstrated that counties in the Amazon region face serious conditions of susceptibility to natural hazards (e.g., floods, landslides, flash floods, droughts) and this is magnified by high levels of socioeconomic inequality. In addition, the Amazon region has a very low capacity to recover and adapt to future environmental and social scenarios because of climate change [13]. All these frameworks require good quality information, and the same type of information is often required irrespective of the type of disaster (e.g., storms, floods, landslides, etc.) [8]. The need for information to enhance preparedness and plans to mitigate the impact of disasters in communities is an important dimension and examples of such initiatives include The Use of Social Work Interventions to Address Climate and Disaster Risk [12]. Traditional approaches to data collection have relied upon availability of data collected via surveys, census, tax returns, etc., but some of these can be time-consuming and expensive.

Another approach that has generated much interest is the use of Earth Observation (EO) via satellites, aircraft, and drones. However, a third approach that has been gaining some prominence in the literature is the use of tools such as Google 'Street View' (GSV) [14]. GSV images are available for several cities, as are open EO data, such as Sentinel 1- radar that supply day and night all-weather EO data [15], or Sentinel 2- optical EO data available every five days under cloud free conditions [16]. GSV has been used to develop different conceptual frameworks and methodologies for different purposes. Wang et al. [17] developed a new machine learning method based on GSV to assess the quality of green spaces in Guangzhou, China. They also examined exposure and access to green spaces associated with socioeconomic inequalities in urban areas, and how neighbourhoods with high socioeconomic status may have better access to quantity and quality of green spaces. Feldmeyer et al. [18] used Open Street Map (OSM) and machine learning to generate socioeconomic indicators where the availability of quality data is limited at specific temporal and spatial resolution. GSV and deep learning have been used to measure the relationships between a Green View Index (GVI) and walking behaviour [19]. Li et al. [20] assessed urban greenery using GSV for monitoring and measuring street greenery that people can see on the ground in different streets. Other examples include virtual tree surveys [21], the use of GSV to identify the elements that affect the probability that individual buildings may suffer flooding in urban areas [22], and the assessment of damage after hurricanes [14]. Hence, GSV has been used for several applications, and proves to be a useful tool for virtual field observation, especially when combined with EO. Giuliani et al. [23] used a combination of free Earth Observation (EO) and crowdsourced (e.g., OSM) EO data to model physical accessibility to urban green spaces in four European cities (Geneva, Barcelona, Goteborg, and Bristol). Unlike other studies [17–22], Giuliani et al. [23] incorporated the use of EO data but measured only one specific indicator associated with green areas.

Using GSV to populate indicators has advantages in terms of cost, but it also has limitations. For example, GSV imagery may not necessarily be up-to-date; in addition, the indicators that can be populated with the tool have to be amenable to a visual assessment (e.g., quantity and quality of infrastructure). EO can also be used in conjunction with GSV, but it too is limited to addressing indicators that can be passively 'seen' using visual wavelengths or the use of 'active' EO such as the use of radar. Image resolution can also be an important factor in EO. Hence, the assessment of urban greenspace has often been a focus with GSV- and EO-based systems in urban spaces, but there are no published examples as yet of using these tools to assess vulnerability to disasters such as flooding. The research set out in this paper addressed that gap in knowledge and sought to develop and apply an alternative method to assess vulnerability in urban area primarily using GSV along with a framework for assessing vulnerability based on the Notre Dame Global Adaptation Initiative (ND-GAIN).

The ND-GAIN approach defines adaptation as 'adjustment to the changing climate that minimize negative impacts on humans and on built and natural systems', and this involves both a mitigation of risk along with an exploration of opportunities [24]. In order to assess vulnerability to climatic hazard events such as flooding, the ND-GAIN framework uses three dimensions:

- Exposure: The size of the population and critical infrastructure (e.g., transport links, health care facilities) which may potentially be exposed to a climatic hazard event.
- Sensitivity: The extent to which a population or infrastructure may be affected by a climatic hazard event. This could be influenced by many factors such as the quality of construction of key infrastructure.
- Adaptive capacity: The ability to respond to the consequences of climate hazards, for example, the presence of emergency services or the ability to bring in support from outside the area affected.

A series of indicators may be chosen or developed to assess the three dimensions, and these can be populated in part using existing datasets such as census data [25] as well as via primary data. The logic at the heart of the ND-GAIN framework is that vulnerability to hazardous events is influenced by all three of these and is maximized when all three are weak. There are nuances, however. For example, a population may have a high level of exposure to flooding events because of geography, but the sensitivity to damage may be low, perhaps because of physical mitigations employed in construction of housing and infrastructure. Similarly, exposure and sensitivity may both be high, but there are good adaptations in place to ensure that affected populations are supported quickly and effectively. The latter could, for example, include early warning systems coupled with rapid evacuation systems. Hence, an appreciation of vulnerability needs to draw upon an understanding of all three of these dimensions. The ND-GAIN approach includes a fourth dimension—readiness—which in essence is the ability to enhance adaptive capacity and includes ability to attract funding as well as the ability of governance structures to make the best use of that funding. There may be various barriers at play that influence these, and one example is linked to the effectiveness of communicative tools to addressing key municipal barriers to climate adaptation [26]. One of the key findings of research based in the Netherlands is that barriers experienced by municipalities are lack of urgency, lack of knowledge of risk and measures, and actions by authorities, which are limiting their adaptation planning and implementation [26]. Enabling factors for 'readiness' can include communication and transportation infrastructure, local laws and regulations, and community-based behaviours during flooding [27].

The research reported here focuses on assessing the vulnerability of relatively rich and poor areas in the cities of Rio Branco (state of Acre) and Belem (state of Para), Brazil, to flood events. The approach is a visual interpretation and does not include machine learning or algorithms to process large numbers of images. The research followed several steps, and these are set out in the methodology section of the paper. Firstly, two cities were selected to explore resilience to flooding, and the choice of Rio Branco and Belem was based upon a series of preliminary experiences and field work visits in Brazil. Within each city, two areas that were especially prone to flooding were selected, one of them regarded as being a poorer area while another was seen as being a comparatively wealthier area. Once cities and areas were chosen and demarcated on maps, GSV was used to populate an indicator framework based upon the ND-GAIN approach set out above. The indicator values for the two cities and areas within the cities were then analysed and conclusions drawn about vulnerability to flooding, the utility of the approach taken, and its potential for further development.

2. Material and Methods

2.1. Study Background—Rio Branco (Acre State) and Belem (Para State) in Brazil

The locations in Brazil of the two cities chosen for this research (Rio Branco and Belem) are shown in Figure 1. Both cities are located within the relatively high-rainfall Amazon region.

The two cities were recommended by experts based at the National Institute of Space Research of Brazil (INPE), The Brazilian Agriculture Research Corporation (EMBRAPA), the National Centre for Disasters and Alerts and Monitoring (CEMADEN), and the Geological Service of Brazil (CPRM), which are responsible for mapping flood hazard risk areas in vulnerable areas in Brazil. All the experts consulted noted the importance and timeliness of the research, as flooding is one of the most severe natural disasters that affects livelihoods in several regions of Brazil. The specific locations within the two cities were based on suggestions provided by experts from CPRM; the cities and locations were known to have a frequency of flooding of at least two or three events in the last ten years. The Brazilian Disaster Risk Indicators (DRIB-Index) proposal by Almeida et al. [28] was also consulted to select the best locations in the Amazon region. The DRIB-Index served as a tool to help assess 32 indicators that include different levels of exposure, vulnerability, and risks in Brazil [28]. DRIB aims to capture and measure four major components: exposure to natural disasters, susceptibility of the exposed communities, coping capacities, and adaptive capacities [28]. The DRIB-Index showed a high level of vulnerability and low capacity to cope and adapt to socioenvironmental changes imposed by disasters and climate changes for several cities in the North (Amazon) and Northeast regions [2] of Brazil. Indeed, some places in the Amazon region were identified by the index as being greatly exposed to multiple hazards such as landslides, floods, flash floods, and droughts [29–31]. The Amazon region also has a very high social-vulnerability index based on indicators associated with employment, social dependency, race and ethnicity, availability of quality sanitation, and housing structures, amongst others [32,33]. However, the DRIB-Index does not use visual tools at the local level to evaluate specific socioeconomic features.

Figure 1. Study area showing two cities in the Amazon region (grey shading) in Brazil: Rio Branco (Acre State) and Belem (Para State).

The city of Rio Branco, the capital of the state of Acre in southwestern Amazon (Figure 1), is an example of an area of almost annually recurrent extreme events (e.g., floods, droughts, and forest fires) [34]. Since 1988, the city has been flooded (River Acre (the main river running through the city) levels exceeding 14.0 m [34]) several times, with 2015 being the most severe in recent history when the River Acre flooded for 32 consecutive days and reached 18.4 m in March that year. The 2015 flood affected 100,000 people, or about one-third of the city's population [34]. Rio Branco's population was estimated at 413,418 in 2020 as compared with 336,038 in the last 2010 census [35]. Rio Branco has 57% of adequate sewage, and 20% urbanization of public roads, which shows some limitations in infrastructure in additional to population growth in the past decade [35]. Belem has an estimated population of about 149,964 in 2020 compared with 139,300 in the 2010 census [36]. In addition to the high population density, Belem has 68% of adequate sewage, and 36% urbanization of public roads [36].

The first author visited both locations for two weeks each in February and March 2020. In this period, the aim was to collect local data from local authorities (e.g., civil defence, local council, urban planning agencies, local environmental authorities). However, access to data emerged as a major challenge, despite visits to over ten local institutions in each location, including federal institutions such as EMBRAPA, federal universities, and the Brazilian Geological Survey (CPRM), and local institutions such as the fire brigade, civil defence, and environmental agency authorities. We assumed that talking directly to personnel in both cities would facilitate data collection. However, it was often mentioned by senior researchers in different federal institutions (e.g., CPRM, EMBRAPA, University of Para) that data availability and access to local institutions are major challenges for local researchers, and there are various reasons for this. For example, it was often mentioned by informants that federal, state, and local authorities are from different political parties and interest in climate change mitigation was at the time not part of the political agenda. Thus, access to data that shows poverty, vulnerability, exposure of minorities within flooding areas was limited or appeared not to be in the interest of local authorities to provide. This issue may have been exacerbated at the time of the fieldwork by political debates and sensitivities regarding forthcoming elections for mayoral posts. It should also be noted that, at the time of fieldwork (and also in 2021), IBGE had postponed the demographic census of 2020 due to the ongoing COVID-19 pandemic, and so the last set of available census data were from 2010 [37].

These factors and the challenging circumstances illustrate the difficulties for many researchers and authorities in accessing 'conventional', timely, and relevant data to assess local vulnerability and resilience. This reinforced the aim of this research to seek alternative, open, remote sensing approaches to assess these characteristics in the local context. Following the approach taken in the ND-GAIN framework summarized above, where vulnerability to events such as flooding is assessed via an appreciation of exposure, sensitivity and adaptation, the aspects chosen for remote sensing evaluation were designed to explore aspects such as the likelihood of floods generating material and nonmaterial losses (e.g., destruction of public and private infrastructure, reduction of accessibility to various locations, disruption of traffic flow, families losing their homes or access to dwellings) [37].

2.2. Choice of Areas in the Two Cities

In this study, two 1 km^2 areas within each city that were known to flood regularly (Figures 2 and 3) were identified. Within each city, one area was selected as being 'poor' and one that was 'rich' on the advice of experts at CPRM based on several technical geological surveys. We used several criteria for selection (e.g., average income, house prices, number of banks, shopping malls) as indicators of poor and richer areas. However, it should be noted here that these are relative rather than absolute categorizations. Other factors such as the number of hospitals, schools, and churches were considered as indicative of the adaptive capacity to provide accommodation, psychological assistance, and shelter available locally to support dislodged people [33]. The following neighbourhoods were chosen for the research:

1. Rio Branco: Cidade Nova and Preventorio—poor area;
2. Rio Branco: City Center, Base and Seis de Agosto—richer area;
3. Belem: Terra Firme—poor area;
4. Belem: Umarizal—richer area.

The suggestions above were confirmed by advice from EMPRABA, local universities, civil defence, and other local institutions. Figure 2 gives an EO image for the poor area in Rio Branco (A), which is subject to both flooding and landslides due to unsustainable land use occupation near the river, and the richer area (B) with high levels of infrastructure (e.g., banks, government buildings, pharmacies) and also proximity to river.

The areas were demarcated using Google Earth Pro (GEP) and measured using its toolbox. The blue lines in Figures 2 and 3 are the streets available on GSV and the user can zoom in to access the street option view, or swipe up and down on the blue line. Users can move the 'yellow man' pegman and can switch between Google Earth (GE) using EO data from unique sources of Google Street View. EO imagery built into GE for every specific location provides the date, geographical coordinates, and elevation at the point of observation. In Google Earth, complete 360 degrees is available at the street level, and the user can move readily up and down the street (Figure 3). Both long and short streets were considered as single units of analysis. As the purpose of this study was to identify the overall vulnerability of a particular area, we used imagery available for throughout the period 2012 to 2020.

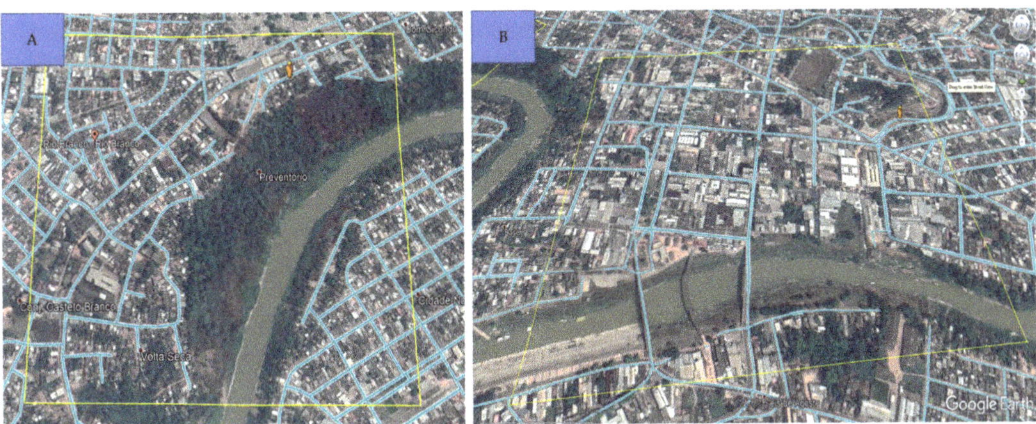

Figure 2. Google Earth EO image (27/05/2020) of selected poor (**A**) and richer areas (**B**) in Rio Branco, Acre State, 2021.

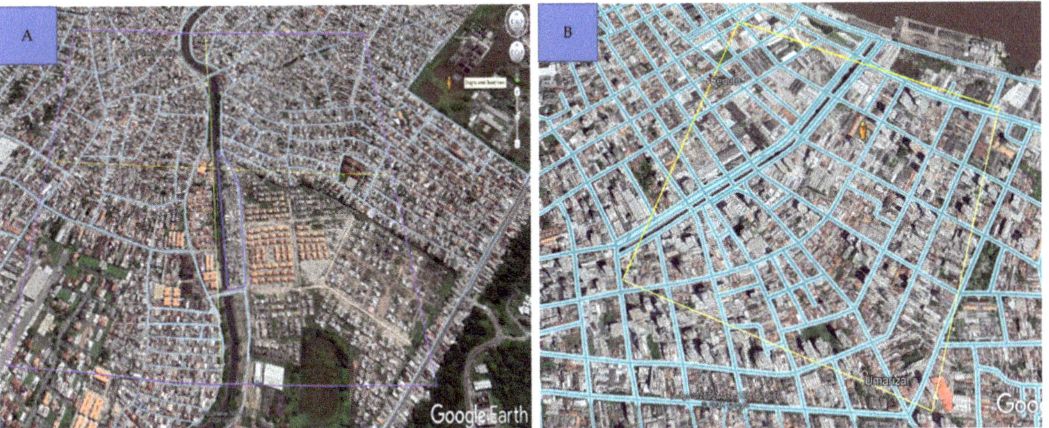

Figure 3. Google Earth EO image (27/05/2020) of selected poor (**A**) and richer areas (**B**) in Belem, Para State, 2021.

2.3. Indicator Framework

The indicator framework for assessing vulnerability and resilience followed the logic set out in the ND-GAIN framework outlined above [10]. Vulnerability to flooding was assumed to have three dimensions:

1. Exposure (E—factors that influence exposure to flooding);
2. Sensitivity (S—sensitivity to flooding events);
3. Adaptation (A—adaptations made to limit incidence or damage from flooding events).

The indicators and variables were then selected based on relevance to flooding and exposure of residents to floods in the selected areas. The initial identification included critical infrastructure and places to support vulnerable communities during flooding or post-flooding (e.g., shelters, churches, sport facilities), or urban infrastructure that could support affected neighbourhoods during a flooding event. We considered an area at high levels of vulnerability if it had few assets available (e.g., weak access to hospitals, schools, poor roads, rubbish in the streets, other sensitivity indicators based on socioeconomic criteria).

We began with the list of 232 indicators of the SDGs, those of the Sendai Framework, and other indicators available in the literature. These were then selected by the authors for those associated with flooding and natural hazards. From these, those that were considered potentially amenable to visual assessment via GSV or EO were further selected. The indicators were also classified in terms of the five major capitals (or assets) of the Sustainable Livelihood Approach:

1. Natural (N);
2. Human (H);
3. Social (S);
4. Financial (F);
5. Physical (P).

Hence, indicators labelled as 'EN' come under the exposure dimension and were considered as 'natural' capitals. The result of this selection process was a framework of 45 indicators we termed the Visual Indicator Framework for Resilience (VFIOR), as set out in Table 1. The framework comprises a mix of indicators assessed by scoring and counting, and lower values (scores and counts) equate to a low resilience or capacity to adapt to flood hazards. Each of the three dimensions (Exposure, Sensitivity, and Adaptation) had differing numbers of indicators and it was not necessarily the case that all five capitals were represented within each dimension. The exposure and adaptation dimensions have indicators that span all five of the capitals, but in the sensitivity dimension the indicators only spanned the human, financial, and physical capitals.

Table 1. Indicators used to measure VIFOR framework.

No.	Indicators Used to Measure Vulnerability	Type of Data Collected	Assumption
	Exposure (factors that influence exposure to flooding)		
1	* EN1: Presence and scale of waterways (extent/size)	Score	More waterways and greater extent can cause greater probability of flooding
2	EH1: Proximity of hospitals to waterways	Score	Proximity leads to greater probability of flood damage
3	EH2: Proximity of clinics to waterways	Score	
4	EH3: Proximity of pharmacies to waterways	Score	
5	EH4: Proximity of schools to waterways	Score	
6	EH5: Proximity of houses near to waterways	Score	
7	EH6: Proximity of business to waterways	Score	
8	EH7: Proportion of business near to ground level	Score	Being near ground level equals a higher chance of flood damage. Some level of elevation above ground.
9	ES1: Proximity of places of worship to waterways	Score	Proximity leads to greater probability of flood damage
10	EF1: Proximity of banks to waterways	Score	
11	EP1: Proportion of business at direct street level	Score	Being at ground level means a higher chance of flood damage. Direct at the street or pavement without elevation.
12	EP2: Proportion of housing at direct street level	Score	

Table 1. *Cont.*

No.	Indicators Used to Measure Vulnerability	Type of Data Collected	Assumption
13	* EP3: Quality of road surface	Score	Indicator of flood damage but also is important in terms of access for emergency vehicles and people wishing to leave
14	EP4 General state of repair of buildings	Score	Indicator of flood damage but also poverty
15	EP5: Presence of soil erosion	Score	Indicator of flood damage
16	EP6: Rubbish in the streets	Score	Indicator of poverty, but also more rubbish leads to a greater chance of drainage systems being blocked
Sensitivity (sensitivity to flooding events)			
17	SH1: Density of housing/construction in the flooding areas	Score	Number of houses per street. Density and intersection with mix land uses.
18	SH2: Sturdiness of dwellings	Score	Quality of materials of houses and buildings.
19	SF1: Cleanliness of streets	Score	Rubbish in the streets
20	SF2: Presence of graffiti	Score	Number of graffiti in public spaces
21	SF3: Unoccupied/boarded-up buildings	Score	Empty buildings or facilities without use
22	SF4: Incidence of decaying buildings	Score	Buildings without use
23	SF5: Value of cars parked on streets	Score	New and expensive cars to damaged and old cars.
24	* SP1: Type of road surface	Score	High quality asphalt to various low-quality pavement layers.
25	* SP2: Overall road width	Score	Size of road width and road markings used, including those across the carriageway.
26	SP3: Quality of pavements/sidewalks to roadsides	Score	Allows cars and people to move during flooding events. Quality of sidewalks, which allows effective mobility of pedestrians.
27	SP4: Presence of water on the streets	Score	Poor drainage that can lead to flooding.
Adaptation (to flooding)			
28	* AN1: Proportion of green areas and vegetation	Proportion (%)	Green spaces help with drainage
29	AH1: Number of hospitals (public and private)	Count	Assets to support health care during a flood event
30	AH2: Number of clinics	Count	
31	AH3: Number of pharmacies	Count	
32	AH4: Number of schools	Count	School premises can provide spaces for people to gather in the event of a flood and be supported.
33	AH5: Number of community public health centre	Count	Supports health care in the event of a flood
34	AH6: Number of universities (public and private)	Count	Physical premises can provide spaces for people to gather in the event of a flood and be supported
35	AS1: Number places of worship	Count	
36	AS2: Number of sport halls	Count	
37	AS3: Number of police stations	Count	Provide support/security during a flood event
38	AS4: Number of fire stations	Count	

Table 1. *Cont.*

No.	Indicators Used to Measure Vulnerability	Type of Data Collected	Assumption
39	AF1: Number of banks (access to local financial support)	Count	Access to cash and financial services
40	AF2: Number of cooperative associations	Count	Supports with information, shelter and community-based cooperation to support socioeconomic recovery
41	* AP1: Presence of flood defences (strengthened banks etc.)	Count	Provide defence for rivers do not burst their banks
42	AP2: Quality of flood defences	Score	Natural and artificial strengthened banks to prevent flooding
43	AP3: Quality of street drainage systems?	Score	Surface water drainage systems to prevent flooding
44	* AP4: Number or bridges (bridges for vehicles)	Count	Provide access and mobility in flooding areas
45	AP5: Durability of bridges (to flooding)	Score	Type of material from small scale wood bridges to large scale concrete bridges. Resilience of infrastructure to avoid bridge collapse during flooding.

Note: Indicators marked with an asterisk (*) are potentially measurable via Earth Observation.

In the Exposure dimension (16 indicators) many of the indicators refer to distance of the community assets (hospitals, clinics, schools, etc.) from a source of flooding. The assumption here is that the farther away an asset is from the source of flooding, then the lower the likelihood is that it would be exposed to damage as a result of flooding. Further included here are indicators that assess whether an asset is at street level or raised (e.g., on the second floor or higher of a multitier building). The Sensitivity dimension has a total of 11 indicators, most of which address the exposure of assets to damage once flooding occurs. The assumptions here are that buildings, roads, and sidewalks that are of poor quality suggest that they may be or have been readily damaged by flood events and/or that they may also be sensitive to further degradation. To some extent these are also indicators of wealth and lack of investment. One of the indicators specifically assessed whether there is standing water on the street, which would suggest a lack of drainage (this depends on when the GSV images were taken). For the Adaptation dimension (18 indicators, the largest number of indicators of the three dimensions), the indicators are mostly counts of important assets such as the number of hospitals (public and private), schools, pharmacies, universities (public and private), clinics, places of worship, sports halls, number of police stations, and number of banks. These are important in terms of health care provision during and after a flooding event, but also of the ability to improve human capital (education), social capital (cooperatives), and financial capital (banks). Further included here are counts of assets such as emergency service stations (police, fire stations) and flood defences (e.g., barriers) and drainage systems. Assessing adaptation in terms of physical entities that can be observed via GSV or EO is admittedly a narrower perspective than envisaged in the ND-GAIN framework. In ND-GAIN, adaptation (or more precisely, adaptive capacity) is seen as an ability to manage flood events and includes the presence of early warning systems and management plans. However, in the VIFOR framework, adaptation was limited to an assessment of structures such as number of facilities (e.g., AS1 to AS2), or specific flood defence systems such as the quality of flood defences (AP2) or the quality of street drainage systems (AP3). There is a relationship here in the sense that the presence of structures such as medical facilities and police and fire stations are important for helping deliver any plans that authorities may have in place, although counting such facilities says nothing about their quality in terms of number of staff available, their training and preparedness, and availability of required equipment. Therefore, a thorough assessment of adaptation would need to go beyond what can be seen and this would necessitate interviews with stakeholders at all levels, including the local community, although there can be constraints here as noted above and especially during a pandemic.

All three dimensions of VIFOR overlap to some extent and there are indicators that could be moved to a different category. For example, under Sensitivity, the indicators that capture building quality (SH2, SF3, and SF4) could also be classified under Adaptation; one of the responses that people could make to improve resilience would be to strengthen their dwellings. Similarly, the quality of the road surface (EP3) in the Exposure dimension could be placed under Sensitivity. It is included under Exposure as it is assumed that a poor-quality road indicates greater damage resulting from flooding, but that can also be applied to the type of road surface (SP1), as paved roads could decay as the result of successive flood events. For this reason, it was decided to weigh all the indicators equally and not apply weights to the three dimensions. Thus, in a sense it does not really matter where an indicator appears in the framework, it is still weighted equally, and the classification is only for ease of use.

It should be noted that some indicators in the VIFOR framework have the potential for assessment via EO. These are demarcated in Table 1 with an asterisk. The presence of waterways (EN1), the proportion of green areas (AN1), flood defences (AP1), and bridges (AP4) are obvious examples, as indeed are indicators of road width (SP2) and road quality (EP3) and material of construction (SP1). There are other indicators that may not be so obviously assessable via EO, such as the number of banks (AF1) and cooperative associations (AF2). These institutions often occupy spaces within larger buildings and provide no visual clues readily tractable by EO to their use.

2.4. Data Collection and Analysis

The indicators in Table 1 were populated via a combination of counts and scoring. For all the indicators, all the streets in the 1 km^2 quadrant (lines in Figures 2 and 3) were 'walked' virtually in GSV. We focused on 1 km^2 due to logistical issues and time constraints to cover all the streets within the boundaries of the two cities. Hence, we selected two areas (rich and poor) for each city based on the recommendations of experts along with the criteria noted above. The number of streets 'walked' in each of the areas is shown in Table 2. Care was taken to ensure that a single asset existing on a junction between streets was not recorded more than once. However, while the streets differed in terms of length, this was not accounted for in the assessment and long streets were treated the same as short streets. Hence, each indicator was not weighted for variation in the length of the street. For those indicators assessed via scores (mainly Exposure and Sensitivity), the scores ranged from 1 to 5, with the polarity set to 1 representing low resilience and 5 representing high resilience (to flood events) and the midpoint score of 3 representing a moderate level of resilience 'performance' for the indicator. For count-type scores (mainly in the Adaptation dimension) these are 'raw' data and no midpoint assessment of a level of resilience is given except for the fact that a higher number is indicative of greater resilience and average values of zero or small fractious per street suggest low resilience.

Table 2. Number of streets assessed for the indicators in Table 1.

City	Area	Number of Streets
Rio Branco	Poor	42
	Rich	29
Belem	Poor	45
	Rich	20

The construction patterns of dwellings are often not designed to withstand flooding, especially in regions with significant inequalities [38]. In Figure 4, we used GSV to classify houses and structures in terms of their likelihood of flooding as described in the following categories: very high, high, high, medium, low, and very low. We assumed that very high (score 5) is the most adaptable structure in the sense of the building and its contents being best able to avoid flood damage (often located in richer areas), and very low (score 1) is the least adaptable structure to cope with a flooding event (i.e., building and contents damage is most likely). In the low and very low scores, we identify houses principally situated at street level, limited paved roads, and high concentration of houses. In high and very high categories, houses are located in buildings situated above street level that are less exposed to floods. EP2 refers to the proportion of housing at street level, or likely to flood during an event. A similar scoring system was used to describe the general state of repair of buildings (Figure 5) in the following categories: very low, low, medium, high, and very high. This refers to the quality of the houses, type of materials, and construction. For example, in Figure 5, scores 4 and 5 represent good quality construction indicating resilient infrastructure when compared with scores 1 and 2. A further example of the scoring system is that for SP3 (Figure 6), which refers to the quality of pavements and sidewalks.

Score 1

Score 2

Score 3

Score 4

Score 5

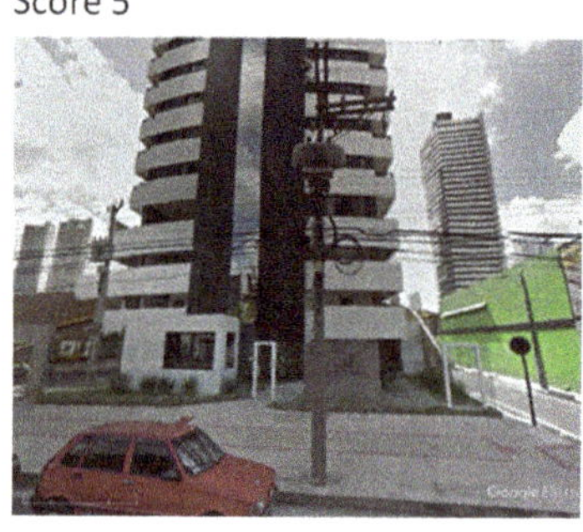

Figure 4. Example of the scoring system showing proportion of houses and structures likely to flood. GSV images from Rio Branco and Belem. EP2: Proportion of housing at direct street level (likely to flood), Score 1 = Very high (80–100%), Score 2 = High (60–80%), Score 3 = Medium (40–60%), Score 4 = Low (20–40%), Score 5 = Very low (0–20%).

Figure 5. Examples of the scoring system showing the general state of repair from buildings. GSV images from Rio Branco and Belem. EP4: General state of repair of buildings, Score 1 = Very low, Score 2 = Low, Score 3 = Medium, Score 4 = High, Score 5 = Very high.

Figure 6. Examples of the scoring system showing quality of pavements/sidewalks. GSV images from Rio Branco and Belem. SP3: Score 1 = Very poor, Score 2 = Poor, Score 3 = Medium, Score 4 = High, Score 5 = Very high.

3. Results

The means and standard deviations (SD) for the indicators as assessed using GSV observation data for the selected streets/city areas are shown in Table 3. Further shown are the number of streets (N) for which the indicator could be assessed. The latter is important; for example, the indicators which assess proximity to a source of flooding were not relevant in some cases where the unit (hospital, clinic, etc.) did not exist within the 1 km^2 grid. There is insufficient space here to describe all the indicators in Table 3, so only a few examples are highlighted. For the two poor areas there are issues with houses and businesses more likely to be located near ground level. The presence and scale of waterways in the poor area of Belem (EN1 average score: 4.11) is higher compared with that of Rio Branco (EN1 average score: 3.61). Indeed, in Belem, many of the businesses and houses are located near canals that are often related to poor technical maintenance of river channels, and disposal of rubbish in the canals also contributes to flooding. In Rio Branco, some houses in poor areas located near the Acre River are built with two floors, while in many of the poorer areas in Belem located near the canals, houses have only one floor. The poor area in Belem has a worse drainage system (AP3 average score 1.62) when compared with the poor area of Rio Branco (AP3 average score: 2.6), which relates to the quality of drainage as an indicator for Adaptation during flooding events. The density of housing is higher in the Belem poor area (SH1 average score: 2) when compared with the Rio Branco poor area (SH1 average score: 2.86). The streets in the poorer area of Rio Branco are cleaner (SF1 average score: 2.76) when compared to Belem (SF1 average score: 2.38), and there is more graffiti in the poorer area of Belem (SF2 average score: 3.27) when compared to Rio Branco (SF2 average score: 3.88). The quality of pavements and sidewalks in Belem (SP3 average score: 1.58) was also lower than found in the poor area of Rio Branco (SP3 average score: 2.33).

Table 3. VIFOR framework scores for the 45 indicators based on GSV observation of streets in the selected areas of Rio Branco and Belem.

| | Rio Branco | | | | | | Belem | | | | | |
| | Poor Area | | | Rich Area | | | Poor Area | | | Rich Area | | |
Indicator	N	Mean	SD	N	Mean	SD	N	Mean	SD	N	Mean	SD
					(a) Exposure dimension							
EN1	42	3.62	0.85	29	3.41	0.98	45	4.11	0.91	20	4.25	1.33
EH1	0	-	-	0	-	-	0	-	-	6	2.17	1.33
EH2	1	2	-	11	2.45	0.82	3	1	0	10	2.4	0.84
EH3	1	3	-	12	2.08	0.67	0	-	-	7	1.14	0.69
EH4	3	3	1	10	2	0.47	3	1.33	0.58	6	2.83	0.75
EH5	41	2.54	1.12	29	2.59	1.12	43	1.33	0.57	19	2.21	0.85
EH6	39	2.69	1.06	29	2.41	1.21	42	1.31	0.6	17	2.12	0.93
EH7	39	2.15	0.74	29	2.34	1.04	43	1.12	0.45	18	2.33	0.77
ES1	13	2	1	9	1.67	1.32	28	1.25	0.52	3	1.67	1.15
EF1	2	3	0	14	2	0.96	0	-	-	5	1.4	0.89
EP1	40	2.25	0.78	29	2.1	0.86	44	1.2	0.46	18	2.44	0.92
EP2	42	2.26	0.8	29	2.17	0.76	45	1.36	0.53	19	3.11	1.52
EP3	42	2.74	0.8	29	3.45	0.69	45	1.96	0.67	20	3.85	0.93
EP4	42	2.71	0.46	29	3.55	0.83	45	2	0.64	20	3.85	1.04
EP5	42	3.1	0.98	29	4	0.8	45	4.07	0.89	20	5	0
EP6	42	3.05	0.62	29	3.45	1.21	45	2.96	0.77	20	3.9	0.64
					(b) Sensitivity dimension							
SH1	42	2.86	0.65	29	3.31	0.66	45	2	0.48	20	3.45	1
SH2	42	2.71	0.64	29	3.38	0.56	45	2.22	0.56	20	3.55	1.19
SF1	42	2.76	0.53	29	3.24	0.87	45	2.38	0.53	20	3.4	1.23
SF2	42	3.88	0.67	29	4.1	1.01	45	3.27	0.86	20	3.7	1.26
SF3	42	3.6	0.63	29	4.14	0.99	45	3.33	0.77	20	4.1	1.07
SF4	42	3.4	0.8	29	3.86	1.13	45	2.38	0.68	20	4.2	0.83
SF5	42	2.76	0.53	29	3.24	0.64	45	2.24	0.68	20	3.6	0.82

Table 3. *Cont.*

| | Rio Branco | | | | | | Belem | | | | | |
| | Poor Area | | | Rich Area | | | Poor Area | | | Rich Area | | |
Indicator	N	Mean	SD	N	Mean	SD	N	Mean	SD	N	Mean	SD
SP1	42	2.79	0.72	29	3.31	0.89	45	2.11	0.49	20	3.7	1.03
SP2	42	2.6	0.63	29	3.38	0.86	45	1.47	0.55	20	3.4	1.14
SP3	42	2.33	0.82	29	3.34	1.14	45	1.58	0.75	20	3.6	1.27
SP4	42	3.81	0.8	29	4	0.85	45	3.67	1	20	4.25	0.97
(c) Adaptation dimension												
AN1	42	2.1	0.66	29	2.28	1.03	45	1.11	0.49	20	2.1	0.97
AH1	42	0	0	29	0	0	45	0	0	20	0.4	0.75
AH2	42	0.02	0.15	29	0.41	0.73	45	0.04	0.21	20	0.65	0.81
AH3	42	0.1	0.48	29	0.41	0.82	45	0	0	20	0.9	2.1
AH4	42	0.07	0.26	29	0.48	0.74	45	0.07	0.25	20	0.5	0.89
AH5	42	0.07	0.26	29	0.07	0.26	45	0	0	19	0.05	0.23
AH6	42	0	0	29	0.03	0.19	45	0	0	20	0.2	0.41
AS1	42	0.43	0.74	29	0.28	0.45	45	1.02	1.23	19	0.21	0.42
AS2	42	0.02	0.15	29	0.24	0.58	45	0.04	0.21	20	0.3	0.57
AS3	42	0	0	29	0.14	0.35	45	0	0	20	0.05	0.22
AS4	42	0	0	29	0	0	45	0	0	20	0	0
AF1	42	0.05	0.22	28	0.79	1.42	45	0	0	20	0.4	0.94
AF2	42	0.02	0.15	28	0.36	0.56	45	0	0	20	0	0
AP1	38	0.03	0.16	29	0.21	0.41	32	0.03	0.18	13	0.38	0.51
AP2	4	0.75	1.5	6	3.33	0.82	10	0.9	0.57	9	4	0.87
AP3	42	2.6	0.77	29	3.07	0.8	45	1.62	0.65	20	3.4	1.05
AP4	21	0.05	0.22	29	0.07	0.26	33	0.33	0.74	11	0.55	0.52
AP5	3	1	1.73	2	5	0	8	1.25	1.16	6	4	0

Note: Means and standard deviation (SD) are based on the streets that were assessed for the indicators. N is the number of streets where the indicator could be assessed.

For the two richer areas, the one in Belem is less exposed to waterways (EN1 average score: 4.25) when compared to the richer area in Rio Branco (EN1 average score: 3.41). The general state of repair of buildings, which includes the quality of the materials and condition of buildings, is similar for the richer areas of Belem (EP4 average score: 3.85) and Rio Branco (EP4 average score: 3.55). Richer areas in Belem include large numbers of high-rise buildings, which may only flood at street level (access points to buildings), but flooding is unlikely to cause damage more widely to the building. The proximity of businesses to waterways in the richer area of Belem (EH6 average score: 2.12) is similar to that for Rio Branco (EH6 average score: 2.41) and suggests that businesses in both areas are relatively close to local waterways. The quality of pavements/sidewalks to roadsides in the richer areas of Belem (SP3 average score: 3.6) and Rio Branco (SP3 average score: 3.34), and the quality of street drainage in the rich of Belem (AP3 average score: 3.4) and in Rio Branco (AP3 average score 3.07), suggest a medium level of vulnerability and the indicators certainly suggest better levels of resilience compared to the poorer areas of the two cities.

The distribution of scores for the Exposure and Sensitivity dimensions to resilience, the two that were assessed entirely via scoring, are shown in Figure 7. The results suggest that the distribution of scores was towards the higher end of the five-point scale for the richer areas (Figure 7b,d) compared with the poorer ones (Figure 7a,c). Indeed, the poorer area of Belem (Figure 7c) had scores that were especially on the low side. Hence, for the Exposure and Sensitivity dimensions, there is a sense here that poorer areas have a lower resilience compared with richer areas. To allow for a direct comparison of resilience across all three dimensions (Exposure, Sensitivity, and Adaptation), the mean score or count per street was calculated for the 45 indicators and the results are presented in Figure 8 as radar (or amoeba) diagrams. Higher values for any of the indicators suggest a greater resilience. The overall pictures of resilience presented in Figure 8 suggest that there is a clear division between Exposure and Sensitivity between poor and richer areas. However, the poor area

Rio Branco (Figure 8a) would appear to have a higher level of Exposure when compared with the poor area in Belem (Figure 8c). Interestingly, the Adaptation dimension is quite low for all four areas but is especially low for the two poor areas.

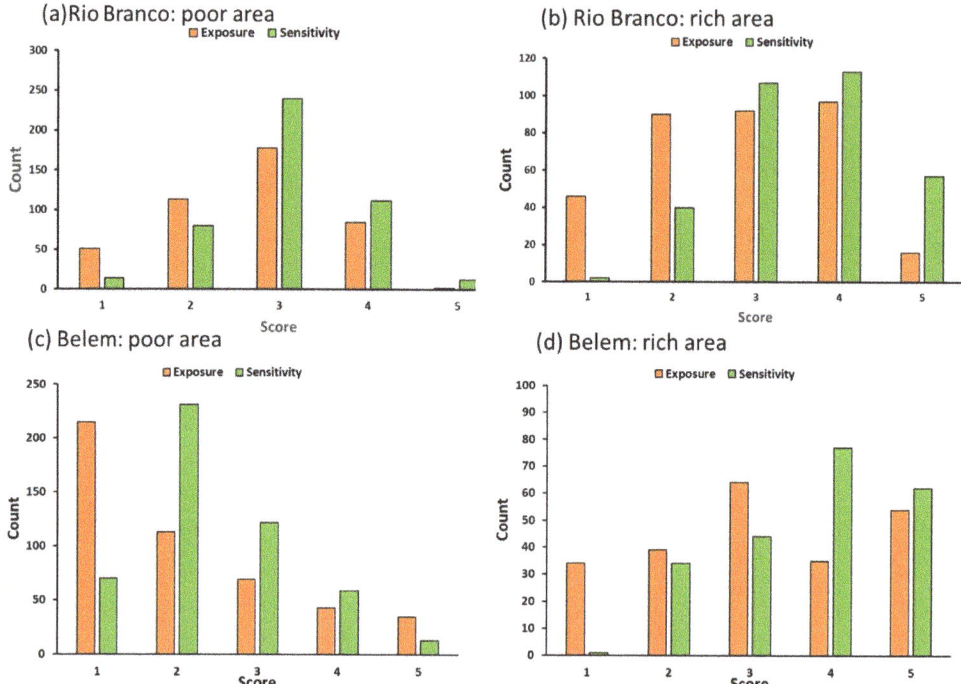

Figure 7. Distribution of exposure and sensitivity scores for selected areas in Rio Branco and Belem. The bars indicate the number of times each score was given for indicators within the exposure and sensitivity categories. (**a**) Rio Branco: poor area, (**b**) Rio Branco: rich area, (**c**) Belem: poor area, (**d**) Belem: rich area.

To explore more fully the differences between cities and areas within cities, cluster analysis was applied to the means in Table 3. The correlation coefficients for the mean indicator values are shown in Table 4. Most of the means are correlated to levels that are statistically significant. The results of the cluster analysis are shown in Figure 9. For the Exposure dimension (Figure 9a), there is no clustering of cities, but the two rich areas do emerge as a distinct cluster. Thus, richer areas have a similar pattern based on the Exposure indicators. However, for the Sensitivity dimension (Figure 9b) the main cluster which emerges is for Rio Branco (poor and rich areas). This suggests a greater degree of similarity for Sensitivity indicators within this city compared with Belem. For the Adaptation dimension (Figure 9c), there is a clear clustering into poor and rich areas, irrespective of city. Rich areas tend to do well with the Adaptation indicators while poor areas do badly, and what emerges is a clear clustering based on wealth.

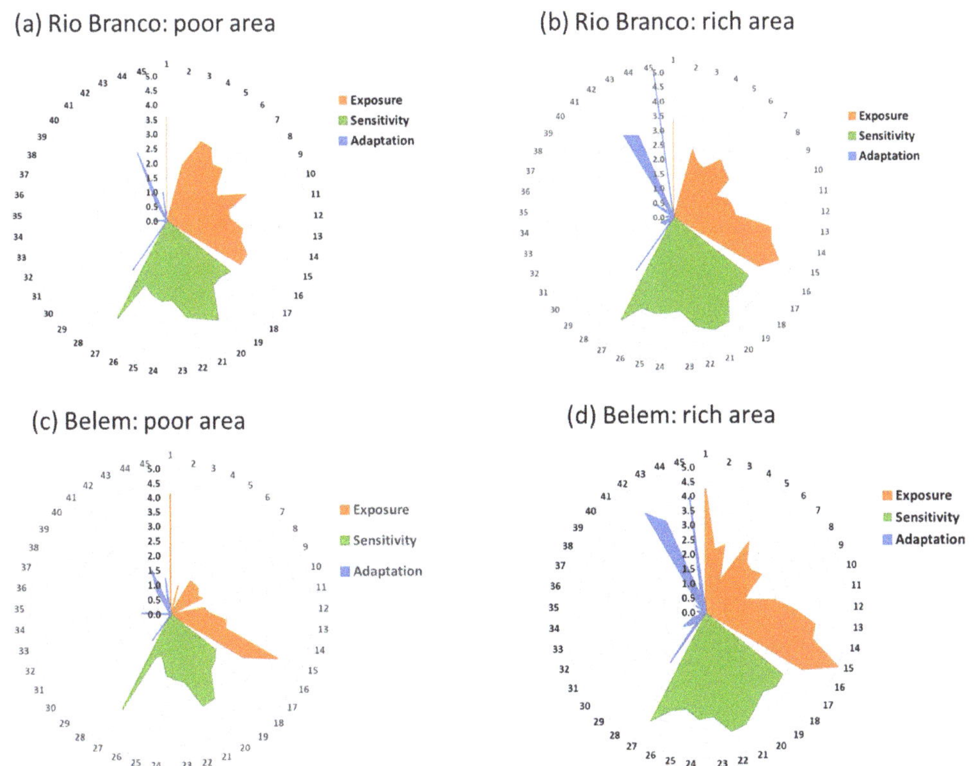

Figure 8. Radar diagrams of the mean indicator values from Table 3. (**a**) Rio Branco: poor area, (**b**) Rio Branco: rich area, (**c**) Belem: poor area, (**d**) Belem: rich area.

Table 4. Correlation coefficients between the mean indicator values shown in Table 3.

	Rio Branco Poor	Rio Branco Rich	Belem Poor	Belem Rich
(a) Exposure				
Rio Branco poor	1	0.52 *	0.83 ***	0.43 ns
Rio Branco rich		1	0.81 ***	0.89 ***
Belem poor			1	0.87 ***
Belem rich				1
(b) Sensitivity				
Rio Branco poor	1	0.93 ***	0.92 ***	0.74 **
Rio Branco rich		1	0.83 **	0.79 **
Belem poor			1	0.70 *
Belem rich				1
(c) Adaptation				
Rio Branco poor	1	0.72 ***	0.89 ***	0.74 ***
Rio Branco rich		1	0.81 ***	0.96 ***
Belem poor			1	0.82 ***
Belem rich				1

Note: Indicator values are mean values for each indicator across the assessed streets. Not significant at 0.05 = ns; * $p < 0.05$; ** $p < 0.01$; *** $p < 0.001$.

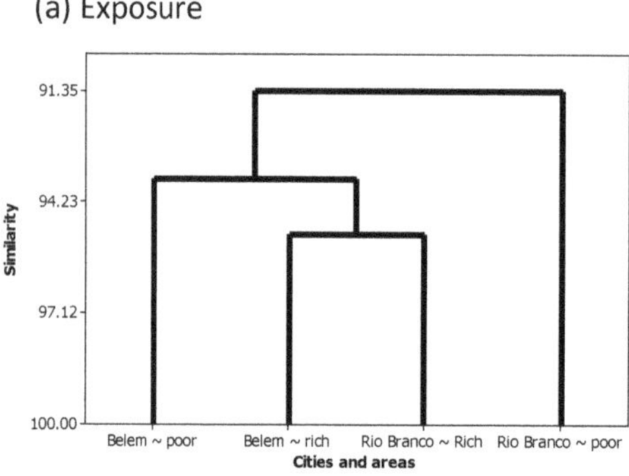

(a) Exposure

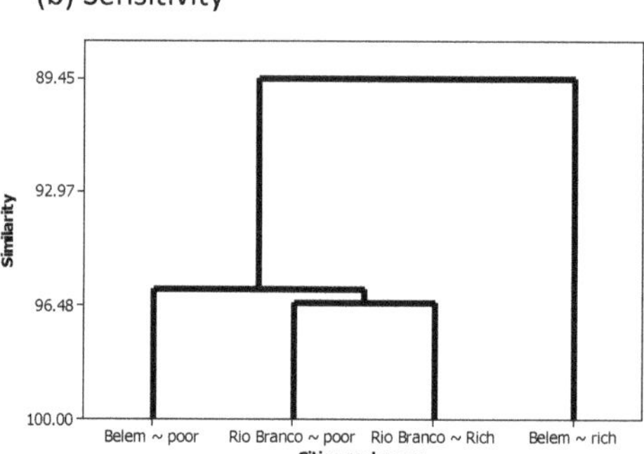

(b) Sensitivity

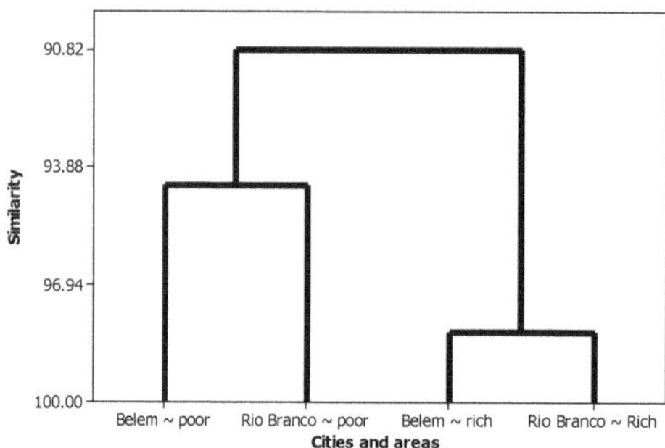

(c) Adaptation

Figure 9. Cluster analysis results for the mean indicator values shown in Table 3. (**a**) *Exposure*, (**b**) Sensitivity, (**c**) Adaptation.

4. Discussion

This research sought to develop and apply a novel method (VIFOR) to assess the vulnerability and resilience to flooding in the urban areas of two cities in Brazilian Amazon primarily using GSV-derived observations. Based on the suite of 45 indicators spanning the dimensions of Exposure, Sensitivity, and Adaptation, it was possible to identify differences in vulnerability and resilience, and these were especially apparent in terms of differences between poorer and richer areas of the cities. Conceptualising resilience in terms of these dimensions has the advantages of being able to explore where the main contributors to resilience (or lack of it) may be coming from, and here the most noticeable differences were with the Adaptation indicators, though this was also reflected in part in the Exposure indicators. Thus, richer areas had more 'assets' that help with resilience, but some of these could themselves be vulnerable to flood damage. Poorer areas had fewer 'assets' that would help with Adaptation. Of course, as with any such framework much depends on the choice of indicators. Here, the focus was upon indicators that could be assessed visually using GSV, with some potentially also assessable via EO. This indicator set focused on a very local scale in two particular regions of two municipalities. Other indicators such as the DRIB-Index focus on the validation of disaster risk reduction models at a higher spatial scale including several municipalities. VIFOR includes a disaster risk analysis at local and microscales. However, a comparative analysis with a different geographical location including other parameters of inequality, natural biome (e.g., outside the Amazon region), or a microregion within the same city may present a different result using the same list of indicators. Regarding the validation of VIFOR and the reliability of the indicators in a different context, the availability of GSV images is obviously essential. The methodology developed here to assess the indicators from GSV imagery can help in providing information when data are not available (e.g., access to institutions, bureaucracy to acquire data, capability for field work and observations). The VIFOR methodology can be improved and is potentially a cost-effective method compared with the use of a field survey in areas where data are not available or for safety reasons (e.g., areas of conflicts, COVID-19 outbreak, crimes). Reasonably up-to-date GSV imagery is also needed and the VIFOR indicators have to be amenable to a visual assessment (e.g., quantity and quality of infrastructure).

Access to high resolution EO data (1 m and below) may help to capture additional features not available from GSV. Image resolution is likely to be an important factor in EO. Hence, the assessment of urban greenspace has often been a focus with GSV- and EO-based systems in urban spaces, but there are no published examples as yet of using these tools to assess vulnerability to disasters such as flooding. However, EO is also limited to populating indicators that can be passively 'seen' using visual wavelengths or the use of 'active' EO such as the use of radar. Machine learning or automatic measurements to process multiple GSV in the same location or extensive areas of the city integrated with EO time-series in real time could improve the quality of the results. The gathering of EO data with multiple tools may also enhance the VIFOR framework by using new algorithms or methods that identify key indicators of our proposed 45 indicators. In the Amazon region, one of the major challenges is to monitor the differences in vegetation and land use characteristics between each biome [39]. The National Space Research Institute of Brazil (INPE) monitors deforestation via the Amazon Deforestation Monitoring Project (PRODES), but only with a minimum unit of 6.25 ha, and it does not include some of the urban areas [39], such as the 1 km^2 area of VIFOR. The improvement of remote sensing products, such as the 50 cm High-Resolution Planet Labs tasking provides rapid daily revisit that help to inform actions [40]. However, private high-resolution data with added value information are expensive and are not affordable to every user. Therefore, the GSV-based approach we demonstrate here provides a simple and ready-to-use tool when funding is not available to pay for the latest technology available. It requires some time and patience to monitor and measure the local area street by street, but it provides helpful information for a specific local context. Similarly, it could help urban planners and policymakers to target further investments and actions to mitigate the effects of flooding in particular streets and locations of the city.

Fundamental to prioritizing disaster mitigation efforts is quantifying flood hazard, exposure, and vulnerability [41]. EO data and remote sensing approaches can directly observe inundation [41], but there are several other indicators that affect lives and livelihoods. This study extends previous research [17,18] by including an extensive list of indicators that take note of dwelling and business locations, conditions, and vulnerabilities, and location-based general socioeconomic resilience features (e.g., observable pharmacies, community centres, etc.). VIFOR is a manual scoring process and, as such, has some limitations. For example, further research may be beneficial in order to adapt and integrate EO data to cover gaps where street view images are not available, although we expect only a subset of the 45 GSV indicators for VIFOR here to be measurable in this way. Overall, we believe that the GSV-based VIFOR approach has several strengths, notably: (i) accessibility and no cost to access GSV, (ii) possibilities to integrate and expand the research with machine learning, (iii) expandability to larger areas, and (iv) inclusion of EO data.

In addition to the selected 45 indicators included in the current VIFOR, its complementation with several other local data sources (e.g., number of health workers, doctors, crime rates) will help to better understand the trends and resilience of a particular location. Future work and new applications will be valuable to test and develop the VIFOR framework in different urban or other settings and could expand this approach further by incorporating the use of social media (e.g., Twitter, Facebook, Instagram) and high resolution EO data.

5. Conclusions

This paper presents the novel Visual Indicator Framework for Resilience (VIFOR) framework based on GSV imagery to assess the vulnerability and resilience of two flood prone urban areas. The framework is useful in locations where data are not available or difficult to access from local authorities. The approach provides a valuable source of information to monitor specific visual indicators, as suggested in our list of 45 indicators.

The existing literature provides several examples of the use of GSV to assess specific indicators. The VIFOR framework extends this by leveraging easy to use and freely available GSV images to enable a multi-indicator assessment of vulnerability and resilience to flooding that is accessible to nonexperts. The framework is likely to be adaptable to several natural and other hazards according to GSV images and local geographical contexts.

Author Contributions: Conceptualization, S.M. (Stephen Morse), M.C., R.M., D.M. and S.M. (Sarah Middlemiss); methodology, M.C., S.M. (Stephen Morse), R.M., D.M. and S.M. (Sarah Middlemiss); formal analysis, M.C., S.M. and R.M.; investigation, M.C.; resources, M.C., S.M. (Stephen Morse), R.M., D.M. and S.M. (Sarah Middlemiss); data curation, M.C.; writing—original draft preparation, M.C. and S.M. (Stephen Morse); writing—review and editing, M.C., S.M. (Stephen Morse), R.M., S.M. (Sarah Middlemiss) and D.M.; supervision, S.M. (Stephen Morse) and R.M.; project administration, R.M. and S.M. (Sarah Middlemiss); funding acquisition, R.M., S.M. (Stephen Morse) and S.M. (Sarah Middlemiss). All authors have read and agreed to the published version of the manuscript.

Funding: The Space Research and Innovation Network for Technology (SPRINT) programme of Research England's Connected Capabilities Fund and Ecometrica Ltd., project award OW131379P2V5.

Institutional Review Board Statement: Not applicable.

Informed Consent Statement: Informed consent was obtained from all subjects involved in the study.

Data Availability Statement: Contact corresponding or University of Surrey authors.

Acknowledgments: The authors are grateful to the SPRINT programme and Ecometrica Ltd. for financial support for this research. We thank Ian James, Sprint Innovation Adviser and Entrepreneur in Residence at the University of Surrey, for his valuable advice and engagement with the project. We are also grateful to the following people and organizations in Brazil for their cooperation and engagement in exploring the cities and districts selected for the case studies: INPE, EMBRAPA, and CPRM.

Conflicts of Interest: The authors declare no conflict of interest.

References

1. World Bank Group. People in Harm's Way: Flood Exposure and Poverty in 189 Countries (English). In *Policy Research Working Paper, No. WPS 9447*; World Bank Group: Washington, DC, USA, 2020. Available online: http://documents.worldbank.org/curated/en/669141603288540994/People-in-Harms-Way-Flood-Exposure-and-Poverty-in-189-Countries (accessed on 25 February 2021).
2. IPCC. Special Report: Global Warming of 1.5C. In *Summary for Policymakers*; IPCC: Geneva, Switzerland, 2018. Available online: https://www.ipcc.ch/sr15/chapter/spm/ (accessed on 25 February 2021).
3. United Nations. *World Cities Report 2020: The Value of Sustainable Urbanization*; UN Habitat: San Francisco, CA, USA, 2020. Available online: https://unhabitat.org/wcr/ (accessed on 25 February 2021).
4. United Nations. The New Urban Agenda. Available online: https://unhabitat.org/the-new-urban-agenda-illustrated (accessed on 24 February 2021).
5. World Bank. *Disaster Recovery Guidance Series: Communication during Disaster Recovery*; World Bank Group: Washington, DC, USA, 2020. Available online: https://openknowledge.worldbank.org/handle/10986/33685 (accessed on 25 February 2021).
6. Gibb, C. A critical analysis of vulnerability. *Int. J. Disaster Risk Reduction.* **2018**, *28*, 327–334. [CrossRef]
7. UNDRR. United Nations Office for Disaster Risk Reduction. Hazard Definition and Classification Review. 2021. Available online: https://www.undrr.org/publication/hazard-definition-and-classification-review (accessed on 9 March 2021).
8. Chen, C.; Noble, I.; Hellmann, J.; Coffee, J.; Murillo, M.; Chawla, N. *University of Notre Dame Global Adaptation Index Country Index Technical Report*; ND-GAIN: South Bend, IN, USA, 2015. Available online: https://gain.nd.edu/our-work/country-index/ (accessed on 24 February 2021).
9. Chen, C.; Doherty, M.; Coffee, J.; Wong, T.; Hellmann, J. Measuring the adaptation gap: A framework for evaluating climate hazards and opportunities in urban areas. *Environ. Sci. Policy* **2016**, *66*, 403–419. [CrossRef]
10. Morse, S. *The Rise and Rise of Indicators: Their History and Geography*; Routledge: Abingdon, UK; New York, NY, USA, 2019.

11. Chen, C.; Hellmann, J.; Berrang-Ford, L.; Noble, I.; Regan, P. A global assessment of adaptation investment from the perspectives of equity and efficiency. *Mitig. Adapt. Strateg. Glob. Chang.* **2018**, *23*, 101–122. [CrossRef]

12. Vivanco, M.L.; Villagrán, A.A.; Martínez, R.V. *Using Social Work Interventions to Address Climate and Disaster Risks in Latin America and the Caribbean*; World Bank: Washington, DC, USA, 2020. Available online: https://openknowledge.worldbank.org/handle/10986/34137 (accessed on 25 February 2021).

13. de Almeida, L.Q.; Welle, T.; Birkmann, J. Disaster risk indicators in Brazil: A proposal based on the world risk index. *Int. J. Disaster Risk Reduct.* **2016**, *17*, 251–272. [CrossRef]

14. Zhai, W.; Peng, Z.R. Damage assessment using Google Street View: Evidence from Hurricane Michael in Mexico Beach, Florida. *Appl. Geogr.* **2020**, *123*, 102252. [CrossRef]

15. ESA European Space Agency. Sentinel 1-Radar Vision for Copernicus. 2021. Available online: http://www.esa.int/Applications/Observing_the_Earth/Copernicus/Sentinel-1 (accessed on 9 March 2021).

16. ESA European Space Agency. Sentinel 2. 2021. Available online: https://sentinel.esa.int/web/sentinel/missions/sentinel-2 (accessed on 9 March 2021).

17. Wang, R.; Feng, Z.; Pearce, J.; Yao, Y.; Li, X.; Liu, Y. The distribution of greenspace quantity and quality and their association with neighbourhood socioeconomic conditions in Guangzhou, China: A new approach using deep learning method and street view images. *Sustain. Cities Soc.* **2021**, *66*, 102664. [CrossRef]

18. Feldmeyer, D.; Meisch, C.; Sauter, H.; Birkmann, J. Using OpenStreetMap Data and Machine Learning to Generate Socio-Economic Indicators. *ISPRS Int. J. Geo-Inf.* **2020**, *9*, 498. [CrossRef]

19. Ki, D.; Lee, S. Analyzing the effects of Green View Index of neighborhood streets on walking time using Google Street View and deep learning. *Landsc. Urban Plan.* **2021**, *205*, 103920. [CrossRef]

20. Li, X.; Zhang, C.; Li, W.; Ricard, R.; Meng, Q.; Zhang, W. Assessing street-level urban greenery using Google Street View and a modified green view index. *Urban For. Urban Green.* **2015**, *14*, 675–685. [CrossRef]

21. Berland, A.; Lange, D.A. Google Street View shows promise for virtual street tree surveys. *Urban For. Urban Green.* **2017**, *21*, 11–15. [CrossRef]

22. Diakakis, M.; Deligiannakis, G.; Pallikarakis, A.; Skordoulis, M. Identifying elements that affect the probability of buildings to suffer flooding in urban areas using Google Street View. A case study from Athens metropolitan area in Greece. *Int. J. Disaster Risk Reduct.* **2017**, *22*, 1–9. [CrossRef]

23. Giuliani, G.; Petri, E.; Interwies, E.; Vysna, V.; Guigoz, Y.; Ray, N.; Dickie, I. Modelling Accessibility to Urban Green Areas Using Open Earth Observations Data: A Novel Approach to Support the Urban SDG in Four European Cities. *Remote Sens.* **2021**, *13*, 422. [CrossRef]

24. ND-Gain. Notre Dame Global Adaptation Initiative. 2022. Available online: https://gain.nd.edu/about/ (accessed on 8 February 2022).

25. Cardoni, A.; Zamani Noori, A.; Greco, R.; Cimellaro, G.P. Resilience assessment at the regional level using census data. *Int. J. Disaster Risk Reduct.* **2021**, *55*, 102059. [CrossRef]

26. Mees, H.; Tijhuis, N.; Dieperink, C. The effectiveness of communicative tools in addressing barriers to municipal climate change adaptation: Lessons from the Netherlands. *Clim. Policy* **2018**, *18*, 1313–1326. [CrossRef]

27. Moftakhari, H.; Shao, W.; Moradkhani, H.; AghaKouchak, A.; Sanders, B.; Matthew, R.; Jones, S.; Orbinski, J. Enabling incremental adaptation in disadvantaged communities: Polycentric governance with a focus on non-financial capital. *Clim. Policy* **2020**, *21*, 396–405. [CrossRef]

28. Almeida, L.Q.d.; Araujo, A.M.S.d.; Welle, T.; Birkmann, J. DRIB Index 2020: Validating and enhancing disaster risk indicators in Brazil. *Int. J. Disaster Risk Reduct.* **2020**, *42*, 101346. [CrossRef]

29. Mansur, A.V.; Brondizio, E.S.; Roy, S.; Soares, P.P.d.M.A.; Newton, A. Adapting to urban challenges in the Amazon: Flood risk and infrastructure deficiencies in Belém, Brazil. *Reg. Environ. Chang.* **2018**, *18*, 1411–1426. [CrossRef]

30. Debortoli, N.S.; Camarinha PI, M.; Marengo, J.A.; Rodrigues, R.R. An index of Brazil's vulnerability to expected increases in natural flash flooding and landslide disasters in the context of climate change. *Nat. Hazards* **2017**, *86*, 557–582. [CrossRef]

31. Alves, L.M.; Chadwick, R.; Moise, A.; Brown, J.; Marengo, J.A. Assessment of rainfall variability and future change in Brazil across multiple timescales. *Int. J. Climatol.* **2021**, *41*, E1875–E1888. [CrossRef]

32. de Andrade MM, N.; Szlafsztein, C.F. Vulnerability assessment including tangible and intangible components in the index composition: An Amazon case study of flooding and flash flooding. *Sci. Total Environ.* **2018**, *630*, 903–912. [CrossRef]

33. de Andrade MM, N.; Bandeira IC, N.; Fonseca DD, F.; Bezerra PE, S.; de Souza Andrade, Á.; de Oliveira, R.S. Flood risk mapping in the Amazon. In *Flood Risk Management*; IntechOpen: London, UK, 2017. [CrossRef]

34. Dolman, D.I.; Brown, I.F.; Anderson, L.O.; Warner, J.F.; Marchezini, V.; Santos, G.L.P. Re-thinking socio-economic impact assessments of disasters: The 2015 flood in Rio Branco, Brazilian Amazon. *Int. J. Disaster Risk Reduct.* **2018**, *31*, 212–219. [CrossRef]

35. IBGE. The Brazilian Institute of Geography and Statistics. Cidade de Rio Branco, Acre. 2021. Available online: https://cidades.ibge.gov.br/brasil/ac/rio-branco/panorama (accessed on 24 March 2021).

36. IBGE. The Brazilian Institute of Geography and Statistics. Cidade de Belem, Para. 2021. Available online: https://cidades.ibge.gov.br/brasil/pa/belem/panorama (accessed on 24 March 2021).

37. IBGE. IBGE Sai em defesa do Censo. 2021. Available online: https://agenciadenoticias.ibge.gov.br/agencia-noticias/2012-agencia-de-noticias/noticias/30350-ibge-sai-em-defesa-do-orcamento-do-censo-2021 (accessed on 24 March 2021).

38. Begum, R.A.; Sarkar, M.S.K.; Jaafar, A.H.; Pereira, J.J. Toward conceptual frameworks for linking disaster risk reduction and climate change adaptation. *Int. J. Disaster Risk Reduct.* **2014**, *10*, 362–373. [CrossRef]
39. Almeida, C.A.; Valeriano, D.M.; Maurano, L.; Vinhas, L.; Fonseca, L.M.G.; Silva, D.; Santos, C.P.F.; Martins, F.S.R.V.; Lara, F.C.B.; Maia, J.S.; et al. March. Deforestation Monitoring in Different Brazilian Biomes: Challenges and Lessons. In Proceedings of the 2020 IEEE Latin American GRSS & ISPRS Remote Sensing Conference (LAGIRS), Santiago, Chile, 22–26 March 2020; pp. 357–362.
40. Planet. High-Resolution Imagery with Planet Satellite Tasking. 2021. Available online: https://www.planet.com/products/hi-res-monitoring/ (accessed on 24 November 2021).
41. Tellman, B.; Sullivan, J.A.; Kuhn, C.; Kettner, A.J.; Doyle, C.S.; Brakenridge, G.R.; Erickson, T.A.; Slayback, D.A. Satellite imaging reveals increased proportion of population exposed to floods. *Nature* **2021**, *596*, 80–86. [CrossRef]

Article

Challenges in Using Earth Observation (EO) Data to Support Environmental Management in Brazil

Mercio Cerbaro [1,*], Stephen Morse [1], Richard Murphy [1], Jim Lynch [1] and Geoffrey Griffiths [2]

[1] Centre for Environment & Sustainability, University of Surrey, Guildford GU2 7XH, UK; s.morse@surrey.ac.uk (S.M.); rj.murphy@surrey.ac.uk (R.M.); j.lynch@surrey.ac.uk (J.L.)

[2] Department of Geography, University of Reading, Reading RG6 6AB, UK; g.h.griffiths@reading.ac.uk

* Correspondence: m.cerbaro@surrey.ac.uk

Received: 27 November 2020; Accepted: 10 December 2020; Published: 12 December 2020

Abstract: This paper presents the results of research designed to explore the challenges involved in the use of Earth Observation (EO) data to support environmental management Brazil. While much has been written about the technology and applications of EO, the perspective of end-users of EO data and their needs has been under-explored in the literature. A total of 53 key informants in Brasilia and the cities of Rio Branco and Cuiaba were interviewed regarding their current use and experience of EO data and the expressed challenges that they face. The research builds upon a conceptual model which illustrates the main steps and limitations in the flow of EO data and information for use in the management of land use and land cover (LULC) in Brazil. The current paper analyzes and ranks, by relative importance, the factors that users identify as limiting their use of EO. The most important limiting factor for the end-user was the lack of personnel, followed by political and economic context, data management, innovation, infrastructure and IT, technical capacity to use and process EO data, bureaucracy, limitations associated with access to high-resolution data, and access to ready-to-use product. In general, users expect to access a ready-to-use product, transformed from the raw EO data into usable information. Related to this is the question of whether this processing is best done within an organization or sourced from outside. Our results suggest that, despite the potential of EO data for informing environmental management in Brazil, its use remains constrained by its lack of suitably trained personnel and financial resources, as well as the poor communication between institutions.

Keywords: earth observation; end-users; environmental management; land use; Brazil

---

1. Introduction

It has been demonstrated that Earth Observation (EO) via satellites provides a useful tool to monitor, report, and verify the management of sustainable land use and land cover (LULC) in tropical forests [1]. For example, the maps generated with EO data enable institutions to monitor deforestation [2,3]. Since 1973, the Brazilian Institute for Space Research (INPE) has been monitoring deforestation in the Amazon forest by using Landsat EO data [4]. Since 1988, INPE has assessed the annual deforestation rates via the Amazon Deforestation Monitoring Project (PRODES), a leading example of operational monitoring to quantify land-cover change via EO data [4–6]. While PRODES provided annual deforestation rates, INPE has also developed the Near Real Time Deforestation Detection System (DETER) to meet the needs of other users that require near-real time detection for early warning and the potential for quick response by government agencies [7]. Information derived from PRODES and DETER is used by law enforcement and other institutions to help support actions on the ground [8]. INPE sends deforestation alerts to institutions such as the Brazilian Institute of Environment and Renewable Natural Resources (IBAMA) [9]. DETER was created in 2004, following demand from IBAMA to provide high-frequency alerts of deforestation instead of relying on annual data from PRODES [9]. The information derived from PRODES and DETER is widely used by different agencies to verify deforestation and land use changes in the Amazon [10–12]. Indeed, deforestation in the Brazilian Amazon, defined as clear cutting and conversion of the original forest cover to other land uses [13], reduced from 7893 km^2/annum in 2016 to 6947 km^2/annum in 2017, but increased to 7536 and 10,129 km^2/annum, respectively, in 2018 and 2019 [14].

This paper builds upon previously published work that illustrated how EO datasets, including the increasing availability of free datasets, has enabled users to generate and use the information available to them [15]. Achard and Hansen [4] suggest that, while unprocessed EO data from a range of satellite sensors may be available for a region of interest, adequate tools are essential to download and preprocess the data, to facilitate user access. Overall, there remains a lack of research to date on the challenges faced by users when trying to access, process, and use EO data for environmental management and other sustainability purposes [9,15]. In the present research, our objective was to address four key questions regarding these challenges:

1. What uses are made of EO data?
2. What are the factors that limit the use of EO data?
3. What are user needs/expectations with regard to EO data?
4. What suggestions do users have for improving EO data use?

"EO data" in these questions span both the raw data and the information that can be derived from those data. Cerbaro et al. [15] noted that some of the major challenges identified in the complex and multifaceted aspects of using EO data were associated with access to, and with the processing of, raw data into usable information. The analysis also revealed novel insights on a lack of inter-institutional communication, adequate office infrastructure and personnel, availability of the right type of EO data and funding restrictions, political instability, and bureaucracy as factors that limit more effective use of EO data in Brazil at present. They used these findings to develop a conceptual model to illustrate the main steps and limitations in the flows of EO data and information for use in the management of land-use change in Brazil. This paper extends that work by reporting the results of a research project undertaken between 2016 and 2017 that was one of the first of its type designed to focus specifically on the challenges faced by end users and their needs/expectations when applying EO data for environmental and sustainability management in Brazil. The paper is structured to present the results obtained regarding each of the four questions, in turn, before arriving at an integrated discussion and conclusions.

2. Materials and Methods

2.1. Research Locations

Initial "scoping" research was undertaken between November and December 2016, whereby the authors conducted 43 interviews with senior scientists at the National Space Research Institute (INPE), policymakers in Brasilia, directors at the Ministry of Environment (MMA), environmental analysts, GIS technicians, senior staff at the Brazilian Space Agency, senior researchers at Brazilian Agriculture Research Cooperation (EMBRAPA), and other public and private institutions in the state of Sao Paulo and in Brasilia. The details of that research are not covered here, but respondents were asked to recommend the most appropriate locations in Brazil, to address the research questions noted above based on examples and experiences of EO data in use. Many of the respondents suggested that the optimum locations in Brazil for such research were the states of Acre and Mato Grosso. Acre was recommended as the best example where EO is already widely used by different stakeholders to inform policymakers. The respondents from EMBRAPA and INPE recommended the state of Mato Grosso as a location where EO data are used to monitor and integrate land-use policies and agriculture over large areas in three different biomes: Amazon, Cerrado, and the Pantanal. In addition, most of the respondents in the scoping study also suggested Brasilia (the capital of Brazil) as a necessary location for the research, because it is where many of the key ministries, in terms of EO data use, are based. Therefore, Brasilia and the cities of Rio Branco (capital of Acre) and Cuiaba (capital of Mato Grosso) were selected as relevant, representative, and informed locations for undertaking the in-depth structured interviews conducted in 2017 (Figure 1).

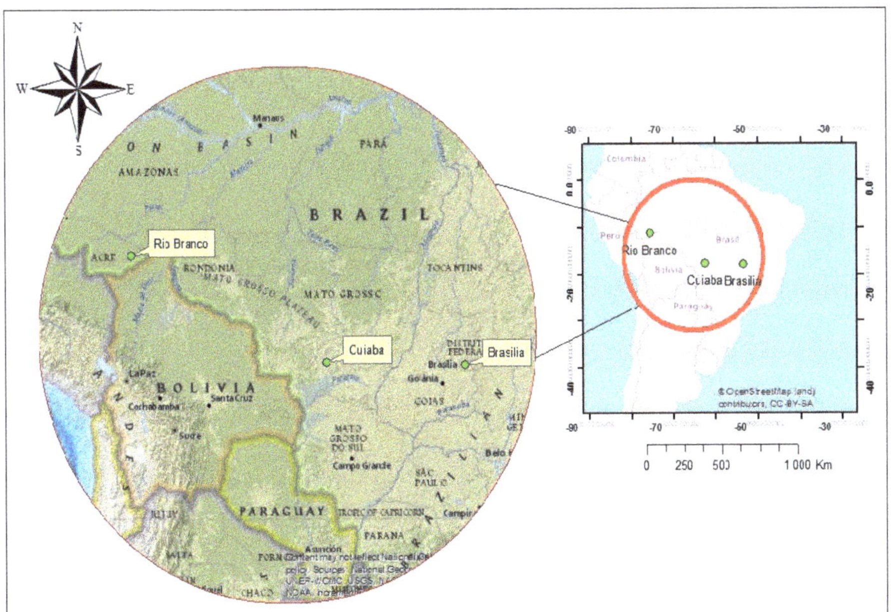

Figure 1. Field work locations: Brasilia (the capital of Brazil), Cuiaba (Mato Grosso) and Rio Branco (Acre). Sources: National Geographic and Open Street Map.

2.2. Data Collection

The primary data reported here are based upon the analysis of 53 semi-structured interviews held between June and August 2017, in Brasilia, Rio Branco, and Cuiaba. The initial respondents were selected based on recommendations from senior researchers at INPE and respondents in the November/December 2016 scoping research. A snowballing approach was used to identify further respondents. The interviews were conducted in Portuguese and translated into English. This number of interviews (53) was considered appropriate for the aim of capturing the breadth of user and provider experience, and the range of organizations was representative of the diversity of main actors engaged in acquiring and using EO data for environmental management in Brazil:

- Federal Government Institutions (26);
- Federal Universities (7);
- State government organizations (16);
- Private institutions (3);
- Non-governmental organization (1).

The questionnaire comprised 15 questions. Two of these were structured quantitative questions (respondents scored on a 1–5 Likert scale a list of factors pre-defined by the research team, based on the initial scoping research in 2016. In addition, thirteen more "open" questions were presented across five more general challenge areas (also identified on the 2016 scoping research) for discussion and comment by respondents. The outline structure of the questionnaire and its relationship with the four key research questions (RQs 1–4) of this study were as follows:

1. EO data use (RQ1);
2. Expectation from providers of EO data (structured questions) (RQ 3);
3. Capacity of institutions to use EO data (RQ 2, 4);
4. Limitations and restrictions on the use EO data (structured questions) (RQ 4);
5. Regional capacity and training for EO users (RQ 4);
6. EO data and public policies in Brazil (RQ 4);
7. Use of EO to help populate indicators (RQ 4).

Within Sections 2 and 4, respondents were presented with a set of potential factors derived from the scoping survey. For capturing respondents 'expectations' from EO data and its use, the predefined list of 7 factors was as follows:

- Low cost (no cost);
- Easy to use systems (minimum need for post-processing);
- Fast delivery;
- Availability of technical support;
- High-resolution images;
- Frequency of coverage;
- Friendly interface (analysis functions and services).

For capturing respondents' perceptions of the limitations/restrictions on EO data and its use, the predefined list of 13 factors was as follows:

- Data management (big data);
- Technical capacity (image processing and data analysis technique);
- Innovation (investment in technology);
- Personnel (human resources, lack of staff, and specialized staff);
- Access (EO images, GIS software's, maps, and web services);
- Institution capacity (long-term planning, management, and business dynamics);
- Infrastructure (IT, computers, and internet);
- Communication (internal and external);
- Political and economic context;
- Bureaucracy (legal frameworks);
- Capacity to implement projects;
- User capacity to combined data from different sources;
- Regional geography (knowledge of local settings and problems, changes in time represented by the imagery, and historic context).

In each of these factors, a score of 5 was equated to "most important" and 1 to "least important.

2.3. Data Analysis

The transcripts of the 53 recorded interviews were subject to content analysis for the thirteen open questions. For the two structured, quantitative questions the Kruskal–Wallis nonparametric test was employed (using SPSS) to identify patterns amongst the amalgamated data from Brasilia, Rio Branco, and Cuiaba. However, not all of the 53 respondents answered the structured questions (mostly due to time constraints), so the question which focused on the expectation of providers of EO data had 47 respondents, while the question which addressed the factors limiting the use of EO data in the respondent institution had a total of 49 respondents.

This enabled the comparison of several treatments and ranking differences between responses to questions [16]. Kruskal–Wallis one-way analysis of variance is the method of statistical choice for non-normally distributed data if there are multiple groups being compared from different populations [17].

3. Results

3.1. Uses Made of Earth Observation Data (RQ 1)

The main reported uses of EO data in this research were for agricultural forecasting, disaster management, land-use planning, environmental management, and monitoring deforestation in the Amazon region. Table 1 shows that a majority of data used were from freely available sources, such as Landsat 8 (30 m resolution), Rapideye (5 m resolution), and Sentinel 2 (10 m resolution).

Table 1. Earth Observation (EO) data use and major providers of EO data.

Earth Observation Data	Number of Users
Landsat 8 (30 m)	23
Rapideye (5 m)	10
Sentinel 2 (10 m)	10
CBERS 4 (20 m)	4
MODIS (250 m)	3
SRTM (30 m and 90 m)	3
RADARSAT (3–100 m)	2
SPOT (10 m)	2
SPOT (2.5 m)	2
ALOS 2	2
GOES	2
ResourceSat	2
FORMOSAT (15, 4, and 2 m)	2
Sentinel 1	1
Provider's Website Used to Download EO Data	**Number or Users**
INPE	25
The United States Geological Survey (USGS)	22
European Space Agency (ESA)	17

Notes: INPE, Brazilian Institute for Space Research. CBERS, China-Brazil Earth Resource Satellites. MODIS, Moderate Resolution Imaging Spectroradiometer sensor of The National Aeronautics and Space Administration(NASA). SRTM, Shuttle Radar Topography Mission. NASA. RADARSAT, Canadian government / industry partnership commercial satellite operator. SPOT is a commercial satellite operator based in Toulouse, France. ALOS 2, Synthetic Aperture Radar (SAR) commercial satellite developed by JAXA (Japan Aerospace Exploration Agency). GOES, Geostationary Operational Environmental Satellite. Joint partnership between The National Oceanic and Atmospheric Administration (NOAA) and NASA. ResourceSat, Indian Earth Observation Satellites program IRS (Indian Remote Sensing Satellite). The EO data is available at not charge to end-users in Brazil via INPE website. Single agreement between India and Brazil governments. FORMOSAT is a commercial satellite operator at the National Space Organization of Taiwan (NSPO).

EO data made available through various government agreements with private sector providers are important. The acquisition of Rapideye (5 m) data by the federal government also enabled users in public institutions to access EO data from such private sector sources. The acquisition of SPOT 6/7 (10 m and 2.5 m; https://earth.esa.int/web/eoportal/satellite-missions/s/spot-6-7) by the state of Mato Grosso and FORMOSAT (15, 4, and 2 m) by the state of Acre are examples of private EO data used in state institutions. SPOT is a commercial satellite operator that offered a range of EO data (2.5, 5, 10, and 20 m), and the main goal is to guarantee a sustainable operational service to customers [18]. FORMOSAT is a program at the National Space Organization of Taiwan (NSPO), and the EO data are available at a cost for end-users, and can be used for applications such as a planning tool for urban development, environment and crop health monitoring, and disaster assessment [19]. Several respondents mentioned that Landsat 8 (30 m) and Sentinel 2 (10 m) meet the demand for monitoring deforestation and land-use applications at a large scale, but not for land-use management in small properties at the national level. Both Sentinel and Landsat EO data are available at no cost for end-users.

However, key respondents complained that all the privately available EO data are over five years old and needed updating at the state level. The demand for new EO data at high spatial resolution for various applications, such as monitoring urban fires in Rio Branco and land-use planning, was mentioned by more than 50% of respondents. In Cuiaba, the state of Mato Grosso purchased the EO data for the entire state in 2008. The final products of the INPE Fire program and the information from PRODES are also widely used in several institutions. For example, the data of PRODES is used by environmental agencies (SEMA) in both Cuiaba and Rio Branco.

All respondents highlighted the importance of EO data to support their respective activities, indicating strong user recognition for its value in environmental management in relation to the first key research question of this study.

3.2. Limitations on the Use of EO Data (RQ 2)

Figure 2 summarizes the mean rank scores at the national level for the set of limiting factors for the following question: *What is limiting your use of Earth Observation?* The details of the five homogeneous subsets identified in the statistical analysis are provided in Table 2. The analysis is based on responses from 49 respondents, with four respondents opting not to respond to the question, mainly because of time constraints during the interviews.

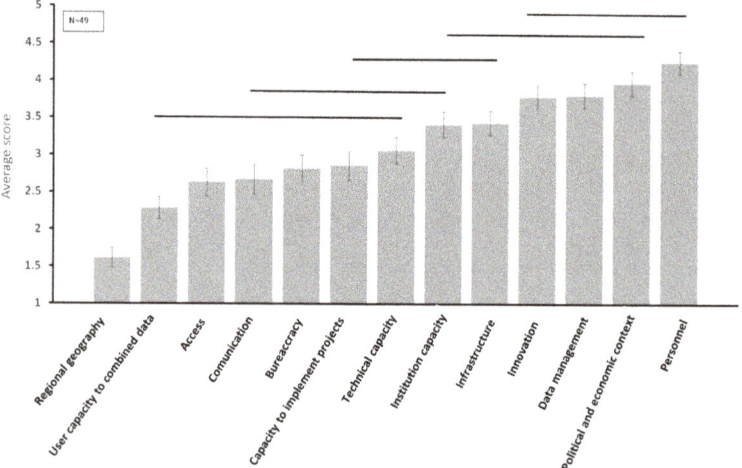

Figure 2. Average scores of factors limiting the use of EO data: 1 = least important, and 5 = most important; error bars = standard error of the mean; horizontal bars represent five homogeneous subsets from the statistical analysis (details presented in Table 2).

Subsets in Table 2 are based on asymptotic significances, with a significant level of $p = 0.05$. The respondents to this question had the opportunity to explain the main reasons why particular factors ranked low or high to complement the numerical rankings given in Table 2. Lack of personnel (average score = 4.24) was seen as the most important individual limitation by respondents, and this was largely associated with a lack of financial resource represented in the political and economic context factor (average score = 3.95). Indeed, the respondents tended to mention the economic context within Brazil as a major factor which limited the use of EO in their institution (both factors ranked high). In general, challenges were also associated with lack of investment in data management and innovation (average scores = 3.8 and 3.78, respectively), which includes investments in IT, internet, and office conditions. Institutional capacity (capacity of institutions to deliver outcomes and use all the information available) was ranked in the middle of the table (average score = 3.41) and was often associated with lack of institutional capacity to implement projects on the ground. These limitations were often associated with the capacity of institutions to act on the ground, based on limited financial resources and personnel. Lower ranked limiting factors included administrative aspects such as capacity to implement projects, bureaucracy, and communication (average scores = 2.86, 2.82, and 2.67

respectively). Interestingly, access to EO data was considered by respondents to be of relatively low importance (average score = 2.63), despite the high demand for high-resolution EO data and limitations to accessing specific types of radar EO data. The capacity of users to combine EO data with other types of information was ranked low (average score = 2.29), which refers to the capacity of users to download EO data and process raw EO data into usable information in the respondent institution. However, end-users often mentioned the need for easy-to-use platforms to allow users with limited technical skills in EO data processing to use large EO datasets in combination with other data (e.g., agricultural statistics, health, income and other indicators). Regional geography (the knowledge of end-users to understand the problems at the regional context, e.g., fire, deforestation, and agriculture) was the single least important limiting factor for the use of EO data (average score = 1.61).

Table 2. Kruskal–Wallis analysis of the responses on the factors limiting the use of EO data.

Factors Influencing the Use of EO Data	Homogeneous Subset					
	1	2	3	4	5	6
Regional geography	125.1					
User capacity for combined data		208.3				
Access to data		253.8				
Communication		261.1	261.1			
Bureaucracy		277.4	277.4			
Capacity to implement projects		284.8	284.8	284.8		
Technical capacity		309.6	309.6	309.6		
Institutional capacity			355.9	355.9	355.9	
Infrastructure				357.7	357.7	
Innovation					405.0	405.0
Data management					408.1	408.1
Political and economic context					430.4	430.4
Lack of personnel						469.8
Test statistic	[a]	10.39	9.33	6.3	9.41	7.48
Significance (2-sided test)		0.065	0.053	0.098	0.052	0.058
Adjusted significance (2-sided test)		0.135	0.133	0.285	0.129	0.177

Notes: Homogeneous subsets (shaded cells) are based on asymptotic significances. N = 49, significance level $p < 0.05$. Each cell shows the sample average rank of factors limiting the use of EO data. [a] Incomputable because subset has only one sample.

The following quotations provide examples of the responses received in the more open elements of the interviews to complement the quantitative data presented above:

I think the use of EO data is systematic (PRODES, DETER, Fire). In my opinion the main gap is that there is information available, but the problem is lack of people to process it. High demand and a few people to do the work. (General coordinator, ICMBIO, Cuiaba)

We hire external services to attend our specific demands here at the Ministry of Environment in Brasilia, and our institutional capacity is related to lack of people. This is a critical problem in government institutions. For example, IBAMA used to have 1600 environmental agents in 2010 and only 900 in 2017. (Senior analyst, Ministry of Environment, Brasilia)

The limitations associated with lack of personnel are often linked with the financial capacity of institutions to hire new personnel and this is reflected in the second highest scoring limitation being the political and economic context.

The user's capacity to combine datasets, communication between institutions, capacity to implement projects, and technical capacity are some of the limitations at the national level (subsets 1 to 3). The major limitations are associated with innovation, data management, political and economic context, and availability of personnel with the required expertise (subsets 4 to 6). For example, financial resources were the critical factor at UCGEO in Rio Branco, initially having 12 staff in 2009 for geoprocessing and having only three in 2017. In Brasilia, the head of EO monitoring at ICMBIO described the problem of only having two permanent staff and the requirement of having to hire staff only on a short-term contract (six months to one year maximum). The work must be complemented with the short-term help of undergraduate students and temporary contracts. This creates problems of quality and control due to high demand of work and constant change. This constant turnover of staff is also a challenge, as time and resources had to be spent training someone for a limited period. A similar situation was evident in other institutions at the national level.

The speed of the internet (infrastructure) was noted to be an important limitation by almost all the users in Brasilia, and at the regional level, it was a main barrier of communication between Brasilia and the state levels. The following illustrate this point:

Internet with poor quality, it takes 12 h to download an EO image. (Environmental Analyst, ICMBIO, Rio Branco)

We need to invest more in system of information and data management. The internet is very slow, old computers and the changes of staff is constant. Investments in technology and to hire new staff is related to the political and economic context in Brazil. For example, one area that requires investment is to integrate land ownership data with different databases. In general, technical capacity is good with a few limitations. This is because we have different training programs in academia or here at IBAMA. Also, EO data image processing and remote sensing is part of our day-to-day life. The main limitation is the speed of the internet here in Brasilia and the regional offices. (Senior director, IBAMA, Brasilia)

When asked about EO restrictions associated with radar images and high-resolution optical data, more than half of the respondents mentioned challenges involved with processing and interpreting radar data.

One of the problems is the capacity of the end-user to use radar images. (Environmental Analyst, ICMBIO, Rio Branco)

Lack of knowledge to use radar images. For example: the high number of clouds in the Amazon region. It would be useful, but we need proper training to process this type of data. (PhD student, University of Acre, Rio Branco)

It is difficult to process EO data from Sentinel 1. It requires specific knowledge. (Senior coordinator, ICMBIO, Brasilia)

The limitations of using EO radar data are also associated with economic limitations to purchase radar data available, but at a cost, in institutions where the end-user is skilled to use and process advanced information, such as IBAMA in Brasilia.

Radar images on band L, 15 day's resolution and 20 m at no cost are essential to detect illegal deforestation in the Amazon. The images are not available for free. (Senior director, IBAMA, Brasilia)

The results show that some of the main challenges in all locations are limitations of end-users with the technical capacity to process and extract the needed information and more than 70% of respondents mentioned challenges involved with processing and interpreting radar data. More than 50% of respondents noted that the problems are associated with both access to radar images and the technical capacity process EO radar data. The need for up-to-date high-resolution EO data of 1 m or lower was one of the main demands noted by respondents, and it was suggested that such high-resolution imagery should be available at no cost for public access.

We need EO data with high resolution and high frequency available for public access. (Environmental analyst, Brazilian Forest Service, Brasilia)

It would be great to have access of high-resolution EO data of 1 m or 2 m. It would help the state of Acre to monitor public land and to monitor fiscalization. For example: problems of land conflicts, landless movements, land invasions and overlapping land areas registered with the different owner. (Senior director, Land Institute of Acre (ITERACRE), Rio Branco)

3.3. Expectations of Earth Observation Data Providers (RQ 3)

In exploring this third key research question we sought responses from users on what they saw as their most important needs/expectations from EO data providers. Figure 3 gives the mean rank scores at the national level for the set of needs/expectations for the following question: *What are your user needs/expectations from providers with regard to EO data?* The details of the three homogeneous subsets identified in the statistical analysis are provided in Table 3.

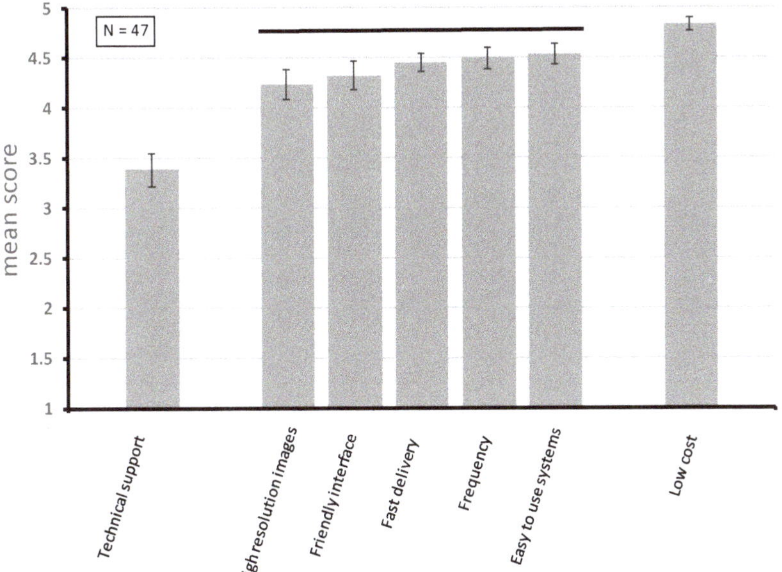

Figure 3. Average scores of the user needs/expectations of EO data providers: 1 = least important, and 5 = most important; error bars = standard error of the mean; horizontal bar represents a homogeneous subset from the statistical analysis (details presented in Table 3).

Perhaps unsurprisingly, low cost was mentioned as the most important factor for the respondents (average score = 4.83, subset 3). The factor products with high frequency and fast delivery was also ranked as very important (average scores = 4.49 and 4.45, respectively), although subset two comprised five of the factors and these could not be separated statistically. Technical support had the lowest average score (3.38); however, this still suggests a degree of importance ascribed to this factor by the respondents. Hence, Figure 3 would suggest that all of these factors were regarded as "important" by the respondents, albeit with low cost emerging as the one having greater emphasis.

Table 3. Kruskal–Wallis analysis of the responses on expectations from EO providers.

		Homogenous Subset		
		1	2	3
Expectations from EO providers	Technical support	**90.3**		
	High-resolution images		160.0	
	Friendly interface		165.0	
	Fast delivery		167.1	
	Frequency		177.4	
	Easy to use systems		180.5	
	Low cost			**214.6**
Test Statistic		[a]	2.076	[a]
Significance (two-sided test)			0.722	
Adjusted Significance (two-sided test)			0.833	

Notes: Homogeneous subsets (shaded cells) are based on asymptotic significances. N = 47; significance level $p < 0.05$. Each cell shows the sample average rank of user needs/expectations. [a] Not computable as subset has only has one sample.

The following quotations provide examples of the responses received in the more open elements of the interviews to complement the quantitative data presented above:

We want products ready to use without the need of additional technical support from EO providers. (Senior coordinator, IBAMA, Brasilia)

EO providers should improve systems to transform complex EO information into simple information for the farmers. (Land analyst, Federation of Agriculture and Livestock of the state of Mato Grosso (FAMATO), Cuiaba)

We need to have access of new EO products that includes both high frequency and high-resolution data. (Environmental analyst, Secretary of the State of Environment (SEMA), Cuiaba)

It would seem that respondents wish to have easy-to-use systems without any additional technical support from the main providers (INPE, USGS, and ESA). The relatively lower ranking of technical support compared to the others is perhaps surprising, but this did not mean that users of EO data saw this as unimportant, as indicated by its scores in the mid-range of the scale. Given that resourcing in terms of staff, IT, fast-internet connection, etc., were regarded as significant limitations in EO use, the services offered by providers were seen as a key ingredient for helping to address that gap by providing information that did not require further in-house processing. The use of EO data via WMS (Web Mapping Services) is a good example of practical solution that IBAMA headquarters provides to regional offices. The service accesses EO time series data in real time, with advanced computers at IBAMA headquarters. This reduces the infrastructure and staffing requirements at the regional level and in the field. The EO data are aggregated with information from PRODES and DETER. In general, respondents were confident that providers generate reliable and accurate information that does not require further verification.

3.4. Suggestions for Improving EO Data Use (RQ 4)

Respondents suggested several ideas for improving EO data use, although these often overlapped with infrastructure problems and institutional challenges. Indeed, the suggestions typically revolved around ways in which the limitations of staffing and other resources noted above could be addressed. For example, the infrastructure planning analyst at the State Secretariat of Planning of Mato Grosso (SEPLAN) in Cuiabá made three suggestions to improve the use of EO data in Mato Grosso:

1. *Policymakers need to see cartography as an essential part of planning.*
2. *The workforce in public institutions needs to be renewed, as there is a lack of people to work in the field.*

3.   *The concept of information management is essential. More transparency of information between all public institutions is required, and there is a need to simplify the use of information for the regular user.*

(Infrastructure planning analyst, State Secretariat of Planning of Mato Grosso (SEPLAN), Cuiaba)

The bureaucracy involved when establishing new partnerships, including with universities, was mentioned as a key barrier by respondents in Brasilia and Cuiaba, as exemplified by the following quotations from respondents:

We need to integrate the work of Universities with other public institutions and the private sector. The knowledge remains at the University and it does not reach the end-user. The problem is bureaucracy to establish partnerships. Internal and external communication with other institutions are major challenges for the University. Political will and conflict of interests are also major issues. (Geography lecturer, Federal University of Mato Grosso (UFMT), Cuiaba)

Universities should participate in more active ways to disseminate knowledge. (Senior director, CONAB, Brasilia)

We need more integration between academia at the state level and centers of information like INPE and federal institutions in Brasilia. Those institutions lack influence at the regional context to influence regional societies. INPE should work more at the state level. (Infrastructure planning analyst, State Secretariat of Planning of Mato Grosso (SEPLAN), Cuiaba)

Respondents often mentioned the need to integrate databases between institutions and improve access to the final information generated by using EO data.

We need to improve the transparency of information in public organizations. One of the main problems is the integration of database between public institutions. We also need to improve communication between institutions. For example, better communication between the State Secretariat for the Environment (SEMA), IBAMA, and the public attorney. Information is power and is associated with politics. (Senior staff, public attorney, Cuiaba)

We expect more transparency of information between public institutions. We have all the data, but we need to improve the links between institutions and the online platforms to share information. For example, between SEMA, IBAMA, INTERMAT, infrastructure departments, and other institutions. (Senior analyst, Secretary of State for the Environment (SEMA), Cuiaba)

We need to improve the quality of the data and how we share the data between institutions. SEMA, the National Institute of Colonization and Agrarian Reform (INCRA), the National Indian Foundation (FUNAI) and improve the flow of information between institutions. We have internal capacity, but one limitation is to manage the large volume of data. One suggestion is to generate more thematic maps to help policymakers. The government budget is linked to our institutional capacity to hire new staff and invest in new technologies such as new data management (big data). We need more integration between public universities and public institutions to integrate our knowledge and resources. (Senior Director, IBAMA, Cuiaba)

Several respondents highlighted that INPE (located in Sao Jose dos Campos) and federal institutions in Brasilia should influence actions at the regional level. The key factors of federal institutions contributions to regional development are associated with the general sense of disconnection between civil servants and policymakers in Brasilia with the reality of problems at the state and regional level. This point was mentioned several times during the interviews in Cuiaba and in Rio Branco.

4. Discussion

This research addresses the important but hitherto poorly understood requirements of, and limitations experienced by, the users of EO data for environmental and sustainable land management. It was conducted by obtaining quantitative and qualitative data from a diversity of relevant actors in Brazil, a country with a long history of interest in using EO data for such purposes. In total, extensive responses were obtained from 53 individual interviews, to provide representative coverage of both governmental institutions, other organizations, and tiers of EO users and enable a number of valuable insights to be drawn on the uses made of EO data, the factors that limit/enhance its use, and, importantly, on the needs of users and their suggestions for improving their use of EO data. While the majority of respondents was from federal or state institutions, this does reflect where much of the experience and responsibility rests for the sourcing and use of EO data and information for environmental and sustainability management in Brazil.

In the discussion below, we have combined the excellent breadth and depth of the freely given responses to the semi-structured represented in the Results section into an overall synthesis narrative addressing the four key research questions posed in the introduction that capture the challenges end-users encountered to extract the right information derived from different EO data. In more than half of the interviews, respondents mentioned that services with "ready-to-use" platforms are available from private providers and government (national level or state) should consider purchasing these specific applications associated with land-use management.

The use of EO data was acknowledged in this research to be indispensable as a means of providing policymakers and environmental and sustainability managers with reliable information e.g., on land use and land-use change. In the present study, the key areas of environmental and sustainability management that users wished to deploy EO data to help address were to monitor deforestation, crop forecast, disasters (e.g., flooding), fire, urban planning, and law enforcement. It is widely used in different sectors, such as government agencies, banks, NGOs, and academia. Potential users include small farmers and non-experts outside the EO community. It was often mentioned the need to create easy-to-use platforms to enable non-experts to understand the information generated from EO data. The main users in this study include civil servants, personnel in research institutions, and government institutions at the federal and state levels.

Lack of skilled personnel and finance emerged as the main limitations identified by users for EO data. The financial limitations restricted both investment in staff and in infrastructure (IT, speed of internet, and big data management), thereby limiting institutional capacity, but they are also a factor restricting investment in field operations and actions that utilize EO data. At the time of the research (2017), Brazil was going through an economic crisis which resulted in fewer funds being available for recruitment by government agencies (political and economic context). Salaries for GIS professionals tend to be low, and there was a lack of long-term career opportunities inside the respondent institutions. Access to EO data at medium resolution (e.g., Sentinel 10 m and Landsat 30 m) was not a limitation for users, but access to up-to-date high-resolution ($\leq$1 m) data from government institutions or via international technical cooperation was an unmet demand.

Two ways emerged in this research whereby the limitations of a lack of skilled personnel and finance could be addressed:

- Improving partnership between institutions to share their data;
- Buy-in processed information from service providers.

It was suggested by some respondents that further cooperation and new partnerships between existing institutions are necessary to overcome limitations on the use of EO data. For example, the results of academic research should be used by policymakers and by institutions that rely on up-to-date information, as well new up-to-date GIS training provided by academics which could improve the technical capacity of regional institutions. In many cases, respondents learned to process EO data alone and technical support from experts could improve institutional capacity to use EO data more efficiently, but the bureaucracy involved in establishing new partnerships was mentioned as a significant barrier and the main reason why institutions do not create partnerships. It was often mentioned that institutions at the federal level, which are generally better resourced, should work closely with institutions at the state level, to improve the use of EO data.

Several ideas to address these barriers to cooperation and transparency were suggested, such as the expectation that INPE should provide high-resolution data with a high frequency of coverage at the national level. These data should be converted into easy-to-use information products. For example, final products or thematic maps that offer multiple information associated with an EO dataset in specific geographical regions could be managed and hosted by INPE. It was suggested that such an "EO portal" would integrate multiple data sources and include information about income, health, agriculture, regional development, and many other indicators associated with EO data to guide informed decisions. The end-user could access in a single online platform EO data and other relevant information according to end-user needs. The second approach, that of purchasing EO-based informational services from private sector providers, was popular and seen by respondents as a useful and trusted means of addressing their in-house limitations.

As noted earlier, there is an important distinction here between EO "data" and "information" derived from those data. In EO terms, data are the raw images, perhaps processed to distinguish factors such as land use. Information is the result of further processing of the data to provide details on the places where (for example) there are changes in the rate and extent of LULC. Hence, service providers may go further than just providing data and also provide high-level information that agencies can make use of. Here the main factor was cost along with a need for high frequency and fast delivery. The information provided need to be easy to use without any additional technical support. However, high-resolution EO data are expensive, and some institutions cannot afford to buy services or products provided by private enterprises. Satellites are valuable tools to monitor land-use change, and they have delivered consistent data on forest change, such as the DETER alert system provided from INPE [2]. The evidence from the interviews suggests that EO data are widely used at the respondent institutions, and additional communication between institutions is necessary to increase the use of the information provided by different providers.

One of the key points identified in the research is how data, information, and knowledge are shared across institutions. Providers of information such as INPE, CONAB, EMBRAPA, IBAMA, and state agencies are users and providers of environmental information derived from EO data. However, access to EO data is not a major limitation in itself, except for the point made about the need for high-resolution EO data, but it is the availability of information derived from the EO data that matters. The problems associated with lack of cooperation and the needs to increase partnerships to improve shared information in different flows of information was mentioned as one of the main limitations by different respondents. This point about the sharing of information has been noted before, although not in the context of EO, and provides a challenge regarding analysis, given that there can be many facets involved. For example, Bruckmeier and Tovey [20] point out the practices of hierarchy of data, information, knowledge, and the interactions amongst social actors involved in generating and using knowledge via complex interactions. Ostrom [21] has suggested the use of multiple-level analytical framework, such as the Institutional Analysis and Development Framework (IAD), to understand the ways institutions operate and interact in terms of information flows, the complexity of such networks, and the changes over time [22]. The IAD framework helps to organize the capabilities of institutions issues related to governance systems and institutional arrangements [21]. However, it can be challenging in practice to determine the patterns of interactions, norms, and strategies between participants within such a complex institutional network [23]. Nonetheless, additional in-depth institutional analysis would enhance the understanding of the flow of EO information between institutions.

Another key point that emerged in the respondent institutions is the skills levels available to use and process all the information available. For example, some respondents (e.g., in IBAMA and

EMBRAPA) mentioned that they are very capable when it comes to using different types of EO data and integration of those data with external sources of information, to provide reliable information for managers. On the other hand, some end-users in small state institutions learned by themselves to process EO data and GIS independently via different online resources. Several respondents mentioned the need for EO data providers to provide thematic maps with "ready to use products" integrating multiple data sources to improve the use of EO data. For example, thematic maps with the best EO data available for specific geographical regions with additional information of socioeconomic factors (economic, unemployment, education, health, poverty, and unemployment) and environmental factors associated with land use and land-use change (agriculture production, deforestation, water quality, and environmental degradation) would be very useful.

Access to raw EO data in not the primary requirement of end-users, but access to online platforms that facilitate access to ready-to-use products and services is critical importance [24]. Vinhas et al. [25] points out that traditional methods where end-users transform data into information are being replaced by ready-to-use Web Map Services (WMS), which significantly facilitate the use of EO information by the wider community. Ready-to-use WMS end-users are not required to transform EO data into information [25]. To extract information, experts use methods that assign a label to each pixel (e.g., grasslands, forest, and pasture), and labels can represent either land cover (observed biophysical cover of the Earth's surface) or land use (description of the socioeconomic activities) [26], but end-users have different knowledge to transform EO data into usable information (e.g., EO terminology, algorithms, software programs, and image processing skills) [27]. The term "data as a service" (DaaS) is defined as the sourcing, management, and provision of data delivered in an immediately usable format to end-users, including non-expert users [25]. Examples include machine-learning techniques and pattern-recognition to process EO data with spatial, temporal, and spectral features of different EO data [28], or machine-learning techniques to solve a wide range of tasks to extract information from multiple EO datasets (e.g., optical and radar) to classify land-cover classes (e.g., crops and forests) [29].

Synthetic Aperture Radar (SAR) is classified as an active system by which the instrument transmits radiation and captures the reflection, while passive systems are designed to only capture radiation from the earth's surface [30]. Passive systems have an important disadvantage in terms of their inability to capture data during unfavorable weather conditions (e.g., rain and clouds) [30], while SAR can "see" through such barriers. Hence, SAR is a high-performing technology for studying the dynamics of forests in tropical forests (e.g., forest classification and detection of clear-cut and burned-out areas), but radar images are not intuitively understandable even by scientists and technicians due to centimetric band classifications [30].

Overall, the results of this research clearly suggest that the challenge, as identified by users in Brazil (mainly in public institutions), is not lack of EO data per se, but rather mainly a lack of institutional resources to translate the data into information and to make use of the information on the ground (e.g., adequately resourced teams to fight fire or illegal deforestation). However, it is also clear that other limitations overlap with this main one and there is no single answer for every user group or institution. A shortage of skilled staff, the speed of the internet (communication between headquarters and regional offices), etc., are recognized problems in several institutions, and this will require further investment from the federal and state government. The federal institutions included in the interviews rely on public funding for new investments, and this seems unlikely in financially challenging times. Nonetheless, a substantial challenge is also associated with lack of cooperation between different government institutions and transparency to provide the environmental information for end-users. What would seem to be required are efficient ways in which the barriers to cooperation and transparency could best be addressed. In a study about forest governance and transparency in public institutions and state agencies in the Amazon region, Bizzo and Michener [31] noted poor governance outcomes (passive transparency) and weak compliance with environmental open-data obligations and policies and with expenditures and procurement practices of government agencies.

Their results are in line with those presented here: End-users complained about the lack of transparent access to environmental information in environmental agencies and government institutions.

The interview with a senior staff member at the public attorney office in Mato Grosso highlighted the power of vested interests. The role of private-sector service providers also requires more attention, particularly regarding the newer high-resolution (<1 m) optical data and what it can offer public-sector agencies. There is clearly a need for more research to untangle the factors at play with these interactions between various public sector institutions seeking to use EO data and between them and the private sector.

To transform EO data into useful information (e.g., maps), end-users require knowledge of technical geospatial approaches, observation, and analysis [32]. It should be noted that the present research did not specifically cover disruptive changes in EO technologies, such as the use of video satellites capable of capturing tens of seconds of Very High Resolution Imagery (VHRI) steady images, such as the Carbonite-2 (Vivid-i) [33], nor were such changes referred to in any substantial degree by respondents. However, it needs to be acknowledged that end-users are increasingly coming from non-EO domains, thus affecting user-centric technology developments to support the visualization of information [32], and that these new disruptive capabilities are already playing an increasingly important role in disaster response and are likely to be used increasingly in environmental and sustainability monitoring of land-use-change detection [33]. VHRI EO data (<1 m resolution) available from optical sensors offer rich spatial details [34]. However, the present study indicates that taking advantage of such advances will be very difficult due to the challenges users in Brazil already face in terms of the lack of skilled personnel able to process existing EO data and potentially without easy access to ready-to-use service platforms for VHRI, etc. Agreements between public institutions and private providers to facilitate purchasing ready-to-use services with new technologies may help the performance of institutions, without adding to the long-term costs of permanent staff. Additionally, further initiatives to improve the use of information, such as the Norway International Climate and Forest Initiative (NICFI) in partnership with Kongsberg Satellite Services (KSAT), Planet, and Airbus, aiming to provide monthly universal access to high-resolution and ready-to-use platform of the tropics [35], may prove to be highly useful in advancing EO usage in support of environmental and sustainability goals.

5. Conclusions

The main conclusions of this research are as follows:

1. The main limitation for wider use of EO data and information is a lack of skilled personnel linked with financial constraints associated with the current political and economic situation of Brazil (lack of government investments). These limit the capacity of institutions to develop the skilled personnel and infrastructure to benefit from the acknowledged gains that EO data and information can bring to their environmental and sustainability management roles.
2. New partnerships and cooperation between institutions at the federal and state levels could be one means of addressing these resource constraints, but there are significant barriers which prevent this, and these warrant further research.
3. Service providers from the private sector are utilized by agencies, and there is potential for more involvement. These service providers are often in a better position to access a variety of platforms and provide the most appropriate EO-based products for the requirements of customers. Further research is required on the interaction between the private and public sector institutions to better understand the needs of both and how they can be addressed.
4. The quantity of EO data is not a real limitation. However, it is also clear that some aspects of the data's "quality" (e.g., resolution, type) and its cost can be.
5. Addressing the widespread demand/expectation identified in this research for "easy to use" EO-based information on "easy to use" platforms will bring substantial benefit for environmental

and sustainability management, especially when combined with complementary information from other sources and databases. This would be of considerable value to the work of diverse public/governmental institutions, other organizations (including the private sector and academic), and for the general public.

Author Contributions: Conceptualization, M.C., S.M., R.M., G.G. and J.L.; methodology, M.C., S.M., R.M., G.G. and J.L.; formal analysis, M.C., S.M. and R.M.; investigation, M.C.; resources, M.C., S.M., R.M., G.G. and J.L.; data curation, M.C.; writing—original draft preparation, M.C.; writing—review and editing, M.C., S.M., R.M., G.G. and J.L.; supervision, S.M., R.M., G.G. and J.L.; project administration, S.M.; funding acquisition, S.M., R.M., G.G. and J.L. All authors have read and agreed to the published version of the manuscript.

Funding: The first author's Ph.D. study program at the University of Surrey, UK, was funded by the UK Natural Environment Research Council's (NERC) SCENARIO Doctoral Training Partnership (NE/L002566/1), which is gratefully acknowledged.

Acknowledgments: We are most grateful to Gilberto Camara and INPE for their assistance in the conduct of the research reported in this paper. We would also like to thank all the individuals and organizations who kindly participated in the research, for their time and cooperation.

Conflicts of Interest: The authors declare no conflict of interest.

References

1. Lynch, J.; Maslin, M.; Balzter, H.; Sweeting, M. Choose satellites to monitor deforestation. *Nature* **2013**, *496*, 293–294. [CrossRef] [PubMed]

2. Popkin, G. Satellite alerts track deforestation in real time. *Nature* **2016**, *530*, 392–393. [CrossRef] [PubMed]

3. Maus, V.; Camara, G.; Cartaxo, R.; Sanchez, A.; Ramos, F.M.; de Queiroz, G.R. A Time-Weighted Dynamic Time Warping Method for Land-Use and Land-Cover Mapping. *IEEE J. Sel. Top. Appl. Earth Obs. Remote Sens.* **2016**, *9*, 3729–3739. [CrossRef]

4. Achard, F.; Hansen, M.C. *Global Forest Monitoring from Earth Observation*; CRC Press: Boca Raton, FL, USA, 2013.

5. Moutinho, P.; Guerra, R.; Azevedo-Ramos, C. Achieving zero deforestation in the Brazilian Amazon: What is missing? *Elem. Sci. Anth.* **2016**, *4*, 125. [CrossRef]

6. Hansen, M.C.; Potapov, P.V.; Moore, R.; Hancher, M.; Turubanova, S.A.; Tyukavina, A.; Thau, D.; Stehman, S.V.; Goetz, S.J.; Loveland, T.R.; et al. High-Resolution Global Maps of 21st-Century Forest Cover Change. *Science* **2013**, *342*, 850–853. [CrossRef] [PubMed]

7. Hansen, M.C.; Shimabukuro, Y.E.; Potapov, P.; Pittman, K. Comparing annual MODIS and PRODES forest cover change data for advancing monitoring of Brazilian forest cover. *Remote Sens. Environ.* **2008**, *112*, 3784–3793. [CrossRef]

8. Laurance, W.F.; Achard, F.; Peedell, S.; Schmitt, S. Big data, big opportunities. *Front. Ecol. Environ.* **2016**, *14*, 347. [CrossRef]

9. Monteiro, M.; Rajão, R. Scientists as citizens and knowers in the detection of deforestation in the Amazon. *Soc. Stud. Sci.* **2017**, *47*, 466–484. [CrossRef]

10. Aguiar, A.P.D.; Câmara, G.; Escada, M.I.S. Spatial statistical analysis of land-use determinants in the Brazilian Amazonia: Exploring intra-regional heterogeneity. *Ecol. Model.* **2007**, *209*, 169–188. [CrossRef]

11. Carvalho, T.S.; Domingues, E.P.; Horridge, J.M. Controlling deforestation in the Brazilian Amazon: Regional economic impacts and land-use change. *Land Use Policy* **2017**, *64*, 327–341. [CrossRef]

12. Rajão, R.; Georgiadou, Y. Blame games in the Amazon: Environmental crises and the emergence of a transparency regime in Brazil. *Global Environ. Politics* **2014**, *14*, 97–115. [CrossRef]

13. Aragão, L.E.O.C.; Shimabukuro, Y.E. The incidence of fire in Amazonian forests with implications for REDD. *Science* **2010**, *328*, 1275. [CrossRef] [PubMed]

14. INPE. PRODES (Brazilian Amazon Rainforest Deforestation Monitoring by Satellite). 2020. Available online: http://www.obt.inpe.br/OBT/assuntos/programas/amazonia/prodes (accessed on 18 August 2020).

15. Cerbaro, M.; Morse, S.; Murphy, R.; Lynch, J.; Griffiths, G. Information from Earth Observation for the Management of Sustainable Land Use and Land Cover in Brazil: An Analysis of User Needs. *Sustainability* **2020**, *12*, 489. [CrossRef]

16. Noether, G.E. *Introduction to Statistics—The Nonparametric Way*; Springer: New York, NY, USA, 1991.

17. Scheff, S.W. *Nonparametric Statistics*; Elsevier: Amsterdam, The Netherlands, 2016.

18. ESA. SPOT-6 and SPOT-7 Commercial Imaging Constellation. 2020. Available online: https://earth.esa.int/web/eoportal/satellite-missions/s/spot-6-7 (accessed on 24 November 2020).

19. ESA. FormoSat -5. 2020. Available online: https://earth.esa.int/web/eoportal/satellite-missions/f/formosat-5 (accessed on 24 November 2020).

20. Bruckmeier, K.; Tovey, H. Knowledge in Sustainable Rural Development: From forms of knowledge to knowledge processes. *Sociol. Rural.* **2008**, *48*, 313–329. [CrossRef]

21. Ostrom, E. Background on the Institutional Analysis and Development Framework. *Policy Stud. J.* **2011**, *39*, 7–27. [CrossRef]

22. McGinnis, M. An Introduction to IAD and the Language of the Ostrom Workshop: A Simple Guide to a Complex Framework. *Policy Stud. J.* **2011**, *39*, 169–183. [CrossRef]

23. Ostrom, E. Do institutions for collective action evolve? *J. Bioeconom.* **2014**, *16*, 3–30. [CrossRef]

24. Mathieu, P.-P.; Aubrecht, C. *Earth Observation Open Science and Innovation*; Springer: Berlin/Heidelberg, Germany, 2018.

25. Vinhas, L.; de Queiroz, G.R.; Ferreira, K.R.; Camara, G. Web services for big earth observation data. *Rev. Bras. Cartogr.* **2017**, *69*, 5.

26. Camara, G. On the semantics of big Earth observation data for land classification. *J. Spat. Inf. Sci.* **2020**, *2020*, 21–34.

27. Arvor, D.; Belgiu, M.; Falomir, Z.; Mougenot, I.; Durieux, L. Ontologies to interpret remote sensing images: Why do we need them? *GISci. Remote Sens.* **2019**, *56*, 911–939. [CrossRef]

28. Marinoni, A.; Iannelli, G.C.; Gamba, P. An information theory-based scheme for efficient classification of remote sensing data. *IEEE Trans. Geosci. Remote Sens.* **2017**, *55*, 5864–5876. [CrossRef]

29. Kussul, N.; Lavreniuk, M.; Skakun, S.; Shelestov, A. Deep learning classification of land cover and crop types using remote sensing data. *IEEE Geosci. Remote Sens. Lett.* **2017**, *14*, 778–782. [CrossRef]

30. Palmann, C.; Mavromatis, S.; Hernández, M.; Sequeira, J.; Brisco, B. Earth observation using radar data: An overview of applications and challenges. *Int. J. Digit. Earth* **2008**, *1*, 171–195. [CrossRef]

31. Bizzo, E.; Michener, G. Forest Governance without Transparency? Evaluating state efforts to reduce deforestation in the Brazilian Amazon. *Environ. Policy Gov.* **2017**, *27*, 560–574. [CrossRef]

32. Sudmanns, M.; Tiede, D.; Lang, S.; Bergstedt, H.; Trost, G.; Augustin, H.; Baraldi, A.; Blaschke, T. Big Earth data: Disruptive changes in Earth observation data management and analysis? *Int. J. Digit. Earth* **2020**, *13*, 832–850. [CrossRef]

33. Elliott, J.R. Earth Observation for the assessment of earthquake hazard, risk and disaster management. *Surv. Geophys.* **2020**, *41*, 1–32. [CrossRef]

34. Zhang, C.; Yue, P.; Tapete, D.; Shangguan, B.; Wang, M.; Wu, Z. A multi-level context-guided classification method with object-based convolutional neural network for land cover classification using very high resolution remote sensing images. *Int. J. Appl. Earth Obs. Geoinf.* **2020**, *88*, 102086. [CrossRef]

35. O'Shea, T. Planet, KSAT and Airbus awarded first-ever contract to combat deforestation. Available online: https://www.planet.com/pulse/planet-ksat-and-airbus-awarded-first-ever-global-contract-to-combat-deforestation (accessed on 20 October 2020).

Publisher's Note: MDPI stays neutral with regard to jurisdictional claims in published maps and institutional affiliations.